2nd EDITION

Productivity Measurement and Improvement

Lawrence S. Aft

Southern College of Technology
Marietta, Georgia

PRENTICE HALL, Englewood Cliffs, NJ 07632

Library of Congress Cataloging-in-Publication Data

Aft, Lawrence S.
    Productivity measurement and improvement / Lawrence S. Aft. — 2nd
ed.
        p.   cm.
    Includes bibliographical references and index.
    ISBN 0-13-728759-3
    1. Time study.   2. Motion study.   3. Industrial productivity.
I. Title.
T60.4.A34   1992
658.5′1—dc20   91–16592
                                                        CIP

Acquisitions editor: *Elizabeth Kaster*
Editorial/production supervision
    and interior design: *Richard DeLorenzo*
Copy editor: *Camie Goffi*
Cover design: *Patricia McGowan*
Manufacturing buyer: *David Dickey*
Prepress buyer: *Linda Behrens*
Supplements editor: *Alice Dworkin*
Editorial assistant: *Jaime Zampino*

 © 1992, 1983 by Prentice-Hall, Inc.
A Simon & Schuster Company
Englewood Cliffs, New Jersey 07632

Printed in the United States of America

10 9 8 7 6 5 4 3 2 1

ISBN 0-13-728759-3

Prentice-Hall International (UK) Limited, *London*
Prentice-Hall of Australia Pty. Limited, *Sydney*
Prentice-Hall Canada Inc., *Toronto*
Prentice-Hall Hispanoamericana, S.A., *Mexico*
Prentice-Hall of India Private Limited, *New Delhi*
Prentice-Hall of Japan, Inc., *Tokyo*
Simon & Schuster Asia Pte. Ltd., *Singapore*
Editora Prentice-Hall do Brasil, Ltdas., *Rio de Janeiro*

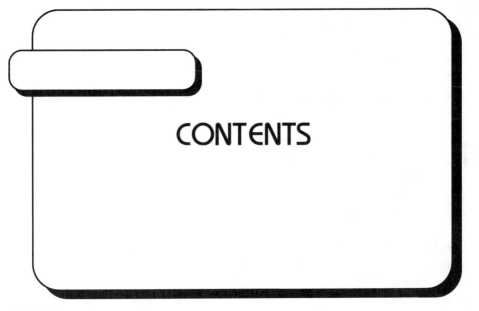

CONTENTS

Contents

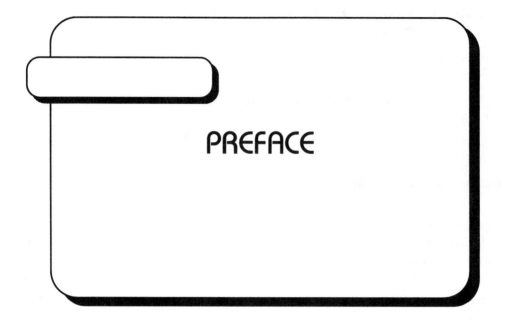

PREFACE

This book has been written to fill the need for an introductory text in the traditional industrial engineering area of motion and time study. Emphasizing the need for productivity improvement, while fully explaining the mechanics of work improvement and measurement, the book is written for the individual who must actually perform the analysis tasks. Being practitioner-oriented, the text provides straightforward descriptions and numerous examples of traditional industrial engineering skills.

The text is also designed to fill the needs of the practitioner in industry, business, or technical education programs. For the practitioner, the book shows some ways to improve worker productivity and how to measure the work performed. For the student, the text is suitable either as an overview of the entire field or as the base for in-depth, applications-oriented courses.

The book is divided into four major sections: An Introduction to Productivity, Work Analysis, Productivity Measurement, and Productivity Improvement. A one-quarter or one-semester survey course, typical of those found in business administration programs, would ideally include the following chapters and topics:

Chapter 1   Productivity
Chapter 2   Methods Analysis
Chapter 5   Time Study
Chapter 10 Human Factors (Ergonomics)
Chapter 11 Incentive Systems
Chapter 12 Alternate Methods to Increase Productivity

The book is especially well suited for in-depth technical education courses of the type found in technical schools or in formal training programs in business and industry. The most effective presentation of this material would be as a two-semester sequence of courses, essentially a methods course and a measurement course, or as a three-quarter sequence of courses which would include a methods course and two measurement courses. Chapters that should be included in a methods course include the following:

Chapter 1   Productivity
Chapter 2   Methods Analysis
Chapter 3   Graphical Productivity Analysis
Chapter 4   Improving Work Methods
Chapter 10 Human Factors (Ergonomics)
Chapter 12 Alternate Methods to Increase Productivity

If a single course in work measurement is to be used, the following chapters should be covered:

Review of:
Chapter 2 Methods Analysis
Chapter 3 Graphical Productivity Analysis
Chapter 4 Improving Work Methods

New presentation of:
Chapter 5   Time Study
Chapter 6   Standard Data Systems
Chapter 7   Predetermined Time Systems
Chapter 8   Work Sampling
Chapter 9   Physiological Work Measurement
Chapter 11 Incentive Systems

A two-course sequence in work measurement would include, in both courses, a review of Chapters 2, 3, and 4. The first course in work measurement would also include Chapters 5 and 11. The second course, following the review, would include Chapters 6, 7, 8, and 9.

Changes Made to the Second Edition

Additions to the second edition include: In *Chapter 1,* material on the relationship of CIM and SPC to the work measurement and productivity improvement process has been added.

Chapter 3 has received a significant revision, with definitions and descriptions of graphical work analysis tools beefed up and brought into line with the ANSI Z94, along with the addition of examples that relate to the service oriented businesses besides manufacturing. The practice section at the end of the chapter doubles the number of problems offered for the student to hone his or her analysis skills on.

In *Chapter 4,* the section on micromotion analysis has been expanded with more indepth descriptions and applications procedures presented in more detail.

Chapter 5 offers an enhanced procedure for conducting a time study, recognizing that many businesses still must rely on the basic stop watch technology. This chapter also features the addition of service sector examples as well as additional manufacturing examples. The statistics of sample size determination are added along with sample size tables. The section on time study difficulties has been expanded.

Chapter 7 continues to illustrate predetermined time systems via the innovative AFTWAYS imaginary predetermined time system. A discussion of MODAPTS has been added.

Chapter 8 presents a new discussion of accuracy, as well as more complete sample size tables for work sampling sample size estimation.

The *annotated bibliography* has been updated with some interesting new readings added to the existing base of standard readings.

The preface would not be complete without a public word of thanks to some of the people who helped make the book possible. Brad Young of Southern College of Technology's Industrial Engineering Technology Department volunteered much of his time as both a sounding board for ideas as well as an able proofreader. Dorothy Ingram of the Southern Tech Library was of tremendous assistance in the research, finding copies of some difficult to obtain references. Patti Futtrell, former Director of Public Relations of Southern Tech, was a most willing and able photographer when it came time to prepare illustrations, as was Curtis Colee in preparing some other illustrations to my specifications. Steve Elliott of Lockheed-Georgia Company and Maurice Chapman of Kennestone Hospital shared some outstanding examples of applications with me. Southern Tech student Lynn Jackson spent many hours working on the computer programs. Special thanks to Dick McGuire and Bob Atkins for their generous donation of additional example problems for the second edition. Finally, a large amount of appreciation goes to my wife Susan, my son Steven, and my daughter Dana, for putting up with me while I prepared this material. Without their patience and understanding, it would have been a much more difficult task.

PART I

Productivity—
An Introduction

1

Productivity

OBJECTIVES

After completing this chapter, the reader should be able to do the following:

- Understand productivity and its importance in competitiveness
- Understand industrial engineer's special relationship with productivity

INTRODUCTION

There is a problem in the land and it can no longer be swept under the rug. For nearly 30 years, our cumulated annual rate of productivity increase has consistently been the lowest among all free industrial nations. Worse yet, the trend is still downward. We are no longer the world's undisputed leader in productivity growth (Birn, 1979, p. 38).

> "Productivity is the buzz word [of the present.] If you want to get peoples' attention, tell them how you can increase their output while decreasing their input. Tell them how to produce more efficiently. Guarantee them a future of progress and comfortable living" (Camp, 1985, p. 80).

Over the last 20 years productivity growth in the U.S. has slowed alarmingly to the point where it is recognized as one of the most significant problems facing our country. Since the mid-1960's, according to *Automotive Industries,* "Japan and West Germany have enjoyed yearly productivity increases of 6.8 percent and 5.3 percent, respectively, whereas the U.S. has had about 2.2 percent." . . . It must be remembered, however, that the American worker started out with a higher productivity base and that other countries are only now catching up. . . . The U.S. is still the most productive nation in the world, although at least four countries will pass the U.S. in the next ten years at the present rate of change. (Otis, 1980, p. 208)

When we examine the manufacturing output per employee, a traditional strength of our industrialized society, we see an equally dismal picture. The U.S., during the 1970s, ranked eighth in the world in terms of growth of manufacturing output per employee. As

Table 1-1 shows, just about every other industrialized Western nation experienced a faster rate of productivity increase than did the U.S. (Kehlbeck, 1978).

TABLE 1-1   MANUFACTURING OUTPUT PER EMPLOYEE

Country	Percent increase	1970s Rank
U.S.A.	1.8	8
Canada	3.0	6
Japan	5.4	2
France	3.4	5
Germany	5.4	2
Italy	6.0	1
Sweden	4.4	4
United Kingdom	3.0	7

Since 1977, U.S. manufacturing output has risen by only 4 percent. During the same period, German output increased 13 percent and Japanese output 29 percent.

As these figures show, the long-time dominance of the U.S. is slipping. "Increased productivity means producing more with the same resources" (*Federationist,* March 1978, p. 1). Decreasing productivity then, must mean producing less with the same resources.

Productivity, once an ignored problem, has emerged as one of the pressing issues of the day. Public awareness has been fueled by three sources: Howard K. Smith's series for the Broadcasting Industry Council to Improve Productivity, the American Institute of Industrial Engineers' designation of January as Productivity Month, and the American Society for Quality Control's initiation of October as Quality Month. The public has become aware that "Productivity is essential for a prosperous economy, a prosperous company, a prosperous employee" (Sellie, 1984, p. 82). This interest has expanded from the historical interest traditionally shown in blue collar productivity into white collar jobs as well (Liker and Hancock, 1984, p. 60).

Productivity has many definitions. It has been defined quite narrowly as the ratio of output in goods and services to the dollar input, both direct and indirect. When output-per-dollar-spent is increased, productivity is improved (Vough, 1979). In a broader sense, "there are two fundamental aspects to the notion of productivity. The first is . . . the productivity approach and the second is productivity measurement" (Riccio, 1976, p. 95). According to Lucius Riccio of the Police Foundation in Washington, D.C., the productivity approach is "the concern for the use of resources and the attainment of objectives without fully appreciating the understanding of the relationship between the two" (Riccio, 1976, p. 96). The productivity approach is a way of understanding the productivity problem and productivity measurement is the actual counting of input and output.

Productivity has three major components: labor, capital, and management. *Labor* or, more specifically, the motivation of the worker, is an often-studied part of the productivity process. Many suggestions, hypotheses, or theories have been proposed over the years to increase the productivity of individual workers. Some have been effective, others have been no more than fads. Concepts tried successfully over the years include employee involvement and incentives. "But again, motivation as the key to productivity (improvement) is still operationally elusive" (Riccio, 1976, p. 97).

Capital is the investment in new technology. In some applications, the introduction of new technology has the potential of significantly increasing productivity. In other instances, the introduction of large capital investment will have no impact on the growth of output for dollar spent.

Management, in its desire to increase productivity, is the third component of productivity. The proper attitude toward improving productivity is set by management and increases in output result only when top management is committed to the program.

"Productivity may, in this sense, be thought of as the effectiveness with which the resource inputs—of personnel, materials, machinery, information—in a plant are translated into customer-satisfaction-oriented production outputs and which today involve all the relevant marketing, engineering, production, and service activities of the plant and company rather than solely the activities of the factory workers, where traditional attention has been concentrated" (Feigenbaum, 1983, p. 339).

Key to this definition of productivity is quality, or quality improvement. The Japanese system, which is so frequently cited as the model to be followed, built its success on quality. "If Western manufacturers are to close the quality gap with the Japanese, there is no better way to begin than by transferring primary responsibility for quality from the QC department to production. Not only is this relatively easy to do, but it also brings quality responsibility back to its natural home" (Schonberger, 1982, p. 35).

WHY IS PRODUCTIVITY A PROBLEM IN THE UNITED STATES?

It has been suggested that the decline in U.S. productivity is directly related to the inefficiency, ineptness, carelessness, or apathy of the American worker. It has been particularly easy to blame the American labor unions and the high standard of compensation they have negotiated for their members for the decline in productivity. In the absence of conclusive data, unions have been a convenient scapegoat. However, when formal studies have been conducted, the results contradict this belief. Dr. Steven G. Allen of the Economics Department of North Carolina State University studied the productivity of construction workers. After intensive research he concluded, "Union workers are more productive" (Allen, 1979, p. 17).

According to a survey by *Productivity,* a monthly newsletter, top managers at 221 concerns blamed a number of factors, "but not workers for weak output" (*Wall Street Journal,* January 6, 1981, p. 1). Writing in the September 1980 issue of the *Atlantic,* James Fallows accredited Robert Reich, Director of the Office of Planning of the Federal Trade Commission, with making the same statement (Fallows, 1980).

Despite evidence to the contrary, many members of our society are convinced, as the AFL-CIO's Bill Cunningham so succinctly stated, ". . . when some people want to increase productivity . . . they just want to increase the workload—specifically somebody else's workload" (*Federationist,* March 1978, p. 1). "What about the productivity of a motel along an interstate highway? Is occupancy the appropriate measure? If so, then

$2 a gallon gasoline will have a more profound effect on the productivity of the employees of that motel than the number of beds changed or meals served in the coffee shop" (Kirkland, 1982, p. 1213).

If workers are not to blame, and the aforementioned studies and reports substantiate that, then just what or who else might be responsible for the state of our nation's productivity? A number of nationally known experts have made some suggestions.

According to Mike Mescon, head of the management department at Georgia State University in Atlanta, "The fault lies at the top. It's the CEO, the chief executive officer, who sets the pattern of productivity or lack of it by what he demands, the habits he lets his subordinates drift into" (*Industrial Distribution,* February 1981, p. 39).

Dr. Phil Hicks, noted industrial engineering consultant and former head of the industrial engineering departments at New Mexico State University and North Carolina A and T University, recently stated that one of the major causes of the declining productivity in the U.S. is management philosophy. Hicks stated, while addressing the 1980 Southeastern Regional Industrial Engineering Conference of the American Institute of Industrial Engineers, that managers who have their eyes on the bottom line of the short term rather than viewing the total picture were constraining the growth of productivity in the U.S. According to Hicks, the proliferation of lawyers, and accountants in top management positions has made the yearly rate of return rather than long-term growth the major objective of many American companies (Hicks, 1980).

The Institute of Industrial Engineers recently surveyed its membership to determine what practicing industrial engineers felt were the major causes of lack of sustained productivity growth in the U.S. The major obstacles to productivity growth, in the view of those surveyed, include the following (Camp, 1985, p. 85):

- 67.3 percent felt management failed to understand how productivity could be improved.
- 61.7 percent felt management failed to authorize sufficient manpower to direct productivity improvements.
- 57.2 percent felt insufficient training was provided for workers.
- 56.0 percent felt management failed to apply proper measurement programs to evaluate productivity improvement.

Management may be to blame but we shouldn't really blame our CEOs entirely for this problem. Although it is not often acknowledged, the leader of an American business firm is probably the most measured creature in captivity. He or she is measured by a monthly statement, a quarterly statement, a semi-annual statement, and an annual financial statement. He or she is asked, in many instances, to present a long range financial plan, but, unfortunately, the boss is rarely ever evaluated on the plan. In fact, he often fails to survive a year if the financial statement doesn't display an optimistic picture quickly. (Walsh, 1980, p. 5)

It is not the purpose of this text to place blame; rather, it is the author's intention to help the reader help his or her organization increase the productivity or the utilization of capital, labor, and management. To accomplish this objective "consideration must be given to increasing the productivity of the enterprise as a whole, along with improving the efficiency of the individuals who are included in the enterprise'' (DeWitt, 1976, p. 24).

The key aspects to this productivity improvement effort is work improvement and work measurement.

WAYS TO IMPROVE PRODUCTIVITY

There are many alternatives available for improving productivity. "A major portion of our energies has been directed at the productivity improvement opportunities which can be taken advantage of by labor-management cooperation'' (Kuper, 1978, p. 208). ". . . clearly the human factor is one of the most important aspects of increased productive output. The worker can no longer be assigned second string status if American management wants to get serious about licking its productivity and quality problems'' (Otis, 1980, p. 210). However, people solutions are not the only solutions available.

According to Jerry Hamlin of the American Productivity Center in Houston, the average company has many potential productivity tools at its disposal. These tools include the following (Hamlin, 1978, p. 224):

- Improved layout, workflow, and materials handling
- Supervisory training
- Work simplification
- Job enlargement and redesign
- Systems analysis
- Attitude surveys
- Incentive plans
- Suggestion programs
- Cost reduction programs.

All of these, when used properly, have their places in industry. "In appraising an organization's potential or capacity for improving productivity, the manager or the analyst must examine the current operations and the current management practices with the objective of trying to decide how they should function in the future'' (Hamlin, 1978, p. 225). "Productivity in this country traditionally has resulted from the use of cheap and abundant energy to add horsepower to manpower, and through the introduction of tools purchased with cheap and abundant capital. If it were any other way—if indeed productivity was the sum of increased human effort—then the building of the Pyramids would be the high water mark for the productivity of mankind'' (Kirkland, 1982, p. 1214).

In the aforementioned survey of its members, the American Institute of Industrial Engineers identified the top priority, as viewed by practicing IEs, for productivity increases to be capital investment for new or automated equipment. (Camp, 1985, p. 82) According to the Organization for Economic Cooperation and Development, only 14.9 percent of GNP was spent on capital improvements in the U.S., while Japan invested 22.6

percent and West Germany 23.1 percent of their respective GNP's in capital improvement. Despite this belief, Joseph Kehlbeck, past President of AIIE, provided this information: "The dollars that the U.S. is spending on fixed capital formation as a percent of Gross National Product is the least of any of its major competitors. The U.S. is running, in 1978, at the 17.4 percent level. Japan is more than double the amount of U.S. Investment as a percent of Gross National Product" (Kehlbeck, 1978, p. 20). Indeed, the modernization of industry, the investment in capital improvements, and the look to the long-term returns will help to improve the productivity picture. Likewise, the application of the tools Hamlin (1978) spoke of will help. However, there is the overriding area of emphasis that continues to emerge. "Productivity Improvement will increasingly depend on factors of human motivation and organization" (Otis, 1980, p. 212).

> Historian Loren Baritz noted in 1960 that companies began social science research, not a minute before they were convinced that this was an effective and relatively inexpensive way to raise production . . . (Gordon, 1979, p. 15).
>
> We must look at the human elements in productivity (Donahue, 1980, p. 14).
>
> European businessmen have been more active in applying the principles of behavioral science to productivity improvement than their U.S. counterparts (Birn, 1979, p. 40).

All of these statements, whether from labor leaders or management executives, express the philosophy that, in terms of the labor component, the best way to increase productivity should be to enlarge and enrich jobs and make them more challenging through the use of enlightened job design (Birn, 1979).

Other areas are also important for productivity improvement. Efforts are expanding into areas that are sometimes referred to as "the factory of the future." Many productivity improvement scenarios are being generated via CIM, Computer Integrated Manufacturing. The philosophies behind this approach include the following (*Modern Materials Handling*, March 1985, p. 60):

1. Ensure early involvement of manufacturing engineering in new product designs to keep products compatible with manufacturing
2. Set up modular manufacturing areas as cost centers
3. Make products for shipment without finished goods storage
4. Closely schedule and monitor manufacturing operations
5. Minimize work-in-process inventory
6. Maximize automation (not necessarily robotics)
7. Require zero defects in manufacturing and vendor parts supply
8. Minimize engineering changes
9. Streamline information handling
10. Train all workers to take on more responsibility.

CIM includes applications that monitor the following types of processes: those that use robots to achieve consistency; those that use cellular manufacturing to improve material flows; and those processes that utilize automated warehousing activities, such as storage and retrieval. The reported results of these implementations have been impressive. The following comments are representative of the successes achieved:

> "We saw a 50 percent increase in accuracy and a 20 percent decrease in labor costs" (Camp, 1985, p. 84).

> "A 56 percent reduction in job change and materials handling" (Dumolien and Santen, 1983, p. 76).

> "New equipment reduced testing time from 2 and one-half hours to 20 minutes" (Camp, 1985, p. 84).

Other areas that need and respond to this increasing attention are employee training and white collar work. According to *Modern Materials Handling* (March 1985, p. 84), "Nissan's first assembly plant in the U.S. blends extensive automation with multi-skilled workers trained to do many jobs and make their own decisions. Worker productivity is unusually high."

A ROLE FOR INDUSTRIAL ENGINEERING

> In order to improve productivity, the manager must first be able to measure it. He must also understand how the use of resources to perform different activities contributes to the overall productivity (Riccio, 1976, p. 97).

> Traditionally, industrial engineers have been involved in various efforts related to measuring manufacturing effectiveness (Hines, 1976, p. 45).

One of the most important roles the industrial engineering function can serve is to describe and measure the current work performed.

> Frederick Taylor, considered the father of scientific management, sought a fair day's work for a fair day's pay by establishing work standards via time studies. During the same period, Frank Gilbreth, and later Lillian Gilbreth, sought the most efficient technique to make measurements via motion studies. After these initial efforts to exercise control over production workers by work measurement developed through stopwatches and clipboards, emphasis shifted toward greater reliance on predetermined time standards, representative of time measured in prior studies of similar motions (Otis, 1980, p. 209).

Before any improvement can be made, it is essential that the present status be known. As improvements are suggested and evaluated, there must be a base for comparison. Without a reference point, there is no way to accurately ascertain whether or not proposed changes really cause an increase in productivity (Hamlin, 1978).

"Work standards, based upon work measurement, will permit management to produce and operate with more satisfactory results. Work measurement may be applied by management to determine how well its employees are performing, not only in production operations, but also in nonproduction, engineering, clerical, and administrative tasks'' (Otis, 1980, p. 293). An important part of the standards-setting procedure is the development of proper work methods. These methods are sometimes called the ''best'' method, although, realistically, proper work methods are better described as effective work methods. Some improvement of methods results from the elimination of obviously sloppy practices, while some come from real ingenuity. When work standards are set, they should reflect a realistic work method.

This case has not always been true. The legitimate work of Taylor and the Gilbreths was sometimes used by charlatans as strictly a way to decrease labor costs. ''Too often the expedient and simple solution to this was an arbitrary adjustment of the standard with an increase in production requirements. Because so many applications were associated with piecerate wage payments, this practice came to be called rate cutting, and timestudy fell into wide disrepute'' (Otis, 1980, p. 293).

However, much of this misuse has disappeared. Organized labor has accepted the concept of work measurement as a valid way to assure a ''fair day's work for a fair day's pay.''* When structured properly, the work standard program can be a most effective tool for measuring current productivity levels.

Irvin Otis of American Motors suggested the following guidelines to assure the successful operation of a standards or productivity measurement program (Otis, 1980). First, and most importantly, is the establishment of realistic work standards. These standards must reflect the work actually performed by the worker, as well as provide an allowance for the unexpected. The standards, regardless of the methods used, must be set by qualified analysts. The analysts must, according to Otis, be trained and must be able to set standards that are consistent from job to job.

Another of the important points Otis made is that work standards should not be changed without reason. Standards should be changed only when a work method changes. If standards are disputed by the worker or the workers' union these disputes must be resolved in a businesslike fashion. Otis maintained that the quicker these disputes about rates were resolved, the better it would be.

''The work standard is an aid to, not a substitute for, effective supervision'' (Otis, 1980, p. 294). Standards provide supervisors with a guide to effective performance. They should be an aid to supervision, not a replacement for it. ''Good supervision remains the only real answer to efficiency. The work standard will help the good supervisor to do a better job. But the poor supervisor nullifies its effectiveness'' (Otis, 1980, p. 294).

WHAT ABOUT JAPAN?

A few years ago, NBC Television produced a documentary entitled ''If Japan Can, Why Can't We?'' This television program addressed the productivity problem. As the statistics shown earlier in this chapter indicate, and as many consumers know, Japan produces, gen-

---

*Some unions, such as the UAW and the ILGWU, maintain their own IE departments.

erally, a higher quality product at a lower cost than does the U.S. An overview of productivity would not be complete without an examination of this point.

According to Malcolm Salter of the Harvard Business School, "They're (the Japanese) less inclined than Americans to want results now, and they aren't obsessed with quarterly earnings statements" (*Wall Street Journal,* October 10, 1978, p. 1). The Japanese are more concerned with other aspects of the organization's performance. For example, when the Japanese company, Matsushita Electronics, purchased Motorola's television assembly plant in Franklin Park, Illinois, a suburb of Chicago, the average television set produced had about 150 defects. Motorola was losing millions of dollars each year from its television operations. Now called Quasar, the plant is approaching breakeven and the defect rate has been reduced to 1.5 defects per set. This statistic compares favorably with the defect rate of Matsushita's Panasonic line produced in Japan, a rate of .5 defects per set (*Quality,* August 1980). What is responsible for this increase in productivity?

One of the first things the Japanese did was to clear out Quasar's top management. Then, after spending $108 million to purchase the company, Matsushita immediately spent an additional $15 million on capital improvements. New assembly procedures were implemented including the replacement of nonstop conveyor belts with belts the workers could stop individually. A worker who needed more time to complete a set was permitted to take that time.* Steps were taken to make the production lines more comfortable. Workers were made more responsible for their work areas. Automated equipment was installed to perform the most repetitive tasks.

In addition to making capital improvements, the management of the Quasar plant actively worked at maintaining more open communications with the workers. The company now meets semiannually with all employees to make a report on the state of the company. Twice each day, a 10-minute meeting is held between the line workers and foremen to discuss production, quality control, and other problems. According to one production worker, "We're made to feel a part of the company." The productivity growth at Quasar, since its acquisition by Matsushita, has been impressive. According to the *Wall Street Journal,* the productivity growth rate averaged 5 percent a year (October 10, 1978).

Although there is no magic formula, and what one company uses successfully may not work for another company, Joji Arai, manager of the Japan Productivity Center, outlined three steps that Japanese firms follow to achieve this type of result (*Quality,* June 1981, p. 17):

1. In the long run, improvement in productivity will increase employment. Before the full effects of improved productivity are apparent, the government and the people must minimize temporary frictions which disturb the national economy. They must cooperate to provide suitable measures, such as the transferring of surplus workers to areas where they are needed to prevent unemployment.

---

*The author's first industrial job was as an assembler at the Motorola plant. The changes instituted by Matsushita could only, in the author's opinion, have a positive influence on the productivity of the workforce and the quality of the product produced. See Chapter 12 for some additional comments about this facility.

2. In developing *concrete* measures to increase productivity, labor and management, conforming to the conditions existing in their respective enterprises, must cooperate in discussing, studying, and deliberating such measures.
3. The fruits of improved productivity must, in correspondence with the condition of the national economy, be distributed fairly among management, labor, and the consumer.

ASSESSMENT

What conclusions can be drawn from these discussions? First and foremost, American industry can maintain a productivity growth if:

- Top management really wants to.
- Top management supports this conviction with the willingness to make the required financial investment.
- Top management communicates the profit potential and rewards of the profit potential to its employees. (*Quality*, August 1980)

The last point implies that management recognizes that its employees are people. "Clearly the human factor is one of the most important aspects of increased productive output. The worker can no longer be assigned second string status if American management wants to get serious about licking its productivity . . . problems" (Otis, 1980, p. 210).

> Job design for motivation: the gradual transfer of responsibilities to the working man, and the creation of more challenging jobs. These are not just gimmicks of psychologists to make people happy. On the contrary, these are most effective tools for hard-boiled profit conscious management. Tools it should put to use to increase productivity . . . (Birn, 1979, p. 43)

Carl Blonkvist, a senior vice president of Booz, Allen, and Hamilton, speaking from his own experiences with Japanese workers, stated "The Japanese (managers) really believe that their assembly line workers are the best experts. We say that, but we really don't believe it" (Atlanta *Journal-Constitution*, January 17, 1981, p. 11A). According to George H. Kuper, the executive director of the National Center for Productivity and Quality of Working Life, the key to improving productivity lies in improving labor-management relations (Kuper, 1978). Lane Kirkland, President of the AFL-CIO, speaking at the 1982 American Productivity Congress, said, "So long as the quest for improved productivity is perceived as either a device to make workers toil harder and longer or simply a means for higher profit, then workers will resist" (Kirkland, 1980, p. 1216).

WHERE TO FROM HERE?

This textbook introduces the topics of work measurement and productivity improvement. The productivity problem has been established. As a result, this book will attempt to lay a foundation in these areas for the practitioner, whether experienced or a novice.

Specifically, the text will examine methods analysis or systematic problem-solving, work methods improvement, and work measurement procedures. However, as important as these traditional industrial engineering methodologies are, the preceding information stresses the importance of motivating workers to perform. The complete examination of productivity improvement must include a look at the traditional and not so traditional ways of motivating workers to perform at their most productive levels. This study includes people-oriented job design as well as the standard use of financial incentives and the not yet so standard use of quality of work life programs. This book will not be a definitive guide, but it will certainly lay the groundwork for the successful application of these principles.

C. Jackson Grayson, Jr., Chairman of the American Productivity Center, painted this picture: "The only way to get real growth in a country is through productivity improvement. I do forecast that if we start to do some of these things . . . we can improve . . . annual productivity growth . . . (to) at least a 3 percent per year increase" (*The Collegiate Forum,* Fall 1980, p. 8). Speaking for organized labor, Lane Kirkland stated,

> I would say "Amen" to what Jack Grayson said . . . productivity, by itself is not an end; it is the means to social, economic, and political freedom . . . work, and pride in work, still have value in America. That is why even rumors of good jobs draw lines of willing workers. As the press reported . . . 4500 people stood in line . . . for 550 jobs at a new downtown Milwaukee hotel. At the same time, it is not surprising that many workers are not interested in, or challenged by, dead-end, boring, repetitive jobs (Kirkland, 1982, p. 1219).

CHAPTER 2

Methods Analysis

OBJECTIVES

After completing this chapter, the reader should be able to do the following:

- Be able to approach any problem in a logical, organized fashion
- Understand the principles of the scientific method
- Understand the relationship between methods analysis and problem solving

INTRODUCTION

According to Charles E. Geisel of the Container Corporation of America and Gerald Nadler of the University of Wisconsin, "A strange phenomenon is overtaking the American organization. We keep adding sophisticated techniques . . . yet the rate of improving productivity . . . is going down!" (Geisel and Nadler, 1978, p. 3). The logical question to ask, after reading this statement, is WHY? Geisel and Nadler have an answer:

> It seems to us that each technique produces specialist professionals who seek status through a separate department in an organization. . . . Such diffuse groups in an organization often fail to relate to or have direct impact where the greatest good is needed . . . IEs must know about and make available all new techniques and concepts, but use them in an integrated fashion only as the organization's needs are served (Geisel and Nadler, 1978, p. 4).

One of the potential strengths of any industrial engineering organization is its ability to identify these needs and help direct the company's resources toward solving these problems. As James L. Hays, President of the American Management Associations, has said, "Much of a manager's success depends on his or her ability to identify the causes of problems and develop workable solutions for resolving them" (Hayes, 1981, p. 71).

In a nutshell, what Hayes said is what will be referred to as *methods analysis*. Although much of the information in the rest of this book describes Geisel and Nadler's "techniques," this chapter will emphasize the most important function of developing a systematic approach to solving problems. This procedure is applicable to just about every situation—from productivity improvement to work design to quality improvement.

THE SCIENTIFIC METHOD

One of the most fundamental ways to approach problems in an organized fashion is the scientific method. This method is a structured procedure to use to help solve problems. It consists of a number of steps that, if followed, should result in a better way of performing some task.

Define the Problem

The first step in the scientific method is to define the problem. This definition or formulation of the problem should be as broad as possible. ". . . detail, restrictions, and the present solution to the problem [should] be avoided during this period" (Krick, 1962, p. 22). Care must be taken during this stage of the process to identify the problem and not a symptom of the problem. "It is the most difficult part of the process because the central problem rarely stands out clearly. It is usually surrounded by murkiness resulting from numerous other secondary problems, symptoms, assumptions, long held beliefs, and so forth" (Cohen, 1980, p. 179).

Consider the case of an organization that is continually losing employees and experiencing a high rate of turnover. This problem has led to an inexperienced workforce, large training costs, and poor customer relations. It would be extremely simple to look at this situation and assume that the organization's major problem was an inexperienced workforce caused by the high turnover rate. A closer examination might reveal that the real problem was payment of low salaries. The low pay caused the high turnover which led to the other problems. The pay was the problem and everything else was a symptom of the problem (Cohen, 1980). The time spent on problem definition is just as important, if not more so, than any other step of the scientific method.

Gather Information

After the problem has been defined, the second step in the process is to gather information. Sometimes called analysis, the information-gathering step has three major components. First, all the known and available material about the problem should be written down. These are the facts. Second, any assumptions that must be made about the problem should be noted. "Omitting a single important assumption can change the nature of your central problem completely, resulting in entirely different conclusions and different actions to be taken to solve the problem" (Cohen, 1980, p. 180).

The third component of the information-gathering process is the development of the limitations or constraints that are to be placed on the solution. There are two types of constraints. The first are the real constraints. "They have been imposed, they are quite genuine, there is no assumption or imagination involved" (Krick, 1962, p. 37). The second constraint is the fictitious constraint. "They are unnecessary and probably unwittingly assumed by the designer" (Krick, 1962, p. 37). This second type of constraint is often based on instinct, with the designer thinking something "just ought to be that way." Sometimes, the fictitious constraint results from the current solution. Just because a certain method is presently being employed does not necessarily mean that this method is justified.

Develop Alternatives

Once the problem has been identified and the analysis performed, the third step in this organized problem-solving methodology is to develop alternatives to solve the problem. At this stage, it is advantageous to generate as many different alternatives as possible. These alternatives should be broadly described and no attempt should be made to evaluate them. The objective at this point in the procedure is to maximize the problem solver's creativity. Five determinants of the designer's ability to be creative have been identified (Krick, 1962). These criteria are: the designer's knowledge of the situation or process, the effort the designer is willing to exert, the natural ability of the designer to be inventive, the method used to generate alternatives, and, finally, chance.

A problem solver should know as much as possible about the problem before attempting to generate alternatives. After all, this step is the second task in the systematic approach to solving a problem. If the designer does not know everything available about the problem, then the designer should not attempt to generate solutions. Effort in this regard cannot be measured, but a reasonable assumption would be that the designer wants to solve the problem, or no effort at all would be devoted to it. Natural ability or aptitude is perhaps the least significant. Again, it is reasonable to assume that a person with no aptitude (or interest) probably would not be involved anyway. The methods used to generate alternatives are varied, but one popular way to generate many alternatives is to use the brainstorming technique.

"Brainstorming is . . . the uninhibited . . . approach to idea getting. It requires green light thinking, makes maximum use of free association as ideas ricochet from one person to another" (*Factory,* May 1956, p. 99). To maximize brainstorming, the designer or leader of the brainstorming session must be able to "state the problem in basic terms, with only one focal point; not find fault with . . . any idea; reach for any kind of idea; and provide the support and encouragement necessary to free restrictive attitudes" (Von Fange, 1959, p. 47). Brainstorming, when handled properly, works. *The New York Times* reported as follows: "An East Coast Electronics Company that regularly brainstorms at every production impasse has cut production time 37 percent on its last six problems . . ." (cited in Osborn, 1963, p. 188).

Brainstorming is not the only technique that promotes creativity. Some others include work simplification, job methods training, suggestion systems, buzz sessions, and even problem-solving conferences. Regardless of the name, the goal is the same—to produce as many alternatives as possible.

Evaluate the Alternatives

The fourth step in the problem solving procedure is to evaluate the alternatives. "In this step you will compare the advantages and disadvantages of your alternative strategies for solving the central problem'' (Cohen, 1980, p. 180). The evaluation involves a narrowing of the alternatives followed by the specification of greater details about the remaining possibilities. This weeding out procedure continues with each succeeding list of survivors being developed in more and more detail. Eventually, a very limited number of alternatives remains. At this point, a final evaluation is made and the ''best'' solution, in light of the original problem definition, is specified. A variety of analysis techniques may be employed to help with this screening procedure. Depending on the nature of the problem, the analysis may focus on cost, timeliness, or any other criteria deemed important by the problem solver. Regardless of the focus of the evaluation, the result will still specify the ''best'' solution.

Implement the Solution

After the ''best'' solution is identified, it is time to implement it. There are two subparts to implementation. The first and most important of these is selling the solution. When the solution to the problem will result in a change from the present method, the proposal will probably run into resistance. This phenomenon, known as *resistance to change,* is quite common. Ideally, every individual in an organization desires the organization to progress. A change that will improve the way the organization functions should mean an improvement for every individual within that organization. However, in reality, people want the organization to succeed as long as it does not conflict with their personal objectives.* ''Because improving productivity always affects what people in an organization do, they should always be involved. People working in the problem area always have a vast amount of knowledge/information/creative ideas at their fingertips, much more than could ever be recorded by any techniques'' (Geisel and Nadler, 1978, p. 6).

The second subpart of implementation is the physical installation of the solution to the problem. This technical portion is easy: A ''high percentage of proposals that are accepted and installed . . . prove notably unsatisfactory, not for technical reasons, but because the people affected by the new methods have resisted and responded in such a way as to render these proposals failures'' (Krick, 1962, p. 67).

Follow Up

The final step in the scientific method of problem solving or methods analysis is the *follow up.* Simply stated, unless the solution is monitored and the results documented, there is no way to judge the effectiveness of the solution. Without follow up it is even possible that the solution, so diligently arrived at, might even be completely ignored. Follow up also provides an opportunity for the analyst to determine that no additional problems have developed, either from the installation of the new solution or from some other change made in another part of the operation.

To summarize, the scientific method described is a systematic way of solving a problem. It methodically defines and analyzes any problems. The six steps are:

*Chapter 4 has a more complete discussion of resistance to change and possible ways to overcome it.

1. Define the problem.
2. Gather information.
3. Generate alternatives.
4. Evaluate alternatives/specify a solution.
5. Implement the solution.
6. Follow up.

At times, it would be beneficial to recycle the entire process. Conditions may change, new technology may develop, or a new procedure may be learned. Thus, any change could lead to the development of an even better solution.

This chapter will conclude with two examples of applications of this logical approach to solving a problem. As the examples will show, every problem solver does not call the steps by the same names, but the steps are nonetheless present.

Example: Ergonomics—Postural Fatigue

In the April 1976 edition of *Industrial Engineering* magazine, Khalil and Ayoub described a "basic, step-by-step job analysis [which] demonstrates how ergonomics* principles can be used to reduce the safety hazards found in a common workplace" (p. 26). Within the article, the authors demonstrated how ergonomic principles were applied to a textile operation. The operation under study was a multioperation panty hose operation. With this operation, the product could be moved from one operation to the next automatically. "The process consists of loading a pair of leg blanks on clamps. The machine rotates the clamps to a scissor station where the leg blanks are cut half way into the panel block. A sewing operation follows to sew the two leg blanks together . . . the only manual operations required during this cycle are the loading and alignment of the leg blanks in the clamps" (Khalil and Ayoub, 1976, p. 26).

The manual operation involved in this example can cause severe problems. Specifically, the problem that will be examined here is postural fatigue. This fatigue is caused by two factors: the continuous standing of the operator and the twisting and bending the operator must do while loading and aligning the panty hose. This fatigue can lead to any or all of the following problems: varicose veins, circulatory problems, back injury, low back pain, tripping, slips, falls, or shoulder pain (Khalil and Ayoub, 1976). The problem is defined as *postural fatigue*. Some possible solutions include modification of equipment, redesign of the workplace, better selection and placement of employees, modification of the task, and modification of the product. An evaluation of each of these alternatives follows:**

Modification of Machine—Apparently not possible; any modification or engineering changes are required to be handled by the manufacturer—a condition that certainly excludes this as a potential solution.

*Ergonomics, or human factors engineering, will be discussed in Chapter 10.

**Reprinted with permission from INDUSTRIAL ENGINEERING magazine, April, 1976. Copyright © American Institute of Industrial Engineers, 25 Technology Park/Atlanta, Norcross, Georgia 30092.

Workplace Layout—Some rearrangements of some or all of the components of the workplace seem possible; however, the important component (the machine) is almost fixed. Thus, a limited success can be expected from modifying the physical layout.

Modification of Task—A well-balanced work/rest schedule may prove to be useful in eliminating some of the physical and fatigue problems. In addition, dividing the task in some basic steps may increase production rate and, at the same time, improve working conditions.

Selection and Placement of Employees—By assigning operators with selected anthropometric characteristics (e.g., height) to the task, some unnecessary motions may be eliminated. In other words, careful selection practices in hiring might be employed to ultimately protect the worker against physical stress and discomfort of the present task (fit the job to the person).

Product Modification—Modify the product (blanks) in such a way that loading can be performed with minimum number of arm (hand) motions. In addition, kinesthetic rather than visual feedback may be employed in checking alignments of blanks.

After analysis of the potential solutions, the optimum seems to be "implementation of a work/rest schedule. Operators should be allowed to rest without leaving the work station . . . Frequent but short rest periods are recommended" (Khalil and Ayoub, 1976, p. 30).

Example: The Yellow Fever Experiment

Joel E. Ross (1981) described a classic case of problem analysis in his book, *Productivity, People, and Profits**

The Spanish-American War of 1898 ended swiftly and with remarkably few American casualties. However, by the year 1900, thousands of soldiers were dying of yellow fever. The job was to find the cause of the fever and eradicate it.

A two-pronged approach was taken to the problem. It was assumed that the disease was communicated either by person-to-person contact or through general unsanitary conditions. First, a program of microscopic research was undertaken. Autopsies were conducted on hundreds of victims. Blood, flesh, and the organs of the bodies were scrutinized microscopically in search of a clue. Clothing, linens, and personal belongings were analyzed with care in hopes of uncovering the cause of "yellow jack." All efforts were in vain.

A second approach involved a massive effort to clean up the dirt and filth of Cuba. This effort resulted in the cleanest and most sanitary Havana in history, but it did not affect the climbing death rate from yellow fever. The massive fatalities increased.

Later that year Major Walter Reed, in whose name the famous Walter Reed Hospital of the U.S. Army was later dedicated, was placed in charge of "the cause and prevention of yellow

*Copyright © 1981. Reston Publishing Company. Reston, Virginia. Reprinted with permission.

fever.'' He reasoned: ''Perhaps we can't find out what causes 'yellow jack,' but maybe we can find out how it is spread.'' The problem was redefined. In other words, if they could keep people from catching yellow fever they would be successful even though they did not know what the basic cause was.

Reed set out to test the popular belief that the disease was spread through contact with other victims. As time passed he became increasingly frustrated because the evidence proved this hypothesis wrong. First, there were nurses who tended the yellow fever victims but were no more suspectible to the disease than any other group. Second, the pattern of disease was entirely random; it was extremely rare for all members of a family to be struck at the same time. Finally, his tentative cause-and-effect relationship was disproven when some of the totally isolated prisoners in the guardhouse contracted the disease, despite the fact that there had been not the slightest contact with outsiders.

One Dr. Carlos Finlay of Havana, considered by many to be a crank, persisted in his theory that yellow fever was carried by mosquitoes. Despite his skepticism, Reed decided to test this possible cause of the problem. A number of volunteers, including two medical doctors, allowed themselves to be bitten by mosquitoes that had previously bitten many yellow fever victims. The results, including the death of one of the doctor volunteers, indicated there might be something in the mosquito theory. Certainly a cause-and-effect relationship had been established, but the cause needed further testing and verification. So began the famous experiment.

Two houses were built. The first was very sanitary and was antiseptically clean. It had double-screened windows and doors that absolutely ruled out the entry of any mosquito. The volunteers in the house, each of whom had been bitten by mosquitoes that had fed on yellow fever victims, lived in spotless comfort, eating carefully prepared meals and having no contact with the outside world. In the second house lived another group of volunteers, none of whom had been bitten by mosquitoes. No mosquito was allowed entrance into this house but the volunteers lived in squalor and filth. They ate from dishes and slept in beds that were contaminated with the filth from the yellow fever wards.

The results of the experiment are now world-famous. The men living in the spotless antiseptic house and who had been bitten by the mosquitoes came down with yellow fever. Those who had lived for twenty days in the squalor and filth of diseased victims, although very uncomfortable, were still healthy.

By the process of problem analysis, Dr. Walter Reed arrived at the single change that produced the effect. Moreover, he tested and verified the cause of the change. He still didn't know the cause of yellow fever but he knew the cause of its transmission. The solution then became an engineering problem of eliminating the breeding places of the Aedes Aegypti mosquito.

SUMMARY

Methods analysis is problem solving. Organized problem solving is more effective than the seat of the pants approach. As the saying goes, ''Sometimes it is difficult to remember that your objective was to drain the swamp when you are up to your rear end in alligators.'' It is sometimes easy to forget that the role of the methods analyst, work designer, or industrial engineer/technologist is really to solve problems that relate to increasing the productivity of the organization to which he or she belongs.

The remainder of this textbook concerns itself with the techniques used in analysis and measurement. It will be very easy to think that these techniques are the objectives themselves rather than means to an end. This message will serve as the final caution to the reader to remember where we are heading.

PART II

Work Analysis

Two important parts of methods analysis are describing the present method and developing alternatives to the present procedure that might improve productivity. *Graphical Productivity Analysis* is the use of charting techniques to describe the activities performed to produce a product or a service, either as presently performed or as they might be performed. Standard charts are described in Chapter 3.

The *operation process chart,* "permits one to obtain a quick overall knowledge of a process" (Lowry, Maynard, and Stegemerten, pg. 33). The *flow process chart,* "analyzes activities more completely than do operation process charts" (*Production Handbook,* p. 11.13). *Flow diagrams* follow the path taken by people or materials while an operation is performed. The *left-hand/right-hand chart* shows, "a detailed breakdown of the operation" (Barnes, p. 99). The *multiple activity chart* shows activity time and idle time and their relationship to each other for several activities.

After the current method of production is described, alternatives must be specified. Chapter 4 describes traditional methods for improving procedures. Whether Therblig analysis is used, or principles of motion economy are described, the objective is the same—to develop a more productive method.

CHAPTER 3

Graphical Productivity Analysis

OBJECTIVES

After completing this chapter, the reader should be able to

- Construct an operation process chart for an existing operation.
- Construct a flow process chart for an existing operation.
- Construct a flow diagram for an existing operation.
- Construct a left-hand/right-hand chart for an existing operation.
- Construct each of the above charts and/or diagrams for a proposed operation.
- Use each of these graphical tools to improve the productivity of the operation charted.

INTRODUCTION

Productivity improvement is important—there is no argument or doubt about that. The industrial engineering analyst, whether a member of an IE department or a worker or supervisor trained in the procedures, has available a set of analytical/graphical tools that help identify portions of work tasks that are eligible or likely candidates for improvement studies. Traditional charting studies, coupled with a very inquisitive attitude, make this type of analysis a valuable tool in the productivity improvement process.

Before looking at the specifics of these traditional procedures, let us discuss the "questioning" attitude. When examining the way a process is completed—any process— a number of questions should enter the analyst's mind. These questions include

WHY?   WHEN?   WHERE?   HOW?   WHO?

The most important of these questions is "why?" Many unnecessary operations are performed when products are produced. Often, there is no justifiable reason for performing an operation other than the very lame excuses, "it seemed right at the time," or "we have always done it this way."

GRAPHICAL ANALYSIS TOOLS

The IE analyst has a number of graphical analysis tools to help with the analytical questions. By using these tools, the analyst can clearly see the answers to these questions by literally drawing a picture of the work being performed or the work being contemplated. This chapter will illustrate the most frequently used of these methods. All of the methods are designed to help the analyst find a better or more productive way of accomplishing the required work, assuming, of course, that the work is really required. All are based on the concept that before work can be made more productive, the requirements must be completely described. Often, just the process of putting a description down on paper will force us to examine a job closely enough to identify some areas that demand immediate improvement.

According to the *Production Handbook,* these graphical techniques are all known as process charts. While certain specific types of charts will be defined on the following pages,* the *Production Handbook*

> recommends that the detailed type of chart should not be followed rigidly since they are all fundamentally process charts and the same theory and method of analysis applies to all. If each chart is made with the definite object of visualizing a process as a means of improving it, it will be built up so that the controlling factors of the work are brought into prominance by the form of the chart. (*Production Handbook,* p. 11.1)

The examination process is often referred to as operation analysis. According to the official Industrial Engineering Terminology ANSI Z94, operation analysis is defined as

> a study of an operation or series of operations involving people, equipment, and the processes for the purpose of investigating the effectiveness of specific operations or groups of operations so that improvements can be developed which will raise productivity, reduce costs, improve quality, reduce accident hazards, and attain other objectives.

Operation Process Chart

The first of the formally defined charts to be examined is the operation process chart. This chart is defined as (ANSI Z94)

> A graphic, symbolic representation of the act of producing a product or providing a service, showing operations and inspections performed or to be performed with their sequential relationships and materials used. Operation and inspection time required and location may be included.

The operation process chart shows all types of production operation. Whether describing a relatively standard production operation such as building a wooden footstool or a non-standard, unconventional operation such as performing heart surgery, a process can be represented with an operation process chart.

The operation process chart uses two symbols:

*The definitions of the charts are taken from ANSI Z94 for *Industrial Engineering Terminology* published by the Institute of Industrial Engineers Industrial Engineering and Management Press, Norcross, Georgia.

An *operation* is designated by a circle. An operation is a subdivision of a process that changes or modifies a part, material, or product and is done essentially at one workplace location (ANSI Z94).

An *inspection* is designated by a square. Inspection is the comparison of observed quality or quantity of product with a quality or quantity standard.

Some general conventions are normally followed regarding the construction of operation process charts:

- Horizontal lines show where a material enters a process.
- Vertical lines show the sequential steps required in processing.
- Purchased material is shown, via a horizontal line, at the point at which it enters the process.
- Flow is from left to right and from top to bottom.

Example: Raincoat Liners

A small apparel company subcontracts the manufacture of men's raincoat liners. The process used, from the time the raw material arrives in the plant until the finished raincoat liners are shipped, is described in the following narrative.

Raw material in the form of rolls of cloth, spools of thread, boxes of zippers, and ribbons of binding material arrives at the back door. These materials are unloaded and stored until needed at the appropriate manufacturing location. The rolls of cloth are first to be processed. The raw material is cut to size in the cutting room. Usually, two dozen liners of a particular size are cut at one time. Each liner consists of three major pieces. One of these is called the *body* and it will serve as the base to which all other parts are attached. The other two parts that are cut are the arms. These are identical and two are required for each assembly.

Once the pieces are cut, the body is taken to the first sewing operation and the arms are taken to their respective assembly locations. At the first sewing station, the binding material is sewn to the body. This material covers the ragged edge left by the cutting. While this procedure is being done, the sleeves are formed from a flat piece of cloth into what appears, loosely speaking, to be a tubular shape. The sleeves are sewn lengthwise with special care being given to make sure that the ends are tucked under so that they look "good."

After the edging material has been attached to the body, the zipper is then affixed. This task is performed at a special sewing machine that uses a jig and fixture designed for this purpose. The operator must check to make sure that the zipper will function—after all, it is a removable raincoat liner. Once the zipper is attached, the arms are sewn to the liner body. After the arms are securely fastened, the garments are cleaned up; that is, loose threads are clipped and the garment is hung up. Finally, the liners are inspected one last time. This company must deliver all liners specified in its contract, but will only be paid

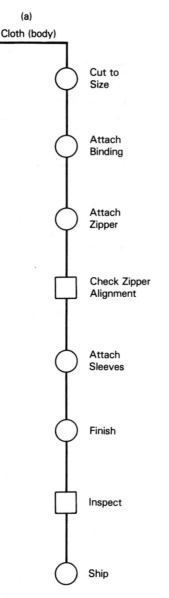

(a)

Cloth (body)

Cut to Size

Attach Binding

Attach Zipper

Check Zipper Alignment

Attach Sleeves

Finish

Inspect

Ship

Figure 3-1(a) Operation Process Chart for Raincoat Liner-Body.

for acceptable liners. It must repair any defectives at its own cost without receiving payment for them.

     Figure 3-1 shows the operation process charts for this relatively simple operation. Figure 3-1(a) shows the operations on the base, 3-1(b) shows the operation involved when sleeves are attached, 3-1(c) shows the addition of the zipper, and 3-1(d) shows the addition of the binder as well as illustrating the entire operation process chart for the raincoat liner.

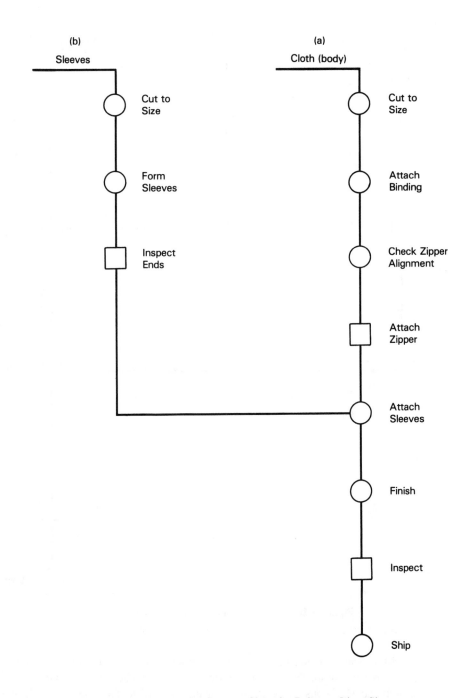

(b)
Sleeves

Cut to Size

Form Sleeves

Inspect Ends

(a)
Cloth (body)

Cut to Size

Attach Binding

Check Zipper Alignment

Attach Zipper

Attach Sleeves

Finish

Inspect

Ship

Figure 3-1(b) Operation Process Chart for Raincoat Liner-Sleeves.

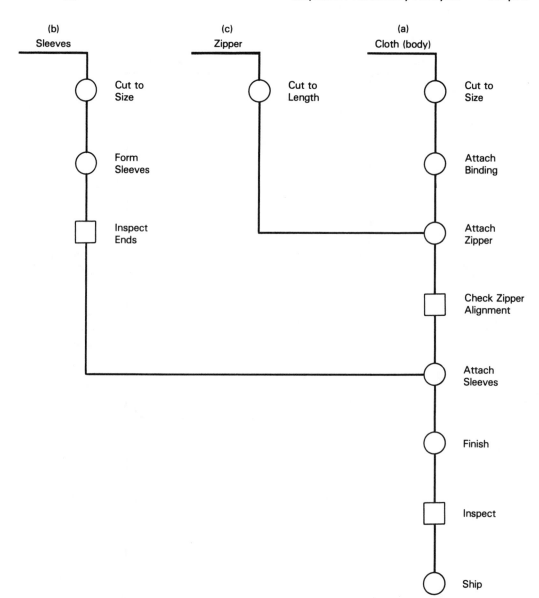

Figure 3-1(c)   Operation Process Chart for Raincoat Liner-Zipper.

More sophisticated products require more sophisticated operation process charts. The following example uses a hypothetical set of manufacturing instruction sheets from the mythical Mr. Mouse Mousetrap Company (MMM) to show more complicated operation process charts.

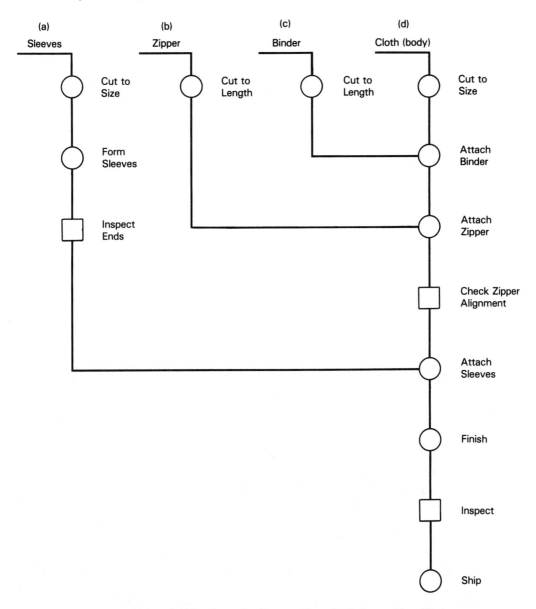

Figure 3-1(d)   Operation Process Chart for Raincoat Liner-Binder.

Example: Mousetrap

The Mr. Mouse Mousetrap Company produces a standard cheese and spring activated mousetrap. This mousetrap is produced by an anything but traditional manufacturing

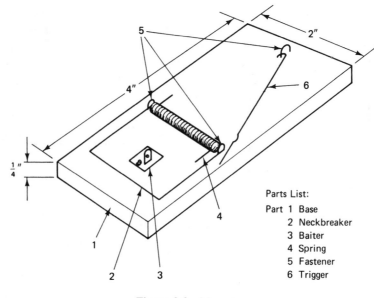

Figure 3-2   Mousetrap.

Parts List:
Part 1 Base
     2 Neckbreaker
     3 Baiter
     4 Spring
     5 Fastener
     6 Trigger

process. As Figures 3-3(a) through (e) show, the manufacturing process uses some unique steps. Figure 3-2 shows a sketch of the mousetrap produced by MMM and Figures 3-3(a) through (d) are summaries of the manufacturing instructions MMM uses to produce the component parts for the mousetrap.

Examination of these charts indicates that the major focus of the assembly is the base. All the other parts are eventually, either individually or as subassemblies, attached to the base. Figure 3-4(a) shows the operations required to make the base. Because the base is the "heart" of the trap, it will appear on the right-hand side of the chart. Figure 3-4(b) shows the addition of the trigger and the staples to the base. Part (c) of the figure shows the addition of the baiter. Figure 3-4(d) adds the neckbreaker spring subassembly. Finally, part (e) of this figure adds the final operations required before MMM's trap is ready for sale to the customer.

Note that the chart shows how all operations feed into the final assembly. The logical flow pattern has been identified. Now the analyst must question the procedure. Had crossed lines been evident, or a "spaghetti" effect been seen, it would have required considerable questioning. It might have been possible or even desirable to market do-it-yourself mousetrap assembly kits. However, someone might have designed a way to build the mousetrap better, so the MMM Company appears to be justified in providing an assembled product to its customers. The overall view of the operations also seems to indicate, as shown by the operation process chart, that MMM has things well in hand.

| PART NAME: Mousetrap Base | | | PART NUMBER: 140 | |
| MATERIAL: Wood Stock | | | PAGE 1 OF 1 | |

Step	Operation	Equipment	Special Tooling	Notes
1	Cut to Length	Saw		
~	Sand Ends	Sander		
3	Affix Trigger	Stapler A		Use 1 staple — Part 154
4	Affix Baiter	Stapler B		Use 1 staple — Part 154
5	Affix Subassembly	Stapler C		Use 2 staples — Part 154
6	Test Activate	Rubber Mouse		
7	Paint	Rubber Stamp		
8	Pack for Final Assembly			
9	To Stock			

SPECIAL INFORMATION:

Figure 3-3(a) Manufacturing Instructions for Base.

Flow Process Chart

A second process chart that is valuable in describing a process is known as the flow process chart. This chart is defined as (ANSI Z94)

> A graphic, symbolic representation of the work performed or to be performed on a product as it passes through some or all of the stages of a process.

PART NAME:	Staple		PART NUMBER: 154	
MATERIAL:	Copper Wire, $3\frac{1}{2}$ Diameter		PAGE 1 OF 1	

Step	Operation	Equipment	Special Tooling	Notes
1	Cut to $\frac{1}{2}$ inch length	Wire Cutter		
2	Bend to Shape	Wire Bender A		
3	To Assembly			

SPECIAL INFORMATION:

Figure 3-3(b)   Manufacturing Instructions for Staple.

In addition to the operation and inspection symbols defined for the operation process chart, the flow process chart uses the following ASME-defined symbols (ANSI Z94):

 Storage is shown with a triangle. It is defined as keeping a product, material, or part protected against unauthorized removal.

| PART NAME: Trigger | | | PART NUMBER: 156 | |
| MATERIAL: Copper Wire $\frac{1}{32}$ Diameter | | | PAGE 1 OF 1 | |

Step	Operation	Equipment	Special Tooling	Notes
1	Cut to 2 inch Length	Wire Cutter		
2	Bend Neck	Wire Bender B		
3	Bend Circle on End	Wire Bender C		
4	To Assembly			

SPECIAL INFORMATION:

Figure 3-3(c)   Manfacturing Instructions for Trigger.

Delay is designated with the symbol shown at the left. A delay is an event that occurs when an object or person waits for the next planned action.

The arrow symbol shows *transportations*. This symbol represents a change in location of a person, part, material, or product from one workplace to another.

	PART NAME: Neck Breaker MATERIAL: Copper Wire, $\frac{1}{32}$ Diameter			PART NUMBER: 157 PAGE 1 OF 1
Step	Operation	Equipment	Special Tooling	Notes
1	Cut to 7 inch Length	Wire Cutter		
2	Bend 1	Wire Bender Number 1		
3	Bend 2	Wire Bender Number 2		
4	Assemble with Spring			Part Number 160
5	Bend 3	Wire Bender Number 3		
6	To Assembly			

SPECIAL INFORMATION:

Figure 3-3(d) Manufacturing Instructions for Neckbreaker.

 The overlapping symbols, as illustrated by the operation and inspection symbols at the left, indicate that these activities are performed simultaneously.

The flow process chart can be as broad or detailed as the analyst desires. It can show an entire process or one step in a process. When flow process charts are used, special care must be exercised to insure that the chart accurately reflects either the work an individual

| | PART NAME: Spring | PART NUMBER: 160 | | |
| | MATERIAL: Copper Wire, $\frac{1}{32}$ Diameter | PAGE 1 OF 1 | | |

Step	Operation	Equipment	Special Tooling	Notes
1	Purchase			Purchase Reg. 2703-11
2	Assemble to Neckbreaker			Part 157

SPECIAL INFORMATION:

Figure 3-3(e)   Manufacturing Instructions for Spring.

performs or the work performed on a product. Sometimes flow process charts are called flowcharts, production process charts, or product analysis charts. After the chart is constructed, activities can be carefully examined and questioned. Often a second chart, showing an improved method, can be constructed and the improvements summarized.

Example—Raincoat Liner

Referring to the raincoat liner described in the operation process chart example, it is possible to construct a flow process chart for manufacturing the liner. Figure 3-5 shows the

	PART NAME: Baiter MATERIAL: Sheet Steel $\frac{1}{2}$" X $\frac{1}{16}$"			PART NUMBER: PAGE 1 OF 1
Step	Operation	Equipment	Special Tooling	Notes
1	Stamp to Size and Shape	Baiter Stamper	4A16 Die	
2	Apply Protective Coating	Tub		Must sit no more than 5 min.
3	To Assembly			
SPECIAL INFORMATION:				

Figure 3-3(f)   Manufacturing Instructions for Baiter.

flow process chart for the liner itself. An examination of the process may lead to questions such as,

- Why is the material for the sleeves stored before it is cut?
- Why is the material for the zippers stored before it is cut?
- Why is the material for the binder stored before it is cut?

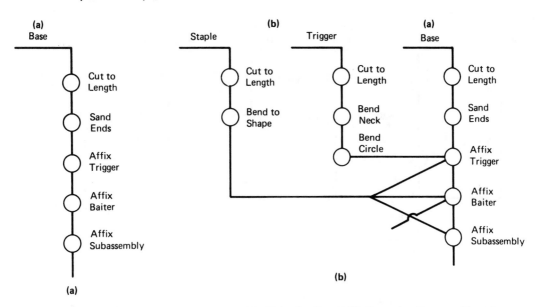

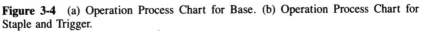

Figure 3-4 (a) Operation Process Chart for Base. (b) Operation Process Chart for Staple and Trigger.

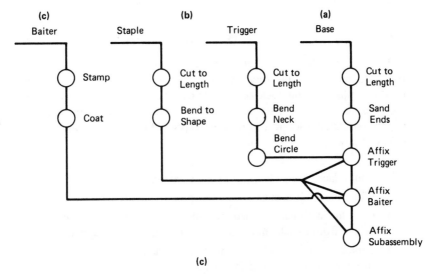

Figure 3-4(c) Operation Process Chart for Baiter.

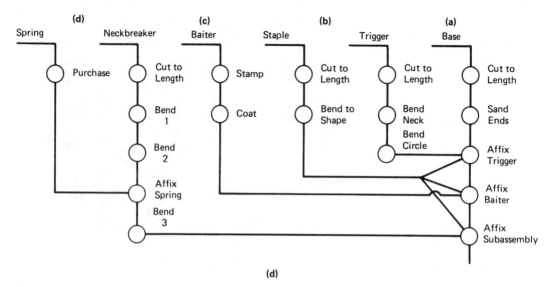

(d)

Figure 3-4(d)   Operation Process Chart for Neckbreacker/Spring Subassembly.

- Why is the material for the body stored before it is cut?
- Why is the zipper alignment checked after it is attached?

The answers to these and other questions should result in an improved method.

Example—Fast Food Preparation

A Mexican-style fast food chain wants to examine the process by which it produces burritos. The process begins after the order for a burrito is received and concludes when the burrito is placed in the sack next to the cash register. The analyst, knowing that the fast food business is really a service business, wants to see if an improved method might be possible so that the service, or really the speed of the service, can be improved. The steps involved in preparing a burrito are

1. Cook places the flour shell in the steamer.
2. While the shell is steaming, cook places a paper wrapper on the counter.
3. After the shell has finished steaming, as denoted by a ringing bell, the shell is removed from the steamer and placed on the paper wrapper.
4. Cook takes a meat scoop from the rack and ladles one scoop of meat onto the shell and returns the scoop to the rack.
5. Cook checks order to see if the burrito will also get a scoop of cheese.
6. If cheese is to be added, cook takes a cheese scoop from the rack and ladles one scoop of cheese onto the meat and then returns the scoop to the rack.

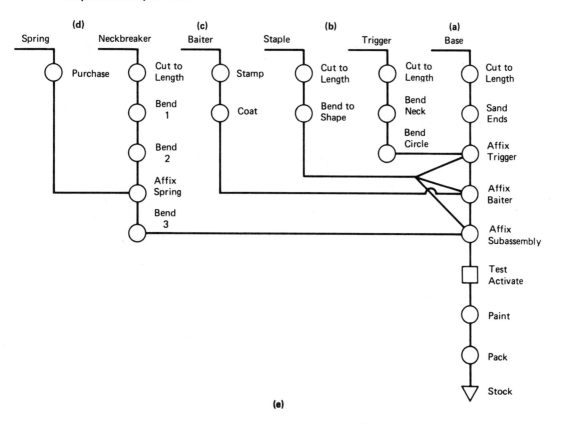

(e)

Figure 3-4(e)   Operation Process Chart for Mousetrap.

7. Cook moves the burrito to the condiment area where sour cream is added, via the dispenser, to the burrito.
8. Cook folds burrito into its normal shape.
9. Cook wraps burrito in its paper wrapper.
10. Cook carries burrito to the order assembly area.
11. Cook finds the order the burrito belongs with and places the burrito into the correct sack.
12. Cook checks the sack to see if the order is complete.
13. Complete orders are carried to the cash register.

As the initial step to improving this existing process, the analyst constructed the flow process chart shown in Figure 3-6. The analyst could have charted either the burrito or the cook. Either chart would have provided useful information. The analyst chose to chart the cook. The analyst also used a pre-printed form. These forms are useful when a charted process is performed at one work station and the operator is being charted. The

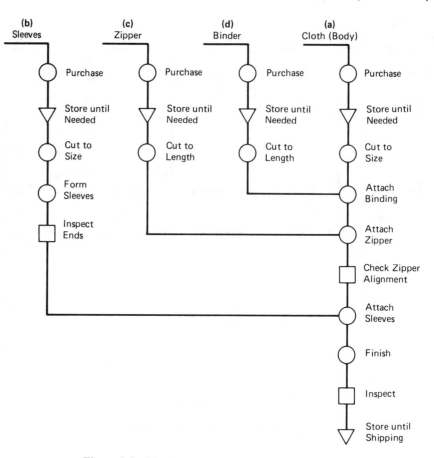

Figure 3-5   Flow Process Chart for Raincoat Liner.

forms can also be used when a product is being charted as it moves from work station to work station. When pre-printed forms are used it is essential that all supporting information on the forms be completed.

After examining the flow process chart some questions came to the analyst's mind:

- Why can't the shells be pre-steamed?
- Why must the cook check for cheese?
- If the cook must check for cheese, why can't it be done when the burrito is steaming?
- Why isn't the sour cream closer to the meat and cheese?
- Why can't the wrapping take place immediately after the sour cream is applied?
- Why are orders assembled away from the wrapping?

FLOW PROCESS CHART

PRESENT METHOD ☒
PROPOSED METHOD ☐

PART NAME: _Burrito_

PROCESS DESCRIPTION: _Cooking_

DEPARTMENT: _Kitchen_

PLANT: _Route 7_

ANALYST: _PLK_

DATE: _12_ / _20_ / _1_____   NOTE: SKETCHES ARE LOCATED ON BACK

SUMMARY		NO.
○	OPERATIONS	10
⇨	TRANSPORTATIONS	3
☐	INSPECTIONS	3
D	DELAYS	1
▽	STORAGES	
	TOTAL STEPS	17
	DISTANCE TRAVLED	15'

STEP	OPERATION / TRANSPORT / INSPECT / DELAY / STORAGE	DESCRIPTION	Distance
1	●⇨☐D▽	Place Shell in Steamer	
2	●⇨☐D▽	Place Wrapper on Counter	
3	○⇨☐■▽	Wait for Shell to Finish Steaming	
4	●⇨☐D▽	Remove Shell from Steamer	
5	●⇨☐D▽	Place Shell on Wrapper	
6	●⇨☐D▽	Scoop Meat onto Shell	
7	○⇨■D▽	Check Order for Cheese	
8	●⇨☐D▽	Scoop Cheese onto Meat	not every order
9	○⬛☐D▽	Carry Burrito/Wrapper to Condiment Area	3'
10	●⇨☐D▽	Add Sour Cream to Burrito	
11	●⇨☐D▽	Fold Burrito	
12	●⇨☐D▽	Wrap Burrito in Wrapping Paper	
13	○⬛☐D▽	Carry Burrito to Order Assembly Area	4'
14	○⇨■D▽	Locate Order	
15	●⇨☐D▽	Place Burrito with Correct Order	
16	○⇨■D▽	Check for Completeness of Order	
17	○⬛☐D▽	Order Taken to Cash Register	8'
	○⇨☐D▽		

Figure 3-6 Flow Process Chart for Burrito Preparation.

• How is the quantity of meat, cheese, and sour cream dispensed controlled?

• How is the quality of the finished burrito checked?

Answers to these and other similar questions might yield an improved method. There may be some real constraints encountered and there may be some artificial constraints found along the way also.

Example: Changing a Tire

Another situation almost everybody has encountered is changing a flat tire. Generally, these incidents occur when least expected and when the driver wants to deal with almost anything else. For example, while driving alone along a back road the driver of a car may suddenly notice that the car is pulling to the right, a good indication that the right tire in the front is low on air.

The first thing the driver will do is pull over to the side and stop the car. Then the driver will put the car in "park" and set the emergency brake. After getting out of the automobile, the driver will verify that the right front tire is indeed flat. The driver will curse and then go to the trunk to remove the spare tire. At this point, the driver will remember that the keys are still in the ignition and will then curse again and retrieve the keys. After opening the trunk, the driver will remove the spare and check its air pressure. Being a farsighted individual, the driver will affirm that the tire has sufficient air pressure to use. Upon checking the spare, the tire is then carried to the right front side of the car. Next, the jack is removed from the trunk and assembled in the approximate area in which it will be used. The car is then raised with the jack and the flat tire is removed. The spare tire is mounted and the car is lowered. The jack and the flat tire are then returned to the trunk. At this point in time, the jack will not be disassembled, but rather it will be thrown into the trunk along with the flat tire. After the equipment is returned to the trunk, the driver is ready to proceed to the next operation, that of fixing or having the flat tire repaired.

The flow process chart for this operation is shown in Figure 3-7. Note that the chart is prepared for the driver of the car. It might also be appropriate to chart the flat tire, the jack, or even the spare tire. The reader should be able to identify some very obvious improvements that will increase the productivity or at least speed up the tire changing process for the driver. One of these obvious suggestions would be for the driver to call the local motor club and let someone else change the tire instead of changing it himself.

Flow Diagram

Another useful tool for graphical productivity analysis is the flow diagram (ANSI Z94).

> A flow diagram is a representation of the location of activities or operations and the flow of materials between activities on a pictorial layout of a process. Usually used with a flow process chart.

PRESENT METHOD ☒			**FLOW PROCESS CHART**		PAGE _1_ OF _2_

PROPOSED METHOD ☐

PART NAME: _Auto_

PROCESS DESCRIPTION: _Tire Changing_

DEPARTMENT: _Outside_

PLANT: _Highway 120_

ANALYST: _A.Ft_

DATE: _4_ / _20_ / _____   NOTE: SKETCHES ARE LOCATED ON BACK

SUMMARY		NO.
○	OPERATIONS	17
ⅾ	TRANSPORTATIONS	12
☐	INSPECTIONS	2
D	DELAYS	1
▽	STORAGES	0
	TOTAL STEPS	32
	DISTANCE TRAVLED	117

STEP	OPERATION TRANSPORT INSPECT DELAY STORAGE	DESCRIPTION	Distance
1	●ⅾ☐D▽	Pull Over and Stop Car	
2	○◆☐D▽	Get Out of Car	3
3	○◆☐D▽	Go to Flat	12
4	○ⅾ■D▽	Check Tire	
5	○◆☐D▽	Go to Trunk	12
6	○ⅾ☐◗▽	Curse	
7	○◆☐D▽	Return to Ignition	10
8	●ⅾ☐D▽	Get Key	
9	○◆☐D▽	Return to Trunk	10
10	●ⅾ☐D▽	Open Trunk	
11	●ⅾ☐D▽	Remove Spare	
12	○ⅾ■D▽	Check Air Pressure	
13	○◆☐D▽	Carry Spare to Flat	12
14	●ⅾ☐D▽	Drop Spare at Flat	
15	○◆☐D▽	Return to Trunk	12
16	●ⅾ☐D▽	Remove Jack	

Figure 3-7   Flow Process Chart for Changing a Tire.

PRESENT METHOD ☒ PROPOSED METHOD ☐		FLOW PROCESS CHART	PAGE 2 OF 2
STEP	OPERATION TRANSPORT INSPECT DELAY STORAGE	DESCRIPTION	Distance
17	O◆☐D▽	Carry Jack to Flat	
18	●◊☐D▽	Assemble Jack	
19	●◊☐D▽	Jack Up Car	
20	●◊☐D▽	Remove Flat	
21	●◊☐D▽	Place Spare on Car	
22	●◊☐D▽	Lower Car	
23	O◆☐D▽	Carry Flat to Trunk	12
24	●◊☐D▽	Place Flat in Trunk	
25	O◆☐D▽	Return for Jack	12
26	●◊☐D▽	Pick Up Jack	
27	O◆☐D▽	Carry Jack to Trunk	12
28	●◊☐D▽	Plack Jack in Trunk	
29	●◊☐D▽	Close Trunk	
30	●◊☐D▽	Remove Keys from Trunk Lock	
31	O◆☐D▽	Return to Car	10
32	●◊☐D▽	Start Car	
	O◊☐D▽		
	O◊☐D▽		
	O◊☐D▽		
	O◊☐D▽		
	O◊☐D▽		
	O◊☐D▽		

Figure 3-7   (Continued)

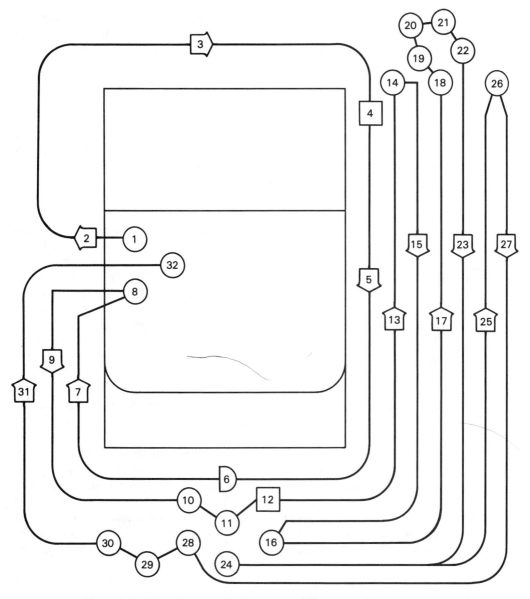

Figure 3-8. Flow Diagram for Tire Change (Numbers refer to the steps used in Figure 3-7).

Example: Changing a Tire

Figure 3-8 shows a flow diagram for the tire-changing episode described in the last example. Notice, where possible, improvements are indicated. The high frequency of trips between the trunk and the flat tire indicate a number of possible areas of investigation. The analyst must question whether or not there is justification for making all of the indicated trips. Perhaps some could be combined or eliminated. Perhaps they are all necessary. Only a complete analysis can answer these questions.

Example: Registration

Most college students dread the event known as registration.* It is, unfortunately, an activity that charitably can be described as "hurry up and wait." The sequence of events, what we should call the flow process chart of the current method used at one institution of higher learning, includes the following activities:

At the indicated time, the student reports to the college gym for registration. After waiting a minimum of 30 minutes past the designated registration time, the student is permitted to enter the gym. At this point, she must report to her department's advisement table and once again wait in line until an advisor is available to help her fill out her schedule card. This card is filled out, at least theoretically, to include those courses that are most appropriate to her education at that time.** After receiving her advisor's blessing and permission to enroll in certain classes, she must then check to see if those classes have sufficient space to actually permit her to register for the course at the time she wants it. If she is given a registration card, she must inspect it to ascertain that it is indeed for the course at the time she wants. After performing this inspection she must sign the card to indicate that she has read the card and is willing to accept it. Typically, she wants five different departments and, just as typically, if she wants five classes, at least one will already be full. In this case, she will have to return to her advisor, wait in line, and then gain approval for a substitute course. If this course is available she will then receive the appropriate course card, which, of course, must be inspected and signed.

After all of the desired course cards have been collected and autographed by the student, she must then proceed to the checkout station where her cards are collected and fees assessed. Also, a part of this registration station is the "operation" of issuing student identification cards and library cards. Upon successful completion of the checkout, the student then takes the fee statement card to the fee station and pays—by check, cash, credit card, or scholarship voucher—for the fees assessed. Once this task is completed, the registration is complete and the student can leave the gymnasium.

Figure 3-9 shows a flow process chart for the registration process. Figure 3-10 shows the flow diagram. The reader is urged to question the method and suggest an improved or more productive way of performing registration. The reader is also cautioned that, despite the fact that some obvious changes are indicated, the likelihood of changing such a process, either marginally or radically, is probably very small!

---

*For the students' information, the faculty dread registration as much, if not more, than the students do.

**Or the courses that are convenient to the student's work or sleep habits.

FLOW PROCESS CHART

Page _1_ of _1_

PART NAME_____

PROCESS DESCRIPTION _Registration_____

DEPARTMENT _____

PLANT _____

RECORDED BY _LSA_____   DATE _4-20_____

	SUMMARY	
		NO.
○	OPERATIONS	7
⇨	TRANSPORTATIONS	7
☐	INSPECTIONS	2
D	DELAYS	5
▽	STORAGES	0
	TOTAL STEPS	
	DISTANCE TRAVELED	285

STEP	Operations Transport Inspect Delay Storage	DESCRIPTION OF _Present_____ METHOD				
1	●○☐D▽	Report for Registration				
2	○⇨☐■▽	Wait for Admission				
3	○◆☐D▽	Report to Advisor	45			
4	○⇨☐■▽	Wait for Advisor				
5	●○☐D▽	Be Advised				
6	○◆☐D▽	Go to Course Cards	20			
7	●○☐D▽	Pull Course Cards				
8	○⇨■D▽	Inspect Course Cards				
9	○◆☐D▽	Return to Advisor	20			
10	○⇨☐■▽	Wait for Advisor				
11	●○☐D▽	Modify Schedule				
12	○◆☐D▽	Go to Course Cards	20			
13	●○☐D▽	Pull Course Cards				
14	○⇨■D▽	Inspect Course Cards				
15	○◆☐D▽	Report to Checkout	60			
16	○⇨☐■▽	Wait for Checkout				
17	●○☐D▽	Have Fees Assessed				
18	○◆☐D▽	Report to Cashier	20			
19	○⇨☐■▽	Wait for Cashier				
20	●○☐D▽	Pay Fees				
21	○◆☐D▽	Leave Gym	100			
	○⇨☐D▽					

Figure 3-9 Flow Process Chart for Registration.

Example: Library Procedures

The following example shows the use of a flow diagram and flow process chart in an actual work situation.*

*Reprinted with permission from *Industrial Engineering,* November 1973. Copyright © American Institute of Industrial Engineers, 25 Technology Park/Atlanta, Norcross, Georgia 30092.

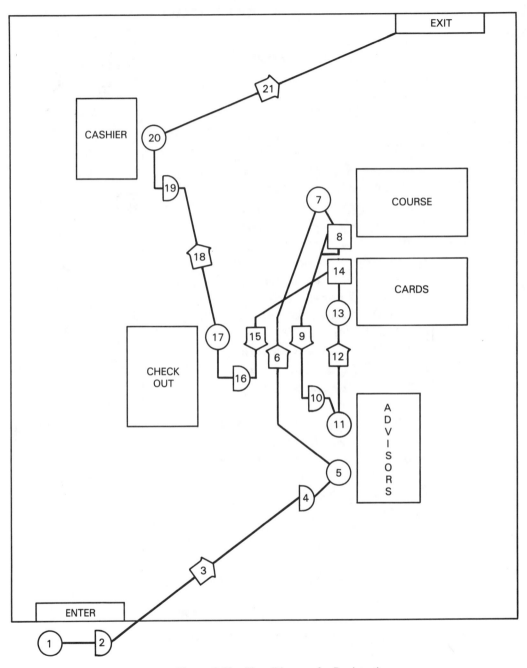

Figure 3-10 Flow Diagram for Registration.

The workroom in the Mountain View Library processes about 8,000 new books per year. All books, once they are received by the library, must pass through this workroom to be prepared for shelving in the library's stacks.

When the library was built, the annual number of new books processed was less than 3,000. The steady increase in the number of new books processed caused a problem with moving production through the workroom. Although the work area was adequate for the original workload, the higher annual volume severely cramped the library's available work area. The layout of the workroom, which could not be significantly changed due to the structure of the library building, is shown in Figure 3-11. The process used to add a book to the collection is described in the following paragraphs.

After a library book is received in the workroom of Mountain View Library, it is moved from the dumb waiter to the check-in area. The check-in process assures that the proper book has arrived and gathers the appropriate paperwork that must accompany the book through its classification process. The check-in also provides for an initial check of call numbers, as well as a chance for duplicate books to be noted. The next step in the process is to classify all books according to the Library of Congress numbering system. No problems exist if the call number is readily available. The number is noted or recorded on the paperwork and the book is ready to move on. Very often, however, especially with new titles, the call number may not be available. If this case applies, the book must be stored until the number is made available. This storage time may be from one day to six months. Once the number is secured, the paperwork is completed and the production process continues. A large amount of shelf space is consumed by books awaiting call numbers.

After call numbers are determined, the appropriate paperwork must be completed and carefully checked. Incorrectly numbered books could very easily be lost. After the verification, the call number is applied to the spine and pocket of the book and a date due slip is attached. When these operations are completed, the book is ready for shelving. Completed volumes are placed in the stacks.

The production process can be summarized in the following steps. They correspond, numerically, to the steps shown on the flow process chart in Figure 3-12 and flow diagram in Figure 3-13.

1. Receive books from outside source(s).
2. Transport books to check-in area.
3. Check books in and see if they duplicate any existing books.
4. Transport books to catalog research area.
5. Catalog books according to Library of Congress classification.
6. If catalog number does not exist, transport book to storage area until the number becomes available.
7. Store books until catalog number becomes available.
8. Transport books back to Step 5 for cataloging.
9. Transport books to verification area.
10. Verify.

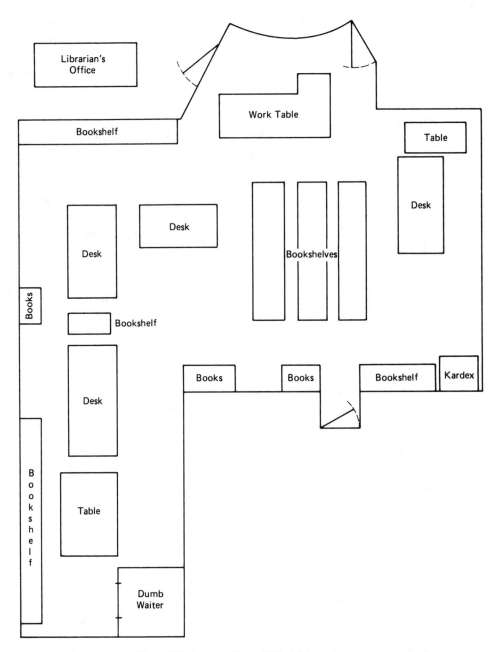

Figure 3-11   Library Workroom. (From "Work Methods Improvement in the Library" by L.S. Aft. *Industrial Engineering*, November 1973, pp. 40–41. Reprinted with permission.)

PRESENT METHOD ☒	**FLOW PROCESS CHART**			PAGE _/_ OF _/_

PROPOSED METHOD ☐

PART NAME: _Book_

PROCESS DESCRIPTION: _Book Cataloging_

DEPARTMENT: _Library_

PLANT: _MV College_

ANALYST: _LSA_

DATE: _4_ / _20_ / ____   NOTE: SKETCHES ARE LOCATED ON BACK

SUMMARY		NO.
○	OPERATIONS	5
⇨	TRANSPORTATIONS	6
☐	INSPECTIONS	1
D	DELAYS	
▽	STORAGES	1
	TOTAL STEPS	13
	DISTANCE TRAVLED	165

STEP	OPERATION TRANSPORT INSPECT DELAY STORAGE	DESCRIPTION	Distance
1	●⇨☐D▽	Receive Books	
2	○⬛☐D▽	Books to Check In	75
3	●⇨☐D▽	Books Checked In	
4	○⬛☐D▽	Books to Research	30
5	●⇨☐D▽	Catalog Books	
6	○⬛☐D▽	To Storage with No Numbers	20
7	○⇨☐D▼	Store	
8	○⬛☐D▽	To Catalog When Number Available	20
9	○⬛☐D▽	To Verification	10
10	○⇨⬛D▽	Verify	
11	○⬛☐D▽	To Call Number Application	5
12	●⇨☐D▽	Apply Call Number	
13	○⬛☐D▽	To Storage	5
	○⇨☐D▽		
	○⇨☐D▽		
	○⇨☐D▽		

Figure 3-12 Flow Process Chart for Processing New Books—Present Method.

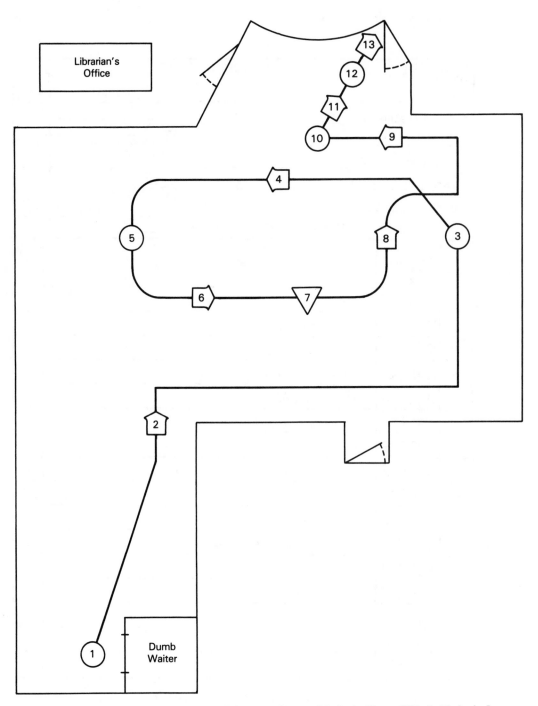

Figure 3-13 Library Flow Diagram—Present Method. (From "Work Methods Improvement in the Library" by L.S. Aft. *Industrial Engineering*, November 1973, pp. 40–41. Reprinted with permission.)

11. Transport books to pocket and call number application area.

12. Install pocket and date due slip.

13. Apply call number.

14. Store books temporarily until they can be shelved.

As shown in Figure 3-13 there is unnecessary movement of books back and forth across the workroom. Also, the main storage shelves constitute a major barrier to the effective flow of materials. The central location of the shelves prohibits them from being as densely loaded as they might be. If the shelves were loaded to capacity, they would block communication in the workroom.

Analysis of the flow process chart suggests the following changes that would increase the productivity of the workroom of Mountain View Library. The rearrangement of furniture, as shown in Figure 3-14, would result in a layout requiring a total of 63 feet less transportation for each book. This layout appears to be the optimum than can be made within the existing framework and fixed structural barriers of the library. The flow diagram for the existing layout, using the 14 activity numbers as described, is shown in Figure 3-15. A much more orderly flow, as well as large savings in the distance books must travel, results from the new layout. It also gives a much less congested appearance to the room and provides considerably more book storage space. And, although a significant amount of book travel distance is reduced, perhaps more important is the extra storage space that can be provided. By moving the major storage shelves out of the center of the room and off to one side, additional storage racks can be added. In this instance, the book storage space was increased by a factor of over 4, or over 400 percent.

The application of the basic analysis tools of flowcharting and flow diagramming has increased the capacity of the Mountain View Library to process new books. This example represents an increase in productivity.

Left-hand/Right-hand Chart

This chart is one (ANSI Z94),

> on which the motions made by one hand in relation to those made by the other hand are recorded using standard process chart symbols.

Although the standard process chart symbols are used for the right-hand/left-hand chart, three special interpretations must be noted. First, the hands do not generally perform inspection. Second, transportations are defined as movements of the hands from one location to another at the workplace. Third, holds are generally considered to be delays.

Example: Pouring Coffee

An activity that many people perform on a frequent basis is filling a cup with coffee. If it is assumed that the coffee is already brewed and in an automatic drip coffee pot with a supply of styrofoam cups, cream, and sugar nearby, the activities that will be performed could be described as follows:

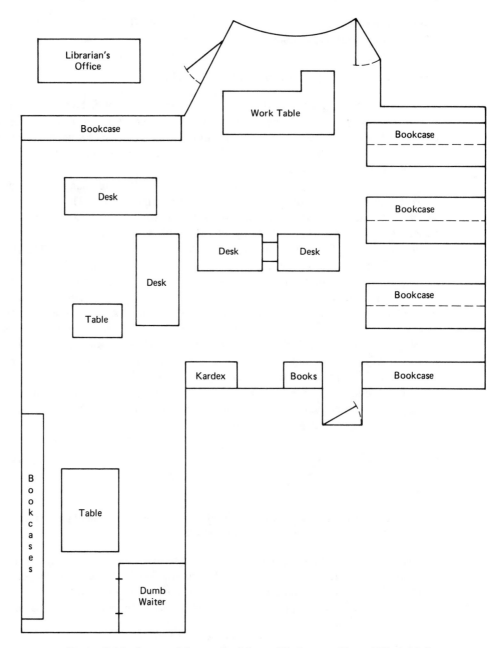

Figure 3-14 Improved Layout for Library Workroom. (From "Work Methods Improvement in the Library" by L. S. Aft. *Industrial Engineering*, November 1973, pp. 40–41. Reprinted with permission.)

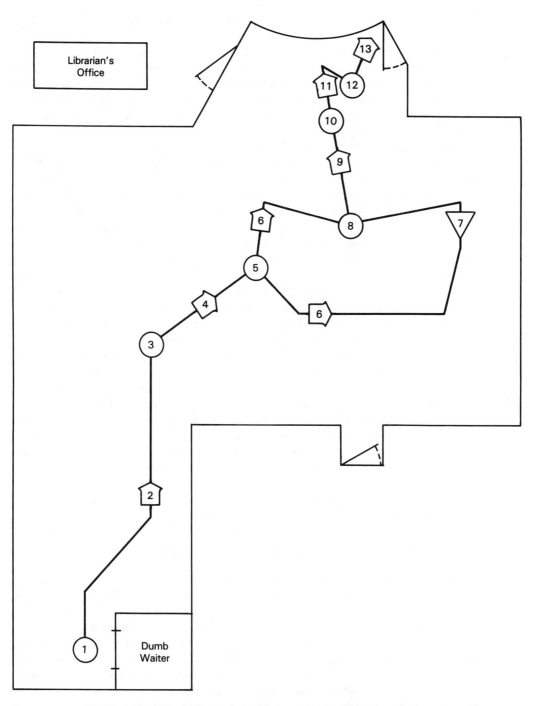

Figure 3-15 Library Flow Diagram—Improved Method. (From "Work Methods Improvement in the Library" by L. S. Aft. *Industrial Engineering*, November 1973, pp. 40–41. Reprinted with permission.)

The right-handed person reaches for the stack of cups with his left hand. Simultaneously, the right hand reaches for the top cup. After the hands grasp the cups, the left hand holds the stack while the right hand removes the top cup. The left hand then sets the stack back down while the right hand brings the cup back to the area right in front of the coffee pot. At this point, the left hand reaches for the cup and grasps it, while the right hand holds the cup. Control of the cup is transferred to the left hand when the right hand releases the cup to reach for the coffee pot. After the hand grasps the pot, it is moved to the cup being held by the left hand and coffee is poured into the cup. As the cup is filled, the person checks to see if the cup is being filled to the desired level. Once the cup has been filled, the right hand returns the coffee pot to the hot plate and then the left hand sets the coffee cup down. Because the operator in this example likes one packet of sugar in his coffee, he reaches with his right hand, after releasing the coffee pot, to the container holding the sugar packets and selects one. This packet is then brought to the left hand which firmly grasps the packet so that the right hand can rip it open. After opening the packet, the right hand releases the packet and the left hand pours the contents into the coffee cup. The right hand then reaches for a stirring stick, picks it up, brings it back to the coffee cup, and stirs the sugar into the coffee. The left hand tosses the empty sugar packet into the trash while the right hand similarly disposes of the stirring stick. The right hand then reaches back for the coffee cup and picks the cup up, thus concluding the "operation."

Although this operation is relatively simple, the LH/RH chart shown in Figure 3-16 should illustrate the use of the chart and should also show the nature of the types of productivity improvements that can readily be made in the process. Remember that the hands of the operator are being charted. Therefore, motions that might appear to be transportations, such as moving the coffee cup from location to location, are really to be viewed as operations. Also, some of the activities that were described, such as checking the level of the coffee in the cup, are not shown at all because they are not operations performed by the hands. On the other hand (pun intended), when the hands reach for an object, they are being transported from one location to another. Any place the delay symbol occurs is an obvious place to start questioning. The reader should be able to suggest a number of potential improvements for the coffee-pouring operation.

Example: Bolt Washer Assembly

A production operation requires an operator to perform a relatively simple assembly. The operator must put together a bolt-washer-nut assembly (see Figure 3-17). The following is a description of the activity as it is presently performed. After examining the LH/RH chart, the reader should be able to suggest a number of productivity improvements.

The bolts, washers, and nuts are found in three containers directly in front of the operator. This operation begins when the operator reaches for a bolt with the left hand. The left hand will pick up the head of the bolt. The operator grasps the bolt and moves it back in front of him. The left hand then holds the bolt by the head as the right hand reaches for and grasps the washer. The washer is brought to the bolt and placed over the end. The right hand, before releasing the washer, slides it down the bolt until it is flush with the head of the bolt. After releasing the washer, the right hand reaches for the nut.

Upon grasping the nut, the right hand then releases the nut and moves aside. The left hand then places the finished assembly aside and reaches for the next nut.

Multiple Activity Chart

The next chart that will be examined is the multiple activity chart. This graphical representation is also known as a man-machine chart, multiple activity process chart, and simo chart. It is defined as follows (ANSI Z94):

> A chart of the coordinated synchronous or simultaneous activities of a work system of one or more workers. Each machine and/or worker is shown in a separate, parallel column indicating their activities as related to the rest of the work system.

This chart is often used to show the most productive way to use multiple operators, machines, or any combination of people and machines.

Example: Operating Semiautomatic Machines

An operator must tend to or operate semiautomatic machines. The operator simply has to load the raw material, turn the machine on, and unload the finished product after the machine finishes its cycle. It takes, for the sake of our example, four minutes for the operator to load the machine and two minutes to unload it after the machine has run. The machine run time or cycle time is eight minutes. The multiple activity chart can be used to determine the best number of machines for the operator to tend. The best number depends, of course, on how *best* is defined. The multiple activity chart will show the activity time and idle time for the operator and the machine(s). If *best* is defined as minimizing operator idle time, the chart can be used to determine the optimum number of machines to be tended. Likewise, if *best* is defined as minimizing machine idle time, the chart can be used to determine that figure as well.

Idle time might be defined as any nonproductive time. In the case of the operator, this time might be when neither loading nor unloading is performed. For the machine, idle time might be the time when the machine is not running, not being loaded, or not being unloaded. A decision about which is more important to minimize—operator idle time or machine idle time—must be made before a study is started. Usually, this type of decision will be made on the basis of cost. If it is less expensive to have machines idle then that direction should be pursued. If it is more cost-effective to have people idle, then that direction or course of action should be followed. The ideal case would be to have no idle time for machines or people.

Let us look first at the case where the operator has one machine to tend. By showing the activity of the operator and machine, we can see a pattern or cycle emerge. This cycle will eventually show in most situations where the work elements have constant times. However, when there are several operators and several machines, each of which have their own unique elemental times, this cycle may not be readily apparent from the multiple activity chart. In this type of situation, we would want to use a more sophisticated mathematical model to answer our questions.

LEFT HAND -- RIGHT HAND CHART

SUMMARY:

SYMBOLS		PRESENT		PROPOSED		DIFFERENCE	
		LH	RH	LH	RH	LH	RH
O	OPERATION	7	14				
⇨	TRANSPORTATION	3	11				
☐	INSPECTION	0	0				
D	DELAYS	16	2				
	TOTALS	26	27				

PROCESS: *Coffee Pouring*
STUDY No: 6
OPERATOR: *JB*
ANALYST: *EB*
DATE 5 / 18 / 82
METHOD (PRESENT) ~~PROPOSED~~
SHEET No. 1 of 2
REMARKS: _____

SKETCH OF:_____   BY: _____

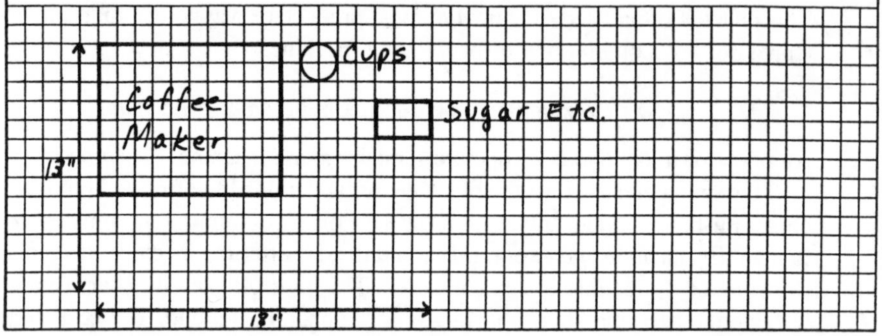

Coffee Maker   O Cups   ☐ Sugar Etc.   13"   18"

LEFT HAND *Present* METHOD		SYMBOL	SYMBOL		RIGHT HAND *Present* METHOD
Reach for Cups		O ⇨ ☐ D	O ⇨ ☐ D		Reach for Top Cup
Grasp Cups		● ⇨ ☐ D	● ⇨ ☐ D		Grasp Cup
Hold Cups		O ⇨ ☐ ◗	● ⇨ ☐ D		Remove Top Cup
Set Cup Down - Release		● ⇨ ☐ D	O ⇨ ☐ D		Bring Cup to Pot
Reach to Cup		O ⇨ ☐ D	O ⇨ ☐ ◗		Hold Cup
Grasp Cup		● ⇨ ☐ D	O ⇨ ☐ ◗		Hold Cup
Hold Cup		O ⇨ ☐ ◗	● ⇨ ☐ D		Release Cup
Hold Cup		O ⇨ ☐ ◗	● ⇨ ☐ D		Reach for Pot and Grasp
Hold Cup		O ⇨ ☐ ◗	O ⇨ ☐ D		Bring Pot to Cup
Hold Cup		O ⇨ ☐ ◗	● ⇨ ☐ D		Pour Coffee into Cup
Hold Cup		O ⇨ ☐ ◗	O ⇨ ☐ D		Return Pot to Hot
Hold Cup		O ⇨ ☐ ◗	O ⇨ ☐ D		Plate

Figure 3-16 Left-hand/Right-hand Chart for Pouring Coffee.

LEFT HAND – RIGHT HAND CHART CONTINUED

STUDY No. __6__   SHEET No. __2__ of __2__

LEFT HAND *Present* METHOD	DIST.	SYMBOL	SYMBOL	DIST.	RIGHT HAND *Present* METHOD
Set Cup Down		○➡□D	●⇨□D		Release Pot
Release Cup		●⇨□D	○➡□D		Reach To Sugar
Idle		○⇨□◗	●⇨□D		Grasp Sugar Packet
Idle		○⇨□◗	○➡□D		Move Sugar to LH
Grasp Sugar		●⇨□D	●⇨□D		Open Packet
Hold Sugar		○⇨□◗	●⇨□D		Release Sugar
Pour Sugar into Cup		●⇨□D	○➡□D		Reach for Stirring Stick
Hold Empty Packet		○⇨□◗	●⇨□D		Grasp Stick
Hold Empty Packet		○⇨□◗	○➡□D		Bring Stick to Cup
Hold Empty Packet		○⇨□◗	●⇨□D		Stir Coffee
Hold Empty Packet		○⇨□◗	●⇨□D		Remove Stick
Dispose Empty Packet		●⇨□D	●⇨□D		Dispose of Stick
Idle		○⇨□◗	○➡□D		Reach for Cup
Idle		○⇨□◗	●⇨□D		Pick up Cup
		○⇨□D	○⇨□D		
		○⇨□D	○⇨□D		
		○⇨□D	○⇨□D		
		○⇨□D	○⇨□D		
		○⇨□D	○⇨□D		
		○⇨□D	○⇨□D		
		○⇨□D	○⇨□D		
		○⇨□D	○⇨□D		
		○⇨□D	○⇨□D		
		○⇨□D	○⇨□D		
		○⇨□D	○⇨□D		
		○⇨□D	○⇨□D		
		○⇨□D	○⇨□D		
		○⇨□D	○⇨□D		

Figure 3-16 (Continued).

LEFT HAND -- RIGHT HAND CHART

SUMMARY:

SYMBOLS		PRESENT		PROPOSED		DIFFERENCE	
		LH	RH	LH	RH	LH	RH
O	OPERATION	2	6				
⇨	TRANSPORTATION	3	4				
☐	INSPECTION	0	0				
D	DELAYS	10	5				
	TOTALS	15	15				

PROCESS: _Bolt Washer Assembly_
STUDY No: ___9___
OPERATOR: ___SR___
ANALYST: ___DB___
DATE _3_ / _12_ / _82_
METHOD (PRESENT) ~~PROPOSED~~
SHEET No. _1_ of _2_
REMARKS: _____

SKETCH OF:_____   BY: _____

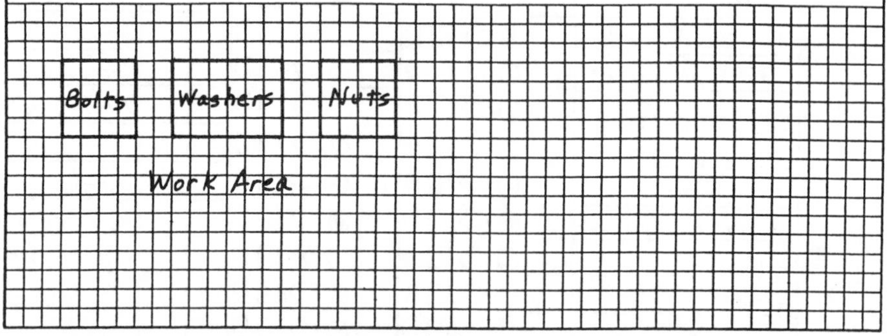

Bolts   Washers   Nuts

Work Area

LEFT HAND _Present_ METHOD	SYMBOL	SYMBOL		RIGHT HAND _Present_ METHOD
Reach for Bolt	O ➡ ☐ D	O ⇨ ☐ ◗		Idle
Grasp Bolt	● ⇨ ☐ D	O ⇨ ☐ ◗		Idle
Move Bolt To Work Area	O ➡ ☐ D	O ⇨ ☐ ◗		Idle
Hold Bolt	O ⇨ ☐ ◗	O ➡ ☐ D		Reach for Washer
Hold Bolt	O ⇨ ☐ ◗	● ⇨ ☐ D		Grasp Washer
Hold Bolt	O ⇨ ☐ ◗	O ➡ ☐ D		Move Washer to Bolt
Hold Bolt	O ⇨ ☐ ◗	● ⇨ ☐ D		Washer onto Bolt
Hold Bolt	O ⇨ ☐ ◗	● ⇨ ☐ D		Release Washer
Hold Bolt	O ⇨ ☐ ◗	O ➡ ☐ D		Reach for Nut
Hold Bolt	O ⇨ ☐ ◗	● ⇨ ☐ D		Grasp Nut
Hold Bolt	O ⇨ ☐ ◗	O ➡ ☐ D		Move Nut to Bolt
Hold Bolt	O ⇨ ☐ ◗	● ⇨ ☐ D		Nut onto Bolt

Figure 3-17   Left-hand/Right-hand Chart for Bolt Washer Assembly.

LEFT HAND – RIGHT HAND CHART CONTINUED

STUDY No. ___9___   SHEET No. _2_ of _2_

LEFT HAND *Present* METHOD		SYMBOL	SYMBOL		RIGHT HAND *Present* METHOD
Hold Bolt		○ ⇨ □ ◗	● ⇨ □ D		Release Nut
Set Aside		● ⇨ □ D	○ ⇨ □ ◗		Idle
Reach for Bolt		○ ➡ □ D	○ ⇨ □ ◗		Idle
		○ ⇨ □ D	○ ⇨ □ D		
		○ ⇨ □ D	○ ⇨ □ D		
		○ ⇨ □ D	○ ⇨ □ D		
		○ ⇨ □ D	○ ⇨ □ D		
		○ ⇨ □ D	○ ⇨ □ D		
		○ ⇨ □ D	○ ⇨ □ D		
		○ ⇨ □ D	○ ⇨ □ D		
		○ ⇨ □ D	○ ⇨ □ D		
		○ ⇨ □ D	○ ⇨ □ D		
		○ ⇨ □ D	○ ⇨ □ D		
		○ ⇨ □ D	○ ⇨ □ D		
		○ ⇨ □ D	○ ⇨ □ D		
		○ ⇨ □ D	○ ⇨ □ D		
		○ ⇨ □ D	○ ⇨ □ D		
		○ ⇨ □ D	○ ⇨ □ D		
		○ ⇨ □ D	○ ⇨ □ D		
		○ ⇨ □ D	○ ⇨ □ D		
		○ ⇨ □ D	○ ⇨ □ D		
		○ ⇨ □ D	○ ⇨ □ D		
		○ ⇨ □ D	○ ⇨ □ D		
		○ ⇨ □ D	○ ⇨ □ D		
		○ ⇨ □ D	○ ⇨ □ D		
		○ ⇨ □ D	○ ⇨ □ D		
		○ ⇨ □ D	○ ⇨ □ D		

Figure 3-17   (Continued).

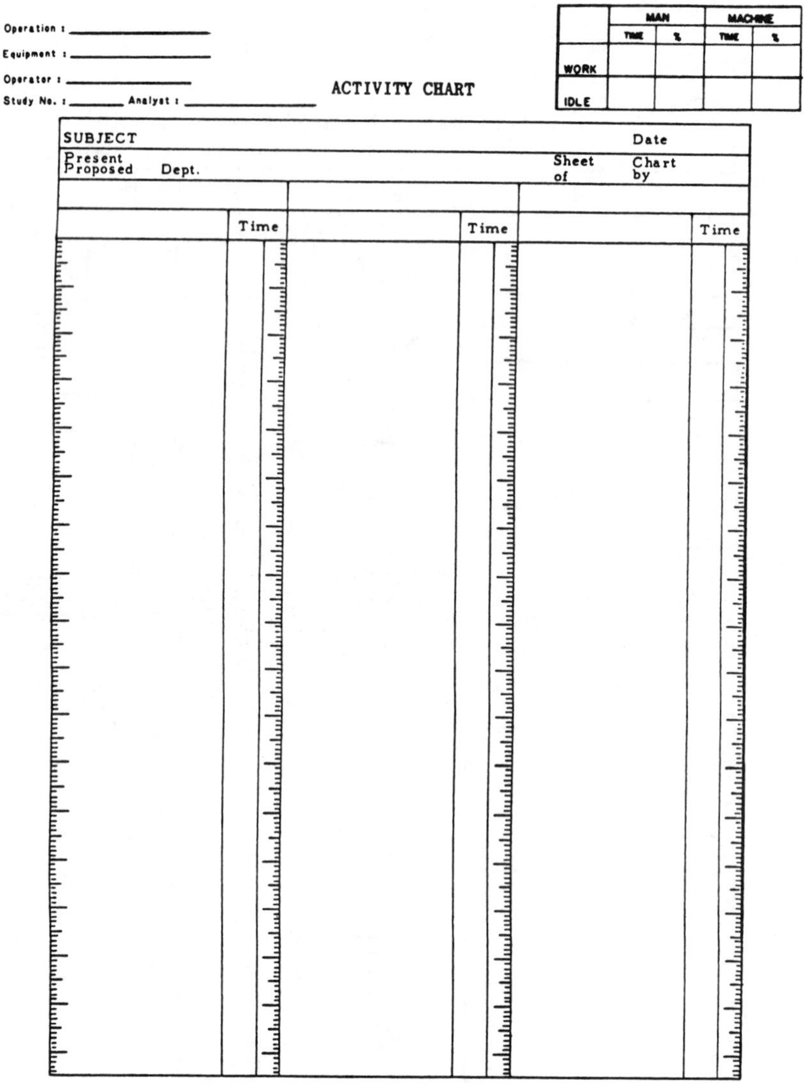

Figure 3-18 Sample Multiple Activity Chart.

However, this problem is elementary. As we construct the multiple activity chart, we start with a blank form as shown in Figure 3-18. Note that the chart shows a time scale and has several columns to list the multiple activities. Figure 3-19(a) shows the building of the chart for this situation. Note that the operator is shown as Activity 1 and the machine is charted where Activity 2 was shown on the original chart. We first note what

	MAN		MACHINE	
	TIME	%	TIME	%
WORK				
IDLE				

Operation : _____

Equipment : _____

Operator : _____   ACTIVITY CHART

Study No. : _____ Analyst : _____

SUBJECT **Semi- Auto Machine**   Date **11-6**

(Present)
Proposed   Dept.   Sheet   Chart
                                                        of   by

	Time	OPERATOR	Time	MACHINE	Time
	2	Load Machine		Being Loaded	
	4				
	6				
	8				
	10				
	12				
	14				
	16				
	18				

Figure 3-19(a)   Sample Multiple Activity Charts for Loading One Machine.

activity the operator performs and mark off the time required for the activity. The first thing the operator does is load the machine. Similarly, we indicate what the machine is doing—in this case, it is being loaded.

Figure 3-19(b) shows the addition of the next activities, but we pick up our chart from the four-minute mark. After loading the machine the operator "runs" the machine

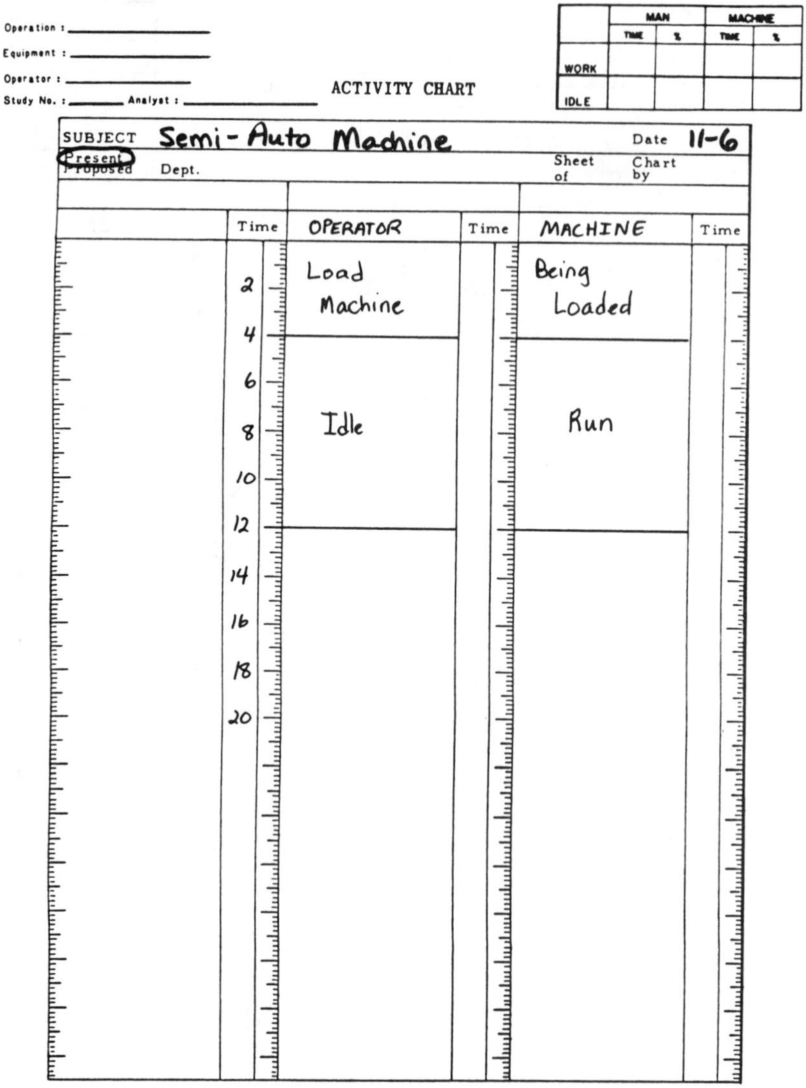

Figure 3-19(b)

and is idle while the machine does its job for eight minutes. Figure 3-19(c) shows the next operations. The machine has finished running and must be unloaded. The operator unloads the machine, taking the required two minutes.

For this example, we have already completed a cycle; that is, all the activities have been completed and we are now ready to begin again in the same sequence and at the same times relative to each other. Figure 3-20 shows three complete cycles for one machine and one operator.

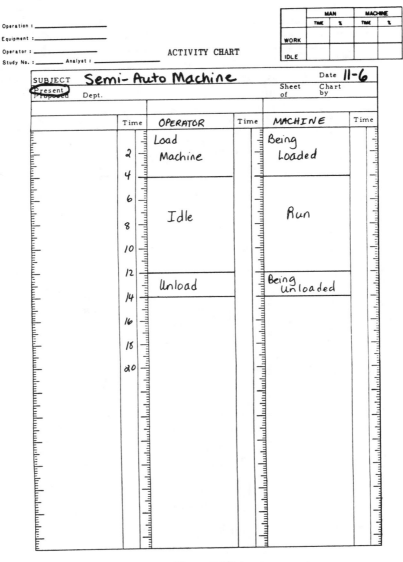

Figure 3-19(c)

An analysis of the complete cycle shows that the entire cycle time for this situation is 14 minutes. The machine is productive 100 percent of the time. We can generalize that for 100-percent machine efficiency or productivity it should be tended by an operator dedicated to its operation. There may be other ways to achieve this utilization, but we can always manage it with one operator and one machine. The operator, however, is not as productive. As can be seen, for eight minutes of every 14-minute cycle the operator is idle. If the operator is not required to monitor the machine's operation

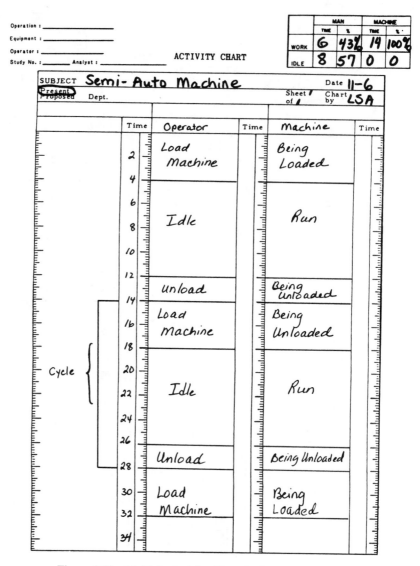

Figure 3-20 Multiple Activity Chart for Identifying Cycles.

or prepare the next batch of material for processing, this time is nonproductive, idle time. This operator, with just one machine to tend, is productive only 6/14, or about 43 percent, of the time.

Although we are getting maximum productivity out of the machine, there seems to be too much nonproductive time for the operator. We should be able to give the operator something else to do. The most logical solution would seem to be to have the operator

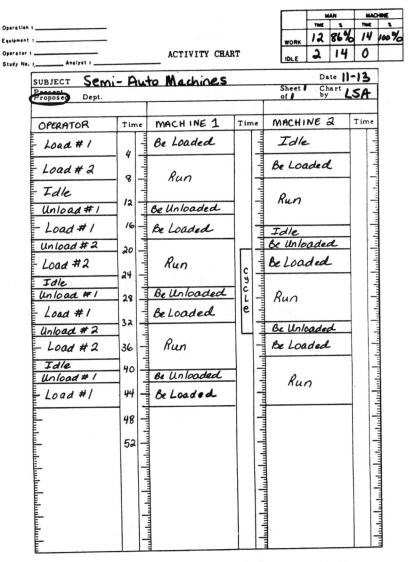

Figure 3-21 Multiple Activity Chart for Loading Top Machines.

tend to a second machine of the same type. Figure 3-21 shows the multiple activity chart for this situation.

Examination of this multiple activity chart shows that once the "steady-state" cycle appears, the cycle time is still 14 minutes. Both machines are productive 100 percent of the time; that is, the only time they are idle is when they are either being loaded or unloaded. However, the operator is idle only two minutes out of every 14. Thus, the

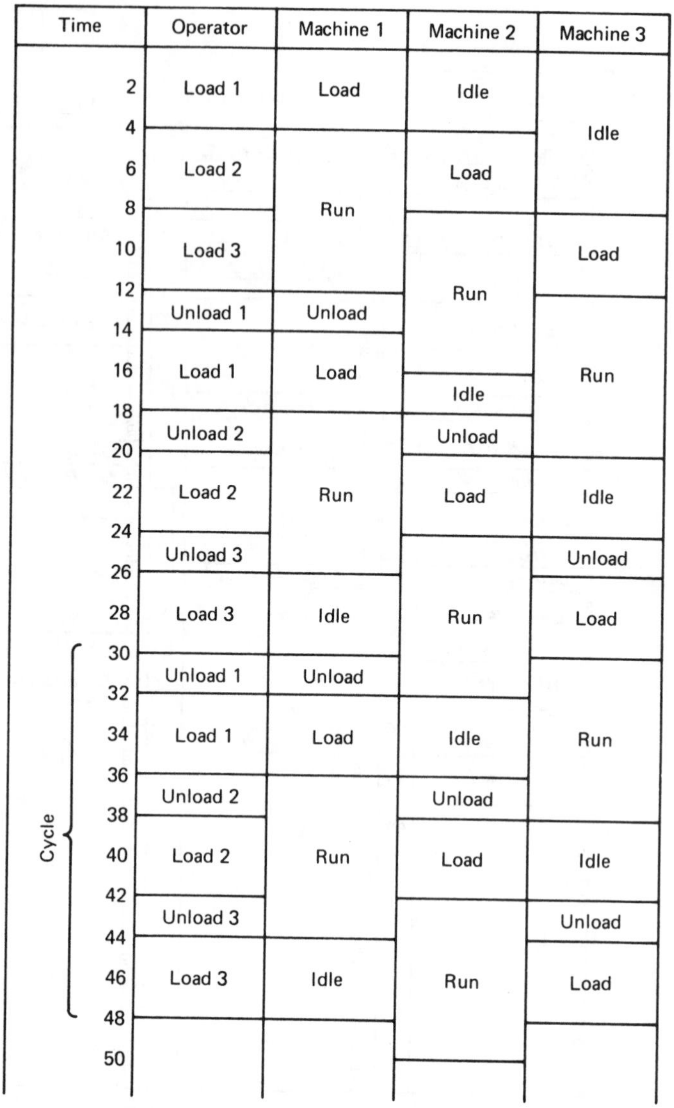

Time	Operator	Machine 1	Machine 2	Machine 3
2	Load 1	Load	Idle	
4				Idle
6	Load 2		Load	
8		Run		
10	Load 3			Load
12			Run	
14	Unload 1	Unload		
16	Load 1	Load		Run
18			Idle	
20	Unload 2		Unload	
22	Load 2	Run	Load	Idle
24				
26	Unload 3			Unload
28	Load 3	Idle	Run	Load
30				
32	Unload 1	Unload		
34	Load 1	Load	Idle	Run
36				
38	Unload 2		Unload	
40	Load 2	Run	Load	Idle
42				
44	Unload 3			Unload
46	Load 3	Idle	Run	Load
48				
50				

(Cycle bracket spans 30–48)

Figure 3-22   Multiple Activity Chart for Loading Three Machines.

operator is productive 12/14, or about 86 percent of the time. We have shown a significant improvement in productivity.

Our next logical question might be, what happens if we give the operator a third similar semiautomatic machine to tend. The multiple activity chart in Figure 3-22 shows that condition. An examination of this chart shows that the operator is productive 100 percent of the time. The cycle time for one operator and three machines has become 18 minutes. Each machine, once its steady-rate cycle has been reached, is idle four minutes every cycle or productive about 78 percent of the time. This situation is obviously a better use of the operator. Surprisingly, this scenario is also a much more productive use of the machines. The following table shows the average time it takes to produce a unit using one, two, and three machines.

NO. MACHINES	CYCLE TIME	TIME/UNIT
1	14	14 minutes
2	14	7
3	18	6

The multiple activity chart shows us that by keeping the operator busy 100 percent of the time with productive tasks, the entire operation will be more productive. Of course the savings in time per unit must be compared with the cost of an additional machine. It will also be influenced by the demand for the product produced by this operation.

Some further questions might follow from this example: What would happen if the operator were given a fourth machine? What would happen if there were two operators and multiple machines? If there were a fourth machine, the idle time for machines would increase. The multiple activity chart in Figure 3-22 shows that the operator is fully occupied with three machines. The addition of a fourth machine would only increase cycle time. (It would be beneficial for the reader to see what happens to the idle time and overall time per unit if this case were true.) If there were two operators and four machines, the best that could be hoped for would be the same times as two machines and one operator. Additional people would affect the situation as multiples of the current utilization as long as each operator performed the loading and unloading activities. If specialization were to be specified, that is, if each person would only load or only unload, then perhaps a better productivity could be achieved.

Example: Assembling Blivets

Sometimes the multiple activity chart can be used to help determine the optimum number of work stations to have on an assembly line to optimize some productivity-related factor such as units produced per hour or dollars spent for labor per unit. For example, the assembly operations required to build a blivet are listed in Figure 3-23. There are 24 different operations, each with a specified required time for completion. An assembly operator gets paid $5.00 per hour. If the Blivet Company wants to minimize labor cost per blivet, the obvious way to accomplish this goal would be to have one employee assemble each blivet. This assembly would require 93 minutes. In terms of assembly labor cost,

93 blivets/minute / 60 minutes/hour = $7.75 per blivet

The next possibility to consider would be two assemblers. Figure 3-24 shows the best balancing of these sequential operations. The cycle time to complete one unit with two assemblers is 48 minutes, translating to a per-unit cost of

Assembly Operation	Time (Minutes)
1	3
2	5
3	4
4	2
5	6
6	8
7	2
8	3
9	5
10	6
11	1
12	7
13	2
14	6
15	5
16	5
17	4
18	2
19	1
20	4
21	3
22	3
23	2
24	4
	‾‾
	93

Figure 3-23 Blivet Assembly Operations and Times.

$$(48/60 \text{ hours})(\$5 \text{ per hour})(2 \text{ workers}) = \$8.00 \text{ per unit.}$$

The cycle time for three assemblers is identified with the multiple activity chart shown in Figure 3-25. The cycle time is shown to be 33 minutes. The labor cost per assembled unit is shown in the following calculation:

$$(33/60)(\$5)(3 \text{ workers}) = \$8.25 \text{ per unit}$$

When four assemblers are used, the multiple activity chart (Figure 3-26) shows the cycle time to be 25 minutes. The per-unit labor cost is

$$(25/60)(\$5)(4 \text{ workers}) = \$8.33 \text{ per unit}$$

With five assemblers, Figure 3-27 shows us that the cycle time is cut down to 20 minutes. The per-unit cost of labor is

$$(20/60)(\$5)(5 \text{ workers}) = \$8.33 \text{ per unit}$$

For six operators on the assembly line, the minimum cycle time, as shown in Figure 3-28, becomes 17 minutes, meaning the labor cost per unit is

$$(17/60)(\$5)(6 \text{ workers}) = \$8.50 \text{ per unit}$$

The minimum time required to assemble a blivet would be for the case when one operator performs only the longest operation. The most time-intensive activity is Operation 6,

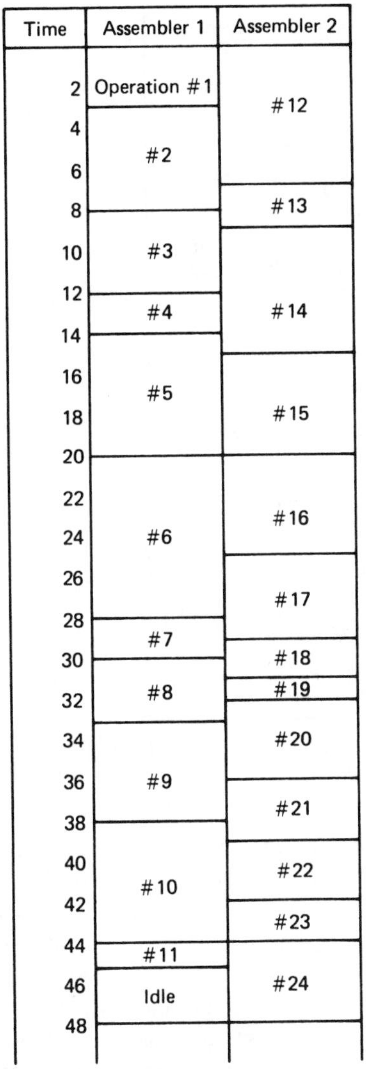

Time	Assembler 1	Assembler 2
2	Operation #1	#12
4	#2	
6		
8		#13
10	#3	
12	#4	#14
14		
16	#5	
18		#15
20		
22		#16
24	#6	
26		#17
28		
30	#7	#18
32	#8	#19
34		#20
36	#9	
38		#21
40	#10	#22
42		#23
44	#11	
46	Idle	#24
48		

Figure 3-24 Multiple Activity Chart for Blivet Assembly (Two Operators).

which takes eight minutes. This situation requires 14 additional operators to balance the remaining operations as closely as possible. Figure 3-28 shows the multiple activity chart for 15 assemblers. The per unit labor cost is

$$(8/60)(\$5)(15 \text{ workers}) = \$10.00 \text{ per unit.}$$

If 15 operators are employed, as shown in Figure 3-29, a blivet can be produced every 8 minutes. According to the chart, operations 2 and 14 appear to limit production time. These bottlenecks, once having been identified by the chart, can be examined for possible improvements. If no methods changes are possible, the addition of multiple work

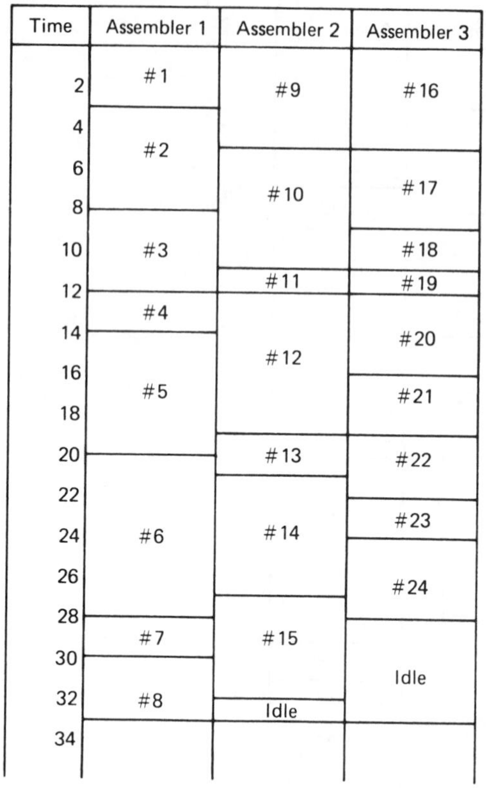

Time	Assembler 1	Assembler 2	Assembler 3
2	#1	#9	#16
4	#2		
6		#10	#17
8			
10	#3		#18
12		#11	#19
14	#4		#20
16		#12	
18	#5		#21
20		#13	#22
22			
24	#6	#14	#23
26			#24
28			
30	#7	#15	
32	#8		Idle
34		Idle	

Figure 3-25. Multiple Activity Chart for Blivet Assembly (Three Operators).

stations can be investigated. Again, costs must be analyzed and the appropriate criteria for decision-making followed. The addition of a second work station for operation 14 might remove that particular bottleneck, but it might create a new one elsewhere. The cost of an additional work station and operator to work at the new position must also be considered.

Although all possibilities have not been examined, it should be obvious at this point that the addition of assemblers only raises the labor cost. However, if the Blivet Company's goal is to minimize production time, this sample information and analysis procedure could be used.

Another factor to consider is first cost or investment. Although the labor cost may be lower with each operator performing all of assembly operations, to match the production quantity per unit time of having multiple assembly operators, the up-front investment in work stations, tooling, and just plain space might be excessive. Another factor that influences the final decision is the number of units required per unit time. If, in this example, the blivet is a subassembly used in another process or assembly, it might be necessary to produce one blivet every nine minutes, or one every 13 minutes, in which case the 90 plus minutes required by one assembler would backlog the rest of the production plant. The available space for assembly facilities might also require fewer ''lines'' with

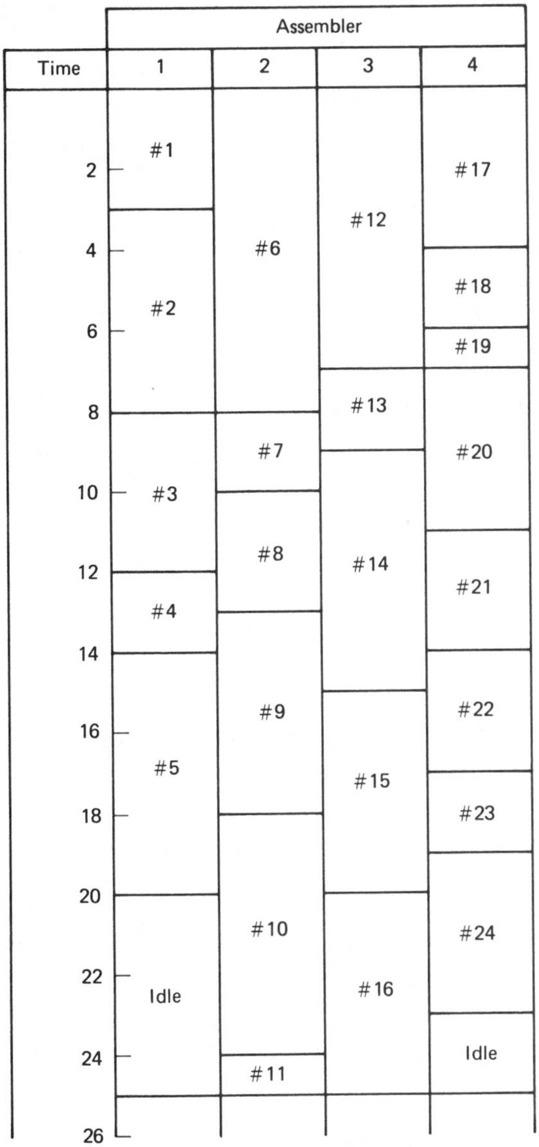

Time	Assembler			
	1	2	3	4

Figure 3-26 Multiple Activity Chart for Blivet Assembly (Four Operators).

more operators per line. One other factor to consider is the skill level required at each step on the assembly line. A single operator would need many different skills and would probably be more difficult to find. When there are multiple stations, workers who are more specialized in their abilities can be used. If these people are not readily available, they are easier to train than people who would be expected to perform many jobs.

The multiple activity chart can also be used to determine the activities of each of an operator's hands. Regardless of the situation, this chart shows the activities of each of

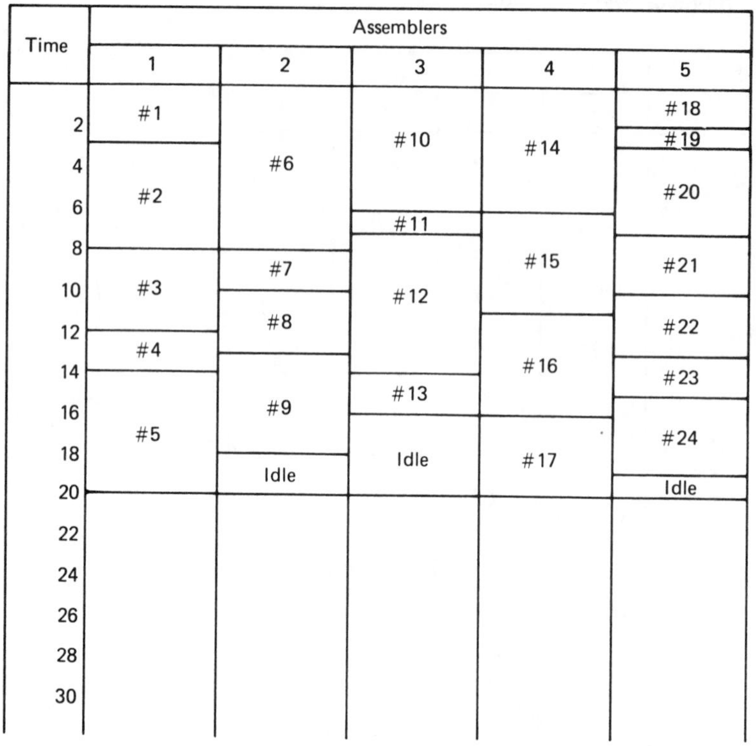

Figure 3-27 Multiple Activity Chart for Blivet Assembly (Five Operators).

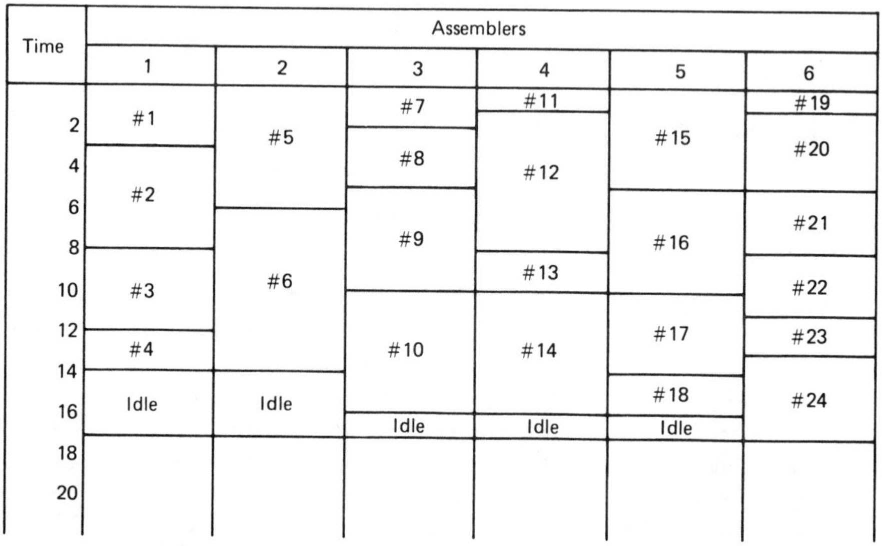

Time	Assemblers				
	1	2	3	4	5
2	#1	#6	#10	#14	#18
					#19
4	#2				#20
6			#11		
8	#3	#7	#12	#15	#21
10		#8			
12	#4			#16	#22
14	#5	#9	#13		#23
16					#24
18		Idle	Idle	#17	
20					Idle
22					
24					
26					
28					
30					

Figure 3-28 Multiple Activity Chart for Blivet Assembly (Six Operators).

Time	Assemblers					
	1	2	3	4	5	6
2	#1	#5	#7	#11	#15	#19
			#8			#20
4	#2			#12		
6		#6	#9		#16	#21
8				#13		
10	#3					#22
12	#4		#10	#14	#17	#23
14					#18	
16	Idle	Idle	Idle	Idle	Idle	#24
18						
20						

several activities: hands, people, or operators and the time required for each activity. Proper construction of the charts, coupled with some simple calculations, can determine the best, most economical, or most productive use of that resource. This knowledge must then be combined with the other factors influencing production before a final solution is suggested.

SUMMARY

Charts tell what is happening in graphical form. In addition, a careful, questioning attitude can help determine why the current method is being used and can help identify potential areas for improvement. These methods and tools can be used effectively to increase productivity.

The following is a summary of the definitions of the graphical analysis tools described in this chapter (from ANSI Z94).

Operation Process Chart: A graphic, symbolic representation of the act of producing a product or providing a service, showing operations and inspections performed or to be performed with their sequential relationships and materials used. Operation, inspection time required, and location may be included.

Flow Process Chart: A graphic, symbolic representation of the work performed or to be performed on a product as it passes through some or all of the stages of a process. Typically, the information included in the chart is quantity, distance moved, type of work done, and equipment used.

Flow Diagram: A representation of the location of activities or operations and the flow of materials between activities on a pictorial layout of a process. Usually used with a flow process chart.

Left-hand/Right-hand Chart: A chart on which the motions made by one hand in relation to those made by the other hand are recorded, using standard process chart symbols.

Multiple Activity Chart: A chart of the coordinated synchronous or simultaneous activities of a work system of one or more machines and/or one or more workers. Each machine and/or worker is shown in a separate, parallel column indicating their activities as related to the rest of the work system.

REVIEW QUESTIONS

1. What does an operation process chart show?
2. What does a flow process chart show?
3. What does a flow diagram show?
4. What does a LH/RH chart show?
5. What does a multiple activity chart show?
6. Define the *standard charting symbols.*

Time	Assemblers														
	1	2	3	4	5	6	7	8	9	10	11	12	13	14	15
2	#1	#3			#7				#13			#17	#20	#22	#24
			#5	#6	#8	#9	#10			#15	#16			#23	
4		#4						#12	#14			#18	#21		
6	#2				Idle	Idel	#11			Idle	Idle	#19		Idle	Idle
		Idle	Idle	Idle			Idle	Idle				Idle	Idle		
8															

Figure 3-29   Multiple Activity Chart for Blivet Assembly (15 Operators).

7. How can each of the following be used to improve productivity?
   (a) Operation process chart
   (b) Flow process chart
   (c) Flow diagram
   (d) Left-hand/right-hand chart
   (e) Multiple activity chart

8. List and describe each of the major questions that should be asked about any operation.

PRACTICE EXERCISES

1. An activity that almost everybody has done at some time in their life is to pound a nail into a piece of wood with a hammer. Using the workplace sketched below, construct a left-hand/right-hand chart of the activity.

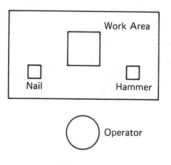

2. At Mountain View College, the following activities are required for a student to successfully graduate once all academic credit has been earned:

   The student must fill out a petition for graduation. This document, which must be filed the quarter preceding the student's graduation, must show the courses the student is currently taking, as well as the courses the student plans on taking his or her final quarter in school. Once the student has filled out this petition, it is submitted to the student's academic department for preliminary approval. The student's department checks the petition for two items: first, to see if all required courses have been completed (or scheduled) and, second, to check that the required number of credit hours for graduation have been earned. Individual curriculum modifications sometimes permit the student to complete the required courses but not to complete the required number of credit hours.

After the academic department approves the petition, the student must take the petition to the cashier's office and pay the graduation fee. After paying the graduation fee, the student takes the petition and receipt to the registrar's office. The registrar of the college checks the petition for four items: the cashier's indication that the fee for graduation has been paid, the student's major department approval, the fact that the student has indeed completed all required courses, and the fact that the student has the correct number of hours to meet the graduation requirements for the degree sought. (Transfer credit may permit a student to receive credit for a particular course but have more or fewer hours credit than the same course at Mountain View.)

After the registrar's office approves the petition, the student's request to graduate is submitted to the college faculty for approval. Each quarter the faculty accepts as candidates for graduation each of those who have met the graduation requirements and followed the above procedure in petitioning for graduation. After the faculty approves the candidates for graduation, the students are free to attend any college course they choose.

(a) Construct a flow process chart for the above process.

(b) Suggest an improved method for accomplishing the above process. Prepare a flow process chart to show the improved method. Document the improvement.

3. Before being named chef at a local restaurant, Gregory was responsible for work-study at some manufacturing companies. (He viewed being chef for a restaurant as a process of manufacturing, consistently, the same meals in the same way.) As part of his strategy to assure consistency in meal preparation. Gregory wants to apply an operation process chart concept to his salad-making operation. He figures he can then have any trainee chef prepare a salad that is up to Gregory's standards because the entire procedure is documented. Use the following information to help Gregory prepare his operation process chart for the salad-making operation.

As prepared at Gregory's restaurant, the typical salad contains several ingredients: lettuce, tomato, pepper, mushrooms, cucumbers, croutons, magic ingredient, and dressing. Although not formally called operations sheets, the ingredients are processed as follows:

LETTUCE

1. Wash
2. Chop
3. Shred
4. Place on plate

PEPPER

1. Wash
2. Slice
3. Remove seeds
4. Chop
5. Place on plate

TOMATO

1. Wash
2. Slice
3. Chop
4. Place on plate
   mixed with lettuce.

MUSHROOMS

1. Wash
2. Chop
3. Place on plate

CUCUMBERS

1. Wash
2. Skin
3. Slice
4. Chop
5. Place on plate

CROUTONS AND MAGIC INGREDIENT

1. Crumble bread
2. Fry bread
3. Add Magic Ingredient
4. Place mixture on salad

DRESSING

1. Check for proper dressing
2. Open container
3. Place dressing on salad
4. Serve salad

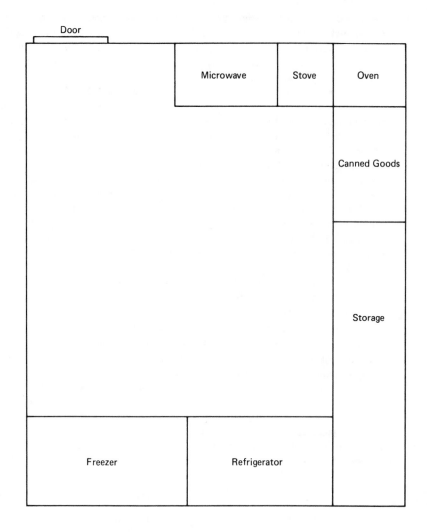

Figure 3-30　Layout of Gregory's Kitchen.

4. The specialty of the house at Gregory's is roast turkey. When Gregory prepares a roast turkey dinner, he performs the following steps in the kitchen. (The layout of the kitchen is shown in Figure 3-30.)

(1) Remove turkey from freezer.

(2) Place turkey in microwave oven.

(3) Defrost turkey with microwave oven.

(4) Place turkey in saucepan.

(5) Apply seasoning to turkey.

(6) Place turkey back in microwave oven and start cooking.

(7) Remove cranberries from refrigerator.

(8) Start cooking cranberries.

(9) Remove vegetables from refrigerator and start cooking.

(10) Check on turkey.

(11) Check on cranberries.

(12) Remove cranberries from stove.

(13) Check on vegetables, continue to simmer.

(14) Grind cranberries into proper consistency for eating.

(15) Cool cranberries in refrigerator.

(16) Check on turkey.

(17) Check on salad.

(18) Check on dessert with pastry chef.

(19) Check on vegetables, add additional seasoning.

(20) Check on turkey.

(21) Remove turkey.

(22) Slice turkey.

(23) Serve turkey.

(24) Serve vegetables.

(25) Serve cranberries.

(26) Bring dinner to customer.

(27) Serve dessert at appropriate time.

   (a) Prepare a flow process chart for the above activities (what will you chart?).

   (b) Prepare a flow diagram for the above activities.

   (c) Suggest an improved method; document it with appropriate charts.

   (d) Suddenly, instead of being faced with the rational judgment of former engineer Gregory, we are faced with new bird chef, Pierre. What influences might Pierre have on the method we are proposing? How would you convince the new bird-cooker that the changes are really warranted?

5. The following describes the activities required for an individual to remove a ball point pen from the left shirt pocket and start writing with it. Prepare a left-hand/right-hand chart of this activity.

   From a rest position with both hands on the table immediately in front of the operator, the left hand reaches for the pen in the left front shirt pocket. This hand grasps the pen and brings it to the right hand. While the LH holds the pen the RH removes the pen cap from the pen. Holding the pen, the left hand rotates the pen 180° for the right hand to stick the cap on the other end of the pen. The left hand then releases the pen completely and the right hand moves the pen-cap assembly to the paper in the work area immediately in front of the operator. The left hand moves to the paper and holds the paper down on the surface so that it will not move when the right hand begins writing. The writing process is then ready to begin.

6. In almost every text on this subject, and this textbook is no exception, the section or chapter on charting includes what is known as the "toaster problem." Loading an old-fashioned toaster, as shown in Figure 3-31, requires both hands. One piece of toast can be loaded at a time. While one hand puts the bread in the toaster, the other hand must keep the panel open until the toasting time begins. Once a piece of bread is toasting it is cooked only on the side facing the electric coil, meaning that while one piece of bread is toasting, a second piece may be loaded into the other side and also have one side toasted. Because this toaster is a "classic," it requires that once a piece of bread has been toasted it must immediately be removed or transferred so the other side can be toasted. Failure to do this procedure would result in burned toast. Bread doesn't have to be transferred or cooked on both sides right away, but it must be removed.

Figure 3-31   A Classic Toaster.

Following are the work elements, their descriptions, and the times required for toasting bread with this classic toaster.

ACTIVITY	TIME	DESCRIPTION
Load Toast	.1 min	Place piece of toast into toaster for toasting. Must be done with both hands—one to place bread in machine, the other to keep door open.
Unload Toast	.1 min	Remove toast from toaster after cooking and place on plate.
Transfer Toast	.05 min	Place removed toast on other side of toaster after unloading and before loading.
Turn Toast	.2 min	Turn toast 180° after removal and before loading. Required to toast same slice in same side of toaster.
Toast	.5 min	Cook bread to perfection on one side. Any time less than .5 minutes is unacceptable and any time over .5 minutes will burn the bread.

What is the minimum time required to cook three slices of bread?

7. A product requires the following 10 operations for assembly. If assemblers get paid $8 per hour and the company can afford a maximum of four assembly lines, how many lines should the company have and how many assembly stations should be on each line?

OPERATION NO.	TIME (HOURS)
1	.14
2	.18
3	.15
4	.11
5	.26
6	.17
7	.16
8	.16
9	.20
10	.14

(a) If the goal is to minimize labor cost?
(b) If the goal is to minimize assembly time?

8. Rolls of lining material, ribbons of binding (border), rolls of zipper tape, boxes of zipper hardware, and boxes of labels are purchased and kept in inventory until needed. The zipper is made at a subassembly operation. The roll of zipper tape is cut to length and the zipper hardware attached at a single operation. The zipper is then tested to verify that it operates properly. The lining material is cut into three body pieces: a back, a left front, and a right front. At the first sewing operation, the right front is sewn to the back and a label inserted in the long side seam. Then, the left front is sewn to the back along the long side seam. The border or binding material is cut to length and sewn to the inside of each armhole, up to the neck opening. The next operation joins the top right and left fronts to the back at the shoulders. Another piece of binding or border material is then cut to length and sewn to the outside edge of the raincoat liner. The zipper is then sewn to the liner and the zipper is check again for proper operation. The entire liner is then sent to final inspection and subsequently loaded onto a freight trailer for delivery to the customer.

Prepare an operation process chart for this assembly operation.

9. A Mexican-style fast food restaurant sells burritos, tacos, refried beans, spanish rice, and nachos. The kitchen layout is shown in Figure 3-32. The process of filling a routine order can be summarized as follows:

The cook gets up out of his chair and walks to the front counter to pick up the next order. He examines the order to make sure he can read it, and then takes it to the food preparation area (work counter). He places the order under a metal clip so he can refer to it as he fills the order. He reads the order carefully and notes that it is for one taco plus side orders of refried beans and spanish rice.

The cook picks up a taco shell and places it in the warmer. He then waits 30 seconds for the shell to get warm. The cook removes the heated taco shell from the warmer unit and holds it in his left hand. He then looks at the order and reads that the customer wants a combination beef/bean taco. So he takes the meat scoop and puts some meat in the taco shell. Then he takes the bean scoop and puts some beans in the taco shell. Next, he again looks at the order and sees that the customer wants hot sauce. So he takes the taco over to the garnishment area where he

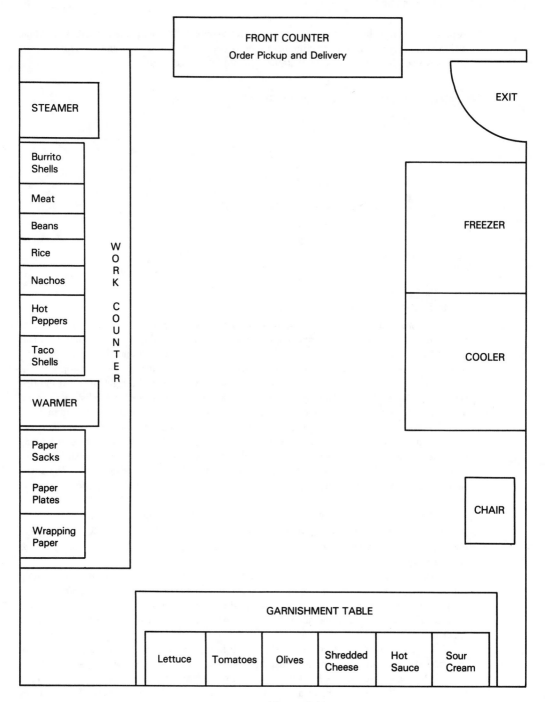

Figure 3-32

throws on a small handful of lettuce, tomatoes, and shredded cheese, in that order. Then he picks up the hot sauce and squirts a generous helping onto the taco. When he finishes, he carries the taco back to the work counter. He then removes a sheet of wrapping paper from the stack and places it on the counter. He places the taco on the paper and wraps it.

He looks at the order again and sees that the customer is going to eat in the restaurant so he picks up a paper plate and puts it on the counter. He then places the taco on the plate.

He looks at the order again and sees the request for a side order of refried beans. So he takes the bean scoop and fills one of the plate's small compartments with refried beans.

He looks at the order again and sees the request for a side order of spanish rice. He takes the rice scoop and fills one of the plate's other small compartments with spanish rice.

He glances back at the order and sees that the has finished his part. He takes the order from beneath the metal clip and picks up the plate with the other hand. He takes both back to the front counter and sets the plate down on top of the order on the counter. He then returns to his chair and sits down.

Prepare a flow process chart and flow diagram for the existing operation. Then improve upon the existing method (and layout), and prepare a second flowchart and diagram for the improved method you develop.

10. U-bolts, clamps, washers, and nuts are located in separate containers in front of a worker. The assembly process begins with the worker reaching for a U-bolt with his left hand. He grasps the bolt at the curved section and moves it back to a position directly in front of him. While the left hand holds the U-bolt in an upright position, the right hand reaches for the clamp, grasps it, and moves it toward the U-bolt. He carefully positions the clamp above the U-bolt and then lowers it into position. The right hand then reaches for a washer, grasps it, moves it to the U-bolt, and lowers it onto the left side of the bolt. The right hand then reaches for a nut, grasps it, moves it to the U-bolt, and assembles the nut onto the left side of the U-bolt. The right hand then reaches for a second washer, grasps it, moves it to the U-bolt, and lowers it onto the right side of the bolt. The right hand then reaches for a second nut, grasps it, moves it to the U-bolt, and assembles the nut onto the right side of the U-bolt. The left hand then places the completed assembly aide.

The process is repeated with the left hand reaching for the next U-bolt. The bolt is shown in Figure 3-33.

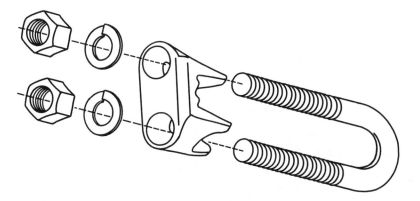

Figure 3-33   U-Clamp Assembly.

Prepare a left-hand/right-hand chart for the existing operation. Then, improve upon the existing method and prepare a second LH/RH chart for the improved method you develop.

11. The following process must be designed so that it produces between 150,000 and 250,000 assemblies per year.

Description	Employee	Machine
Get bottom piece with LH, load into station fixture.	.40	
Get right piece with RH, place in fixture against bottom.	.25	
Get left piece with LH, place in fixture against bottom.	.25	
Get top piece with RH, place on top of right and left pieces.	.20	
Activate machine with both hands-station rotate 1/4 turn.	.10	.10
Station "B" cycle time to process parts.		.20
Remove completed assembly with both hands.	.40	
Inspect assembly by inserting into GO-NOGO gauge with RH.	.35	
Stack pack assembly into case with RH.	.10	
REPEAT PROCESS		

NOTES:

a. Release, Pickup, or Give Piece from one hand to the other = .02 minute.

b. Station B is reserved for machine processing—no loading or unloading required.

c. Stations A, C, and D can be used for loading, unloading, or both.

Each multi-head machine costs $54,000 and has a three-year life with a zero salvage value. Each employee is paid an annual salary of $11,000 per year. The plant operates one shift

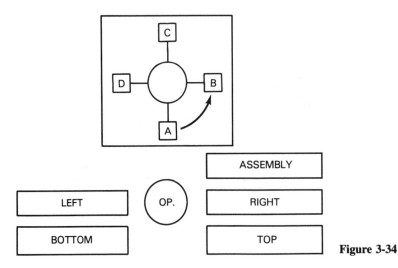

Figure 3-34

per day and has 10% non-productive time (432 minutes productive time per day). The initial load and final unload times each day are included in the 10% non-product time figure. The equipment is used 50 weeks per year, but the employees are paid for 52 weeks (includes two weeks holiday pay). No fractional parts can be produced during a day (example: 27.6 parts = 27 parts). All cost per unit figures should be rounded to three decimal places.

Given the above information, find the best solution. Use any chart or charts that will illustrate why your solution is optimal. Document your solution with cost per unit calculations. The funds are available to purchase any number of machines, or to hire any number of employees (or any combination of the two). This exercise is a cost minimization problem, so search for the lowest overall total cost per piece (equipment plus labor).

12. Below is a list of the assembly operations and times (minutes) required to assemble a bicycle.

Job	Description	Time	Follows
A	Attach Reflector to Rear Fender	.40	—
B	Affix Decals to Rear Fender	.25	A
C	Affix Decals to Front Fender	.30	—
D	Assemble Seat to Seat Pipe	.30	—
E	Attach Reflector to Rear Wheel	.20	—
F	Attach Reflector to Front Wheel	.20	—
G	Assemble Sprocket to Pedal Rod	.15	—
H	Place Pedal Assembly in Frame	.25	G
I	Assemble R & L Pedal to Pedal Assembly	.50	H
J	Assemble Rear Fender to Frame	.35	B
K	Place Chain over Sprocket	.10	I
L	Assemble Rear Wheel to Frame and Chain	.75	E,J,K
M	Adjust Rear Wheel and Chain Tension	.20	L
N	Assemble Chain Guard to Frame	.40	M
O	Assemble Front Fender to Frame	.35	C
P	Assemble Front Wheel to Frame	.60	F,O
Q	Assemble Handlebars to Front Pipe	.40	—
R	Assemble Headlight to Bracket	.50	—
S	Assemble Headlight Asm. to Front Pipe	.35	Q,R
T	Assemble Handlebar Assembly to Frame	.80	P,S
U	Slide R & L Grips onto Handlebars	.60	T
V	Attach Seat Assembly to Frame	.50	D
W	Assemble Kickstand to Frame	.40	P,N
X	Lubricate	.15	U,V,W
Y	Inspect	.85	X

Production requirements are between 300 to 315 bicycles per 8-hour day. Deducting 10% non-productive time, this output yields a 432-minute work day. An operator earns $5.00 per hour (working or idle). Saturday work is at a rate of time and one-half. Balance this assembly line and show your recommendations on a multiple activity chart. Indicate the total number of operators, daily line output, and labor cost per bicycle in your solution.

13. A customer has placed an order for 50,000 units per year for the next three years. In addition, the customer has agreed to purchase an additional 100,000 units per year if they are available (total of 150,000 units per year). Either of the following machines could be used to manufacture this new product.

	MACHINE A	MACHINE B
MACHINE DATA:		
Cost Each ($)	12,000	10,200
Life (Years)	3	3
Depreciation per Year ($)	4,000	3,400
EMPLOYEE DATA:		
Pay Grade Required	C	E
Rate per Hour ($)	5.00	6.00
Wages per Year ($)	10,400	12,480
OPERATING DATA:		
Run Cycle (minutes)	1.8	2.0
Load Machine (minutes)	0.9	0.5
Unload Machine (minutes)	0.3	0.3

The plant operates one shift per day and has 10% non-productive time (432 minutes productive time per day). The equipment is used 50 weeks per year, but the employees are paid for 52 weeks (includes two weeks holiday pay). No fractional parts can be produced during a day (example: 27.6 parts = 27 parts). All cost per unit figures should be rounded to three decimal places.

Prepare a multiple activity chart for the best combination of labor and equipment for both machines (A and B). Document your solution with cost per unit calculations for each alternative. The funds are available to purchase any number of machines, A or B (or any combination of the two). This exercise is a cost minimization problem, so search for the lowest overall total cost (equipment plus labor).

14. The Morse Rubber Products Company manufactures products made wholly or partly from rubber. One of their major products is a marine bearing. These bearings are manufactured in a variety of sizes. Ninety-eight percent of these bearings are made by the same process. This process is described below.

The bearings consist of a brass sleeve and the molded rubber liner. Each of these parts is first considered separately.

Naval Brass Sleeve

The engineering print calls for a particular size naval brass tubing with a specified minimum I.D. and wall thickness. The purchasing department seldom has difficulty in obtaining the correct I.D., but has difficulty in obtaining the proper wall thickness. In the interests of lowering cost, they usually purchase "mill run" tubing which has a heavier wall thickness. This tubing is usually bought in small lot quantities (possibly fifteen sticks, where a stick is a ten-foot length of tube). These lots are delivered to the Shipping/Receiving department where they are off-loaded. Paperwork is processed for the receiving action and the tubing is taken to the "saw building" via fork truck.

The bundle of tubing is placed inside the building and then broken. The individual tubes are placed in racks by hand. They remain there until they are needed to fill a production release.

The machine shop foreman receives a production order for sixty xxxxxx bearing sleeves. He gives the routing sheet to the saw operator. The operator goes to the "saw building" and prepares to cut the tube into the desired lengths. This set-up activity includes such activities as turning on the lights and heat, starting the compressor, positioning the roller conveyor, etc. Once the set-up is complete he manually obtains the correct number of tubes for the initial order. He then adjusts the saw for the correct speed, feed, and cut. The first sleeve cut from each "stick" undergoes a check activity and the saw is adjusted as required. "Sticks" are cut until the required number of sleeves are produced. The operator allows the cut sleeves to fall into a wire basket. Near the end of the production run, the operator will "clean up" around the area and get ready to close the activity down. On completion, he goes to the main plant, obtains a fork truck and returns. He carries the newly cut sleeves to the machine shop via the fork truck. Here he places the basket of sleeves near the machine (usually a lathe) which is scheduled to perform the next activity.

The foreman next assigns a machine operator to the lathe. He (the operator) will perform the indicated operation on the material in the queue. Note: each basket of sleeves contains a route sheet which calls out the needed activities.

The operator will usually do the following:

- Position the basket near the work table.
- Deburr the I.D. on both ends of the sleeve.
- Position sleeve to lathe and set for turn operation.
- Turn sleeve to correct diameter.
- Remove from lathe.
- Check O.D.
- Place sleeve in basket.

The wire baskets of turned sleeves is moved via fork truck to a staging area. Here it waits for a "tank load" of parts to be degreased. When this process is completed, the baskets of sleeves are moved to the paint room where they await having their "insides" sprayed with a bonding agent.

After two coats of bonding agent are sprayed on the I.D., a sleeve is staged for the molding operation, where raw rubber compound is molded to the sleeve. Usually, the molding operation molds two bearings at a time.

Rubber For Marine Bearing

Raw rubber, carbon black, oil, and assorted polymers are received at the receiving dock. These materials are unloaded by fork truck and are given a simple visual check. They are then forwarded to the material preparation area where they are stored prior to mixing. Marine bearing rubber is mixed in lots according to the formula set forth by the chemist. Immediately following mixing, the batch is "rolled" and "hung." Next it is placed on pallets and marked awaiting use in the 6" extruder. Here it is formed into $24'' \times 36'' \times \frac{1}{2}''$ sheets and placed on large carts awaiting preform preparation.

The preform operator secures a small cart full of sheets (about 10) and moves them about 40 feet to the clicking press. Here he positions the sheets to the press platen and cuts formed

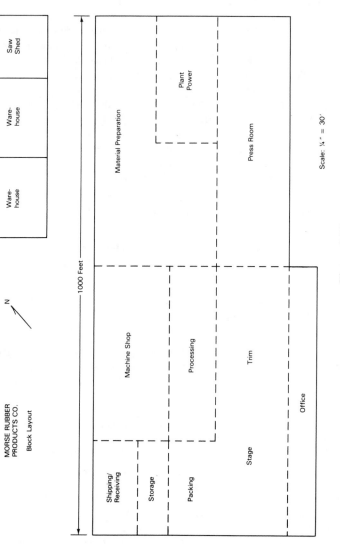

MORSE RUBBER PRODUCTS CO.

Block Layout

Ware-house

Ware-house

Saw Shed

1000 Feet

N

Shipping/ Receiving

Storage

Packing

Stage

Machine Shop

Processing

Trim

Office

Material Preparation

Plant Power

Press Room

Scale: ¼" = 30'

Figure 3.35

pieces. These pieces are assembled and weighed. Then they are placed on a shelf awaiting pickup by the pressman, who moves them to the press. He then assembles the sleeves and preforms them to the mold and places them into the press. The unit cures under heat and pressure for a specified period.

Bearing. The mold is removed from the press and opened to remove the cured bearings. They are stacked in a wire basket (the same type used for the sleeves). When the lot is complete, the baskets are moved to the trim department to await clean up. The excess rubber is trimmed and the bearing is ground to length on a stand grinder. The unit is placed in its wire basket awaiting movement to the machine shop.

The small units (bearings) await turning to their final diameter on the centerless grinder. The larger bearings (over 3″) are turned to the correct diameter on the lathe. Once a bearing is of proper diameter, it is placed in a staging area awaiting movement to the packing area.

The baskets are placed in a staging area to await packaging in cardboard boxes. The bearings are moved to a work area where they are cleaned and identified. Boxes are made and the bearings are packed into the box and labeled. These boxes are then placed on a cart awaiting movement to storage.

The completed bearings remain on the shelves until sales orders require their removal, packing and shipping to the customer.

Prepare appropriate charts for the processes described in this exercise.

15. The hardware package for assembling a do-it-yourself bookcase consists of the following items:

Hardware	No.
Angle Brackets	8
15/16″ Screws	16
9/16″ Screws	16
3/4″ Screws	4
7/8″ Bolts	2
Hex Nuts	2
Magnetic Catch	1
Strike Plate	1
Hinge Bracket	1
Hinge Bushings	2
Assembly Tool	1
Plastic Bag	1

All of the above hardware (including the Assembly Tool) is put into the plastic bag by hand by an operator seated at a workbench. When the bag is full, the operator seals the bag with a sealing machine located to one side of the workbench. The filled, sealed bag is then put into a 18″ × 22″ tote pan for future use in the next operation.

All hardware is kept in 5″ × 9″ trays on the workbench in front of the operator. The 6″ × 8″ plastic bags are in stacks of 50, and are also located on top of the workbench. The floor model sealing machine is 14″ × 14″, with the sealing device about 38″ from the floor.

Prepare a workplace layout for this operation, showing where all the hardware bins are located on the bench. Then, prepare a left-hand/right-hand chart for your proposed assembly method (which should correspond to your layout).

16. The following items are sold as a standard blood gas sampling kit to hospitals and physicians:

No.	Item	Quantity
1	Box	1
2	Blister	1
3	Syringe	1
4	Needle—Small Gauge	1
5	Needle—Medium Gauge	1
6	Test Tubes	4
7	Antiseptic	1
8	Chemical A	1
9	Chemical B	2
10	Chemical C	1
11	Ice Pack	1
12	Cotton Balls	2
13	Bandaid	1
14	Instructions	1
	Identification Stamp	

The LH picks up the box and both hands fold in the bottom. The LH holds the box while the RH picks up a blister and positions it into the bottom of the box. The RH then picks up items 3 through 10, one at a time, and positions them into their individual locations within the specially-designed blister. The RH then picks up the ice pack and places it in the box, covering all other items. Next, the RH picks up items 12 through 14, one at a time (except it picks up the cotton balls at once), and places them loosely on top of the ice pack. The RH then closes the lid to the box, picks up an identification stamp, stamps the box, and sets aside the stamp. The LH then puts the full box into a case.

1. Using the principles of motion economy, improve this assembly method by designing an optimal workplace layout.

2. Describe your improved assembly method in a brief, concise narrative with supporting reasons for its design.

3. Prepare a LH-RH chart for your improved assembly method.

CHAPTER 4

Work Methods Improvement

OBJECTIVES

Upon completing this chapter, the reader should be able to

- Define *methods improvement*.
- Explain how methods improvement relates to productivity improvement.
- Understand the use of and be able to apply the principles of motion economy to the work improvement process.
- Understand some basic principles regarding work design that reflect the capabilities and limitations of people and machines.

INTRODUCTION

In the preceding chapter, we learned how to chart various work activities. While studying the operation process chart, flow process chart, flow diagram, left-hand/right-hand chart, and the multiple activity chart we indicated that these graphical representations of work being performed could be used to improve productivity. By questioning the inclusion of every line on each chart, we identified those activities that should or could be improved. Every time an element of a job can be eliminated, combined, or shortened, the worker will be able to produce more in the same time period. It is hoped that the improvement will make the work easier to perform as well. This type of improvement translates to an increase in productivity.

METHODS IMPROVEMENT DEFINED

Traditionally, the industrial engineering analysis function has been concerned with how people perform their jobs. Two areas have received considerable attention. The first is determining the best way to perform a job. This determination is commonly called *methods analysis*. The second is measuring the time required for a job to be completed using the best method. This measurement has most often been concerned with time study. This chapter will concentrate on the first of these areas, methods analysis, while subsequent chapters will illustrate some of the methods used to measure the time required to perform certain types of work.

Since the founding of industrial engineering, practitioners have always been very concerned about the methods of performing work. Frank Gilbreth applied his ideas about methods to jobs as diverse as bricklaying and surgery.*

Methods improvement is the process whereby a task is analyzed and possible changes are identified that will either increase the productivity of the worker, make the work easier to perform, or both. Analysis is usually performed by identifying the current method, questioning the need for each step in the process, and devising an improved way to perform portions of the task that are really necessary. The final portion requires the analyst to convince the worker and the worker's supervisor that the improved method is really better.

The first portion of the analysis process is relatively straightforward. Determination of the present method can be accomplished by completing one of the standard charts we learned how to use, by recording the job on film or videotape, or by any other similar "permanent" record that shows all activities performed.

Second, a list of questions, such as those suggested in the preceding chapter, can be prepared to use regarding each activity. Some possible questions include: WHY, WHEN, WHERE, WHO, IS IT REALLY NECESSARY, and so forth.

The third portion, developing a better way, can be more difficult. This portion involves creativity and common sense. No matter how much of an improvement anyone makes, an underlying axiom of this analysis is that there is always a better way. The classical industrial engineering way of identifying improved work methods is through the use of motion analysis. Gilbreth, through the use of Therbligs (which we will discuss later in this chapter), was able to identify the microscopic elements of work being performed. These elements were then classified as effective or ineffective. Some guides have evolved, in more general terms, over the years, called the *principles of motion economy.* They are really just common sense recommendations.

Although this third portion is one of the most important parts of the work improvement process, it is thought by some experienced industrial engineers to be one of the most difficult principles. It is a skill or talent that must be individually nurtured for an individual to be successful. Guides such as Therblig analysis and motion economy principles provide very valuable starting points, but complete analysis of existing jobs should not culminate with the use of these guides.

The final part of the methods improvement process is convincing the workers and supervisors involved that the new method is really an improved method and not just a result of management's desire to make life more difficult. Overcoming resistance to change is perhaps one of the trickiest parts of the analyst's job. Most people really don't like to change what, when, and how they do whatever it is they do.

The remainder of this chapter will be devoted to a more thorough examination of these aspects of work improvement. The charting techniques discussed in the

---

*Spriegel and Myers have compiled *The Writings of the Gilbreths* (Irwin) which, as the name implies, is a serious examination of the work of the Gilbreths. For a lighter look at the Gilbreths, *Cheaper by the Dozen* by Gilbreth, is recommended.

previous chapter form the basis for describing the present method of performing work. The following sections examine the remaining portions of what is known as methods improvement.

THE BEST METHOD

Often, the best way to utilize a person in the manufacturing process is not to use the person at all. In other words, the best use of the human is no use of the human.* All production operations should first and foremost be subject to the question WHY? Why is this operation being performed? Why is this operation necessary? Why is this operation being performed at this time? If the answers to these questions indicate that it really is necessary to perform an operation, then the following questions should be asked:

Where is the person used in the process? People do many things well, decision-making being an excellent example. There are many things that machines can do better than people, such as highly repetitive tasks. Machines are much more consistent in the performance of these tasks and machines don't get bored. To use our resources most productively, we must check to see that we are making the best use of the person. We should not expect a person to perform tasks that a machine could do better.**

How is the person used in the process? The following generalities aid in the design of the manufacturing processes pertaining to the work that is preferred for people, procedures, layout, and equipment:

1. The work procedure relates to the distribution of work to part of the body, sequence of motions, and parts of the body to use.
2. The layout of the work involves placing those portions of the work to be done by the individual in the most accessible locations.
3. Equipment design uses a number of human factors guides to optimize human performance in the operation of controls, the use of tools, and the analysis of information (Neibel, 1976).

Can the operation be eliminated? The question to be asked is: Can the operation be done away with completely? Is it really necessary? Will the work have to be undone by the next user of the product? Elimination of unnecessary activities increases operator productivity.

Can the operation be combined with others? Can the work element be combined with another activity that either precedes or follows the activity? Combined motions usually don't require stops or starts and the lack of deceleration or acceleration time will make a contribution toward increased output.

*The emerging science of robotics is the best illustration of this statement.

**The chapter on human factors or ergonomics will examine this concept in more detail. Numerous checklists have been developed, based on physiological and anthropometrical studies, that suggest whether a person or machine is more qualified to perform certain general types of tasks.

Can the sequence of activities be changed? Sometimes the order of activities can be altered to increase productivity. A logical sequence of activities—logical by space arrangement, for example—can be more productive than a haphazard sequence of operations.

Can the activities be simplified? People tend to be more productive on simple motions. The more complex the motions, the more difficult it will be for an individual to perform. Productivity can be increased when the operator gains skill performing certain operations. Generally, due to the "learning curve" effect, the simpler the motion, the better it can be performed.

THERBLIG ANALYSIS

Frank and Lillian Gilbreth maintained that every motion could be subdivided into a series composed of several of the 17 motions they identified. They called these basic motions *Therbligs*. (Therblig is almost Gilbreth spelled backwards.) The Gilbreths believed that every task was a compilation of the following motions (their original symbol appears after each definition):

REACH: The motion of the empty hand moving from one location to another. Reach begins when the hand (or body member) begins to move without a load and ends when the hand (or body member) touches the object or stops moving. (TE)

MOVE: The motion wherein an object is transported from one location to another. Move begins when the hand starts to move with an object and ends when the hand either arrives at its destination or when movement stops. (TL)

RELEASE: The motion wherein the hand relinquishes control of an object. Release begins when the hand starts to relax its control over the object and ends when the hand loses contact with the object. (RL)

GRASP: The motion wherein the hand gains control of an object. Grasp begins when the hand touches the object and ends after control is secured. (G)

PREPOSITION: The motions wherein the operator prepares an object for future use. Preposition begins when the hand causes the object to begin to be placed into position and ends when the object is in the correct position for later use. Preposition occurs in a location other than where the object will be used. (PP)

USE: The motion of manipulating a tool for the function for which it was intended. Use begins when the hand starts to manipulate the tool and ends when the hand stops manipulating the tool. (U)

ASSEMBLE: The motion wherein the operator places two or more objects together so that they are more productive in their combined state than in their separate conditions. Assemble begins when the hand causes the objects to begin to go together and ends when the objects are together. (A)

DISASSEMBLE: The motion wherein the operator takes two or more objects apart. Disassemble begins when the hand causes two objects which were together to begin to come apart and ends when the objects are separated. (DA)

SEARCH: The motion wherein the hand tries to find an object. Search begins when the hand starts to hunt and ends when the hand has found the location of the object. (SH)

SELECT: The motion of the hand finding an object within a group of objects. Select begins when the hand touches several objects and ends when the hand has located the object desired. (ST)

POSITION: The hand motion of aligning, orienting, or changing the position of an object. Position begins when the hand starts lining up or orienting the object and ends when the hand has aligned or oriented the object. (P)

INSPECT: The motion of comparing an object to a standard. Inspect begins when the hand begins to feel or the eye begins to view the object and ends when the object has been felt or viewed. (I)

PLAN: The process of determining which activity to perform next. Plan starts when the hand is idle and ends when the course of action is decided upon. (PN)

UNAVOIDABLE DELAY: Delays that are part of the normal work method and beyond the control of the operator. Delay begins when the hand becomes idle and ends when work resumes. (UD)

AVOIDABLE DELAY: Delays that are not part of the normal work method. Avoidable delay starts when normal method is deviated from and concludes when normal method is resumed. (AD)

REST: The lack of motion while an operator intentionally stops work due to fatigue. Rest begins when motion ceases and ends when work is resumed. (R)

HOLD: The motion wherein the hand acts as a fixture controlling an object while the other hand operates on the controlled object. Hold starts when the held object ceases movement and stops when the object moves.

To convert the Therbligs to the standard process chart symbols used with the left-hand/right-hand chart, the following conventions are usually used (Mundel, p. 315):

Used with Operation Symbol:	Grasp	Disassemble
	Position	Search
	Preposition	Select
	Use	Release
	Assemble	Inspect
Used with Transportation Symbol:	Reach	Move

Used with Delay Symbol: Hold

Used with Storage Symbol: Avoidable Delay
                               Unavoidable Delay
                               Rest
                               Plan

As stated earlier, Gilbreth indicated that certain of the Therbligs were necessary for work. They are called effective Therbligs. Effective Therbligs include the following:

Reach   Preposition
Move   Use
Grasp   Assemble
Release   Disassemble

When effective Therbligs are present in the motions performed, productive work is being performed. Ineffective Therbligs are those which describe work that is generally not productive. The analyst should strive to improve the job whenever any ineffective Therbligs are present. These Therbligs are identified when a job is analyzed with a left-hand/right-hand chart. The ineffective Therbligs include the following:

Search   Unavoidable Delay
Select   Avoidable Delay
Position   Rest
Inspect   Hold
Plan

When the motions identified in a left-hand/right-hand chart are described in terms of Therbligs, ineffective Therbligs present the ideal place to begin the work improvement process.

Example: Bolt-Washer-Nut Assembly

To illustrate the use of Therbligs in the work improvement process, the following assembly operation will be analyzed. The assembled product that is shipped to customers from the Wash Me Not Washer Company is the bolt-washer-nut assembly shown in Figure 4-1. The work area is shown in Figure 4-2.

The procedure for assembling the final product is as follows: The operator begins the task by reaching, with her left hand, for a bolt from the tray containing bolts. Upon reaching the tray, the operator selects a bolt, grasps it, and moves it to the work area in front of her. The operator then reaches with her right hand for a washer from the tray containing washers. A washer is selected and is brought to the work area in front of her. She then positions the washer so that it will fit on the end of the bolt and proceeds to assemble the washer and bolt. This task completed, the operator then reaches to the next

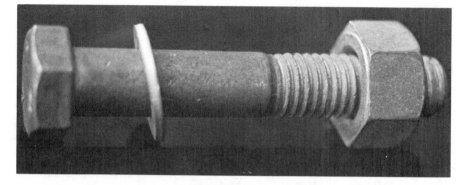

Figure 4-1   Bolt-Washer-Nut Assembly.

tray, selects a nut from the pile, and brings the nut to the bolt. Then she places the nut on the bolt. After this task is completed, the left hand sets the finished assembly aside and begins the cycle again.

The analyst's first inclination is to prepare a left-hand/right-hand chart of the operation. This chart is shown in Figure 4-3. Figure 4-4 shows the Therbligs that correspond to the LH/RH chart. Some may require additional discussion, such as the avoidable delay used for the activity of the right hand on Line 1. While the left hand is performing some task, the right hand should be doing something. It is obviously not productive for the operator to work only with one hand.

Lines 11 and 17 show positioning Therbligs. The rationale here is that, to perform the ensuing assembly, the operator must place the washer in exactly the same plane orientation as the bolt.

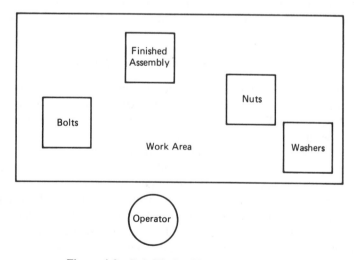

Figure 4-2   Bolt-Washer-Nut Assembly Area.

LEFT HAND -- RIGHT HAND CHART

SUMMARY:

SYMBOLS		PRESENT		PROPOSED		DIFFERENCE	
		LH	RH	LH	RH	LH	RH
O	OPERATION	3	10				
⇨	TRANSPORTATION	3	4				
□	INSPECTION	0	0				
D	DELAYS	14	6				
	TOTALS	20	20				

PROCESS: _Bolt - Washer - Nut Assembly_
STUDY No: _1_
OPERATOR: _OLDS_
ANALYST: _LSA_
DATE _12_ / _4_ / _82_
METHOD (PRESENT) ~~PROPOSED~~
SHEET No. _1_ of _2_
REMARKS: _____

SKETCH OF: _____ BY: _____

See Figure 4-2

LEFT HAND Present METHOD	Therblig	SYMBOL	SYMBOL	Therblig	RIGHT HAND Present METHOD
Reach for Bolt	TE	O ➡ □ D	O ⇨ □ ◗	UD	Idle
Select Bolt	ST	● ⇨ □ D	O ⇨ □ ◗	UD	Idle
Grasp Bolt	G	● ⇨ □ D	O ⇨ □ ◗	UD	Idle
Move Bolt To Work Area	TL	O ➡ □ D	O ⇨ □ ◗	UD	Idle
Hold Bolt	H	O ⇨ □ ◗	O ➡ □ D	TE	Reach for Washer
Hold Bolt	H	O ⇨ □ ◗	● ⇨ □ D	ST	Select Washer
Hold Bolt	H	O ⇨ □ ◗	● ⇨ □ D	G	Grasp Washer
Hold Bolt	H	O ⇨ □ ◗	O ➡ □ D	TL	Move Washer to Work Area
Hold Bolt	H	O ⇨ □ ◗	● ⇨ □ D	P	Align Washer
Hold Bolt	H	O ⇨ □ ◗	● ⇨ □ D	A	Place Washer on Bolt
Hold Bolt	H	O ⇨ □ ◗	● ⇨ □ D	RL	Release Bolt
Hold Bolt	H	O ⇨ □ ◗	O ➡ □ D	TE	Reach for Nut

Figure 4-3 LH/RH Chart for Bolt-Washer-Nut Assembly.

STUDY No. ___1___   SHEET No. _2_ of _2_

LEFT HAND Present METHOD	Therblig	SYMBOL	SYMBOL	Therblig	RIGHT HAND Present METHOD
Hold Bolt	H	○ ⇨ □ ◗	● ⇨ □ D	ST	Select Nut
Hold Bolt	H	○ ⇨ □ ◗	● ⇨ □ D	G	Grasp Nut
Hold Bolt	H	○ ⇨ □ ◗	○ ➡ □ D	TL	Move Nut to Work Area
Hold Bolt	H	○ ⇨ □ ◗	● ⇨ □ D	P	Align Nut
Hold Bolt	H	○ ⇨ □ ◗	● ⇨ □ D	A	Place Nut on Bolt
Hold Bolt	H	○ ⇨ □ ◗	● ⇨ □ D	RL	Release Nut
Move Assembly To Drop	TL	○ ➡ □ D	○ ⇨ □ ◗	UD	Idle
Release Assembly	RL	● ⇨ □ D	○ ⇨ □ ◗	UD	Idle
		○ ⇨ □ D	○ ⇨ □ D		
		○ ⇨ □ D	○ ⇨ □ D		
		○ ⇨ □ D	○ ⇨ □ D		
		○ ⇨ □ D	○ ⇨ □ D		
		○ ⇨ □ D	○ ⇨ □ D		
		○ ⇨ □ D	○ ⇨ □ D		
		○ ⇨ □ D	○ ⇨ □ D		
		○ ⇨ □ D	○ ⇨ □ D		
		○ ⇨ □ D	○ ⇨ □ D		
		○ ⇨ □ D	○ ⇨ □ D		
		○ ⇨ □ D	○ ⇨ □ D		
		○ ⇨ □ D	○ ⇨ □ D		
		○ ⇨ □ D	○ ⇨ □ D		
		○ ⇨ □ D	○ ⇨ □ D		
		○ ⇨ □ D	○ ⇨ □ D		
		○ ⇨ □ D	○ ⇨ □ D		

Figure 4-3 (Continued).

LEFT-HAND	THERBLIG		RIGHT-HAND
1 Reach for bolt	TE	AD	Idle
2 Select bolt	ST	AD	Idle
3 Grasp bolt	G	AD	Idle
4 Move bolt to work area	TL	AD	Idle
5 Hold bolt	H	AD	Idle
6 Hold bolt	H	TE	Reach for washer
7 Hold bolt	H	ST	Select washer
8 Hold bolt	H	G	Grasp washer
9 Hold bolt	H	TL	Move washer to work area
10 Hold bolt	H	P	Align washer
11 Hold bolt	H	A	Place washer on bolt
12 Hold bolt	H	RL	Release bolt
13 Hold bolt	H	TE	Reach for nut
14 Hold bolt	H	ST	Select nut
15 Hold bolt	H	G	Grasp nut
16 Hold bolt	H	TL	Move nut to work area
17 Hold bolt	H	P	Align nut
18 Hold bolt	H	A	Place nut on bolt
19 Hold bolt	H	RL	Release nut
20 Move assembly to "drop"	TL	AD	Idle
21 Release assembly	RL	AD	Idle

Figure 4-4 Left-hand/Right-hand Chart for Bolt-Washer-Nut Assembly Showing Therbligs.

Beginning with Line 20, the right hand is again idle, which is again believed to be an avoidable delay. The right hand could be doing some other productive work rather than just waiting for the left hand to finish its task.

Figure 4-5 shows the ineffective Therbligs; they are circled. The large number of circles shows that there is a potential for a large improvement in job methods and hence, a potential for a large improvement in the operator's productivity.

By asking the following questions, the work method can be improved:

LEFT-HAND	THERBLIG		RIGHT-HAND
1 Reach for bolt	TE	(AD)	Idle
2 Select bolt	ST	(AD)	Idle
3 Grasp bolt	G	(AD)	Idle
4 Move bolt to work area	TL	(AD)	Idle
5 Hold bolt	(H)	(AD)	Idle
6 Hold bolt	(H)	TE	Reach for washer
7 Hold bolt	(H)	(ST)	Select washer
8 Hold bolt	(H)	G	Grasp washer
9 Hold bolt	(H)	TL	Move washer to work area
10 Hold bolt	(H)	(P)	Align washer
11 Hold bolt	(H)	A	Place washer on bolt
12 Hold bolt	(H)	RL	Release bolt
13 Hold bolt	(H)	TE	Reach for nut
14 Hold bolt	(H)	(ST)	Select nut
15 Hold bolt	(H)	G	Grasp nut
16 Hold bolt	(H)	TL	Move nut to work area
17 Hold bolt	(H)	(P)	Align nut
18 Hold bolt	(H)	A	Place nut on bolt
19 Hold bolt	(H)	RL	Release nut
20 Move assembly to "drop"	TL	(AD)	Idle
21 Release assembly	RL	(AD)	Idle

Figure 4-5   Ineffective Therbligs for Bolt-Washer-Nut Assembly.

- Can motions be eliminated?
- Can any of the operations be combined?
- Can the sequence of the operations be changed?
- Can the work itself be simplified?

If the answer to any of these questions is yes, then the analyst must develop an improved method. Although this job sounds easy, and in fact involves really nothing more than some organized common sense, it can be a difficult process. It is one thing to identify a problem; it is quite a different matter to design a way around the problem. In an effort to help, industrial engineers have developed a list of suggestions that help to guide thinking regarding methods improvement. Before we look at these "principles of motion economy" however, let us return briefly to this example.

When we started examining the work method, we indicated that our first inclination was to prepare a left-hand/right-hand chart. However, even before we do this preliminary analysis task, it is most prudent to look at the first of our questions and ask, "Is it really necessary to assemble this component?" Just because a company has always performed an operation, there is no reason to believe that it is necessary. A first step in answering this question is to examine how the product will be used by the customer. Quite often, the first thing a customer will do with the product is disassemble it. The moral of this story is, rather than repair an unnecessary method, it is better to eliminate it.

PRINCIPLES OF MOTION ECONOMY

A number of suggestions have been made as to how to improve a worker's productivity. A simple way of accomplishing this goal, at least simple in terms of handling the suggestions about improving worker productivity, is to list the suggestions, guidelines, principles, or laws according to the general category to which each relates. Neibel, Barnes, and many others have suggested some general classifications. Barnes' (1980) commonly-used categories include:

- *Use of the Human Body:* This classification deals with the most effective use of the body when it is required in a production situation.
- *Workplace Arrangement:* This classification deals with the most effective layout of the work area.
- *Design of Tools and Equipment:* This classification deals with appropriate design of equipment for the ease of operation by the operator.

Within each of these classifications, a large number of guidelines have been developed. Following is an explanation of these guidelines, principles, or laws. These principles are not unique to this text. Taylor and the Gilbreths were the first to formalize these principles.* They have been written about, documented, and researched through the years until the following "rules," which are indeed principles, have evolved. When followed, they should help the analyst do a better, or more productive, job of work design and work improvement. Specifically, the principles described below are directly attributable to Barnes (1980), Konz (1979), or Neibel (1976), as they appear in their original writings.

Use of the Human Body

These principles of motion economy, as listed and described below, are concerned with the most productive use of the human in the production process.

*Gilbreth first published an article about motion study in *Industrial Engineering* in 1910. For a discussion of the response to this article see, for example, the already mentioned *The Writings of the Gilbreths.*

1. *Both hands should begin and end their motions at the same time.* Both hands should not be idle at the same time. It is fairly easy, actually a matter of routine, to start and stop the hands simultaneously. This rule involves planning a sufficient amount of work to keep the operator productive. Ideally, there should be no idle time at all; however, if some is necessary to balance the starting and stopping times of the hands, it should be kept to a minimum and both hands should never be idle at the same time.

2. *Hand motions should be made simultaneously. They should be made symmetrically away from and toward the center of the body.* Motions, when performed in this manner, will permit the operator to work faster. The same thought process and muscular commands will direct both hands. In terms of Therbligs, ineffective Therbligs will no longer apply.

3. *The motions of the hands should be performed to take advantage of the natural rhythms set up by the hand motions of the body.* Once this rhythm has been established, the operator will fall into the pattern of the motions being performed. This rhythm, this 1-2-3, 1-2-3 pattern, is valuable in any repetitive task. The more identifiable the rhythm, the easier the job of the designer to harness the feeling and make the job more productive.

4. *Whenever possible, momentum must be used to overcome resistance to movement.* Objects, while they are being transported from one location to another, develop momentum. This law is a basic tenet of physics. It is much easier to use this natural momentum to direct an object rather than to start out "in the hole" by having to expend considerable energy to overcome the initial resistance to movement or expend even more energy to stop the object that is in motion and abruptly change its direction. Using a toss release for an object wherein the momentum that is created by moving the object from location of assembly to release point is not lost, but is instead used to help the object gain its resting place, is an example of the principle. Expending significant amounts of physical or muscular energy unnecessarily is not a productive way of working.

5. *Motion paths should follow continuous curved paths when the motion requires reaches or transports of objects.* Curved motions follow the natural path of travel that parts of the body circumscribe. Although straight lines may initially seem better because they are shorter, there are some limitations to using the direct motion path of travel. Foremost among these limits is that when a sharp change of direction is involved, the hand must literally come to a stop before changing directions. When the hand moves at a constant speed, as in a curved path, there is no acceleration or deceleration. When a straight move approximates this curved path it can be done only with a considerable number of starts and stops. This slowing down and speeding up makes the operator less productive.

6. *The lowest classification of basic motions should be used.* There are five major classifications of motions, the lowest of which requires the least amount of time and effort to complete:

   a. *Class I* motions involve work that can be performed by the fingers acting alone.

b. *Class II* motions involve work that can be performed by the fingers and the wrist.

c. *Class III* motions involve work that can be performed by the fingers, wrist, and lower arms.

d. *Class IV* motions involve work that can be performed by the fingers, wrist, lower arms, and the upper arms or shoulders.

e. *Class V* motions involve work that can be performed by the entire body, including the fingers, wrist, lower arm, upper arm and shoulder, and other body motions.

Whenever possible, work should be designed to use the lowest motion classifications possible. Each successive motion should use the same motion class. Although there will be situations when higher classifications are used, it is generally believed that the simpler the motions, that is the lower the class of motion used, the simpler the job will be to perform. It follows that these jobs will be more productive.

7. *Ballistic-type motions are easier and hence, more productive than other types of motions.* For example, requiring hands to arrive at approximate locations rather than requiring definite locations will be more productive.

8. *Work should be designed so that the operator can use the preferred hand whenever possible.* A right-handed person develops certain habits regarding the use of the hands during certain motions. The operator almost instinctively attempts to have the normal hand perform these motions. Motion sequences must recognize that operators are creatures of motion habits. It is more productive to take these habits into account and to use these natural preferences, rather than forcing the issue for the sake of symmetry or rhythm. This principle forces the analyst to consider the common sense aspect of designing a job.

9. *Outside stimulations, such as flashing lights, should vary inversely with the internal stimulations that the job itself creates.* The motions within a job should provide the necessary cues to stimulate the operator on to the next step in the job. When this stimulation is not present, or is not possible due to the nature of the work, an outside stimulus might be helpful. Outside and internal stimuli should complement each other, not compete with each other.

10. *If the hand must handle heavy loads over extended periods of time, the thumb and middle finger should be used.* The strongest fingers of the hand are the middle finger in combination with the thumb. The other three fingers are not capable of performing heavy work over a prolonged period of time without excessive fatigue.

11. *When gripping tools, the portions of the fingers closest to the palm should be used.* More force can be applied in this area than over the extended distance that would be required by using the outer ends of the fingers. Additionally, the fatigue in the fingers and hand will be greater if the extreme regions are used.

12. *When the hands perform twisting motions, the elbows should be bent.* Fatigue is much greater when the elbows are not bent. Additionally, if the elbows are not bent, the operator must be at a greater distance from the work. Finally, rigid elbows

reduce the force that can be applied. Although it is not recommended to use the operator as a tool, when it is required, it is important to keep the exertion minimal and the output as high as possible.

13. *The feet are not capable of productively operating pedals when the operator is in a standing position.* Not only is it awkward for an operator to lift a foot, it is also very slow. All other activities almost come to a stop while the operator thinks about the task at hand, or foot. In other words, it is not productive. When an operator sits, the feet can be used relatively easily to depress controls. Also, foot operations can and should be performed simultaneously with hand motions as part of the normal sequence of activities.

Workplace Arrangement

These principles of motion economy relate to the physical layout and environmental conditions present at each work station. Proper attention to this area can help increase productivity.

1. *Tools and materials should be placed in fixed locations at the workplace.* This arrangement permits the operator to develop a sense of habit about a job, which in turn will lead to the rhythm that is so important. When tools and materials are in the same place all the time and the search and select Therbligs don't have to be used, the job will be more productive. Reliable placement implies that the effective Therblig, preposition, has been used in the design of the job.

2. *Gravity bins and drop delivery should be used whenever possible.* These motions require that the hands move only to approximate locations and take advantage of the momentum that the parts and hands have built up. Careful placement of parts in exact locations can slow down a process significantly. The use of the position Therblig is eliminated.

3. *Tools and materials should be located in the normal working area.* Most people have a normal reaching span that is limited by arm length and other physical characteristics. The arcs made by the arms as they rotate about the shoulder help to define this normal working area. Anthropometric studies, such as those done by Rosenthal, resulted in the work station dimensions shown in Figure 4-6. By having the work performed within these dimensions, the operator will not expend too much energy performing routine aspects of the job.

4. *Good posture will increase worker productivity.* Bad posture can tire the worker much more quickly than good posture. Tired workers are not as productive as rested workers. Good posture is encouraged through the use of comfortable chairs, arranged so the worker can sit or stand.

5. *Machines should be located so that they are available at the point of use.* Not only do the machines have to be at the proper work station, but their physical orientation should be such that the material will flow directly into the process and not have to be rearranged.

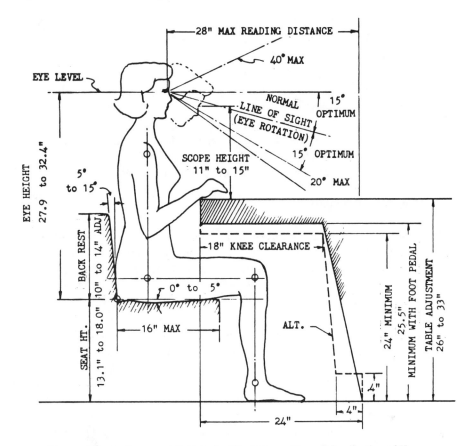

Figure 4-6 Anthropometric Data for Work Area. (From ''Application of Human Engineering Principles and Techniques in the Design of Electronic Production Equipment'' by M. Rosenthal. *Human Factors,* Vol. 15, No. 2, 1973, p. 140. Copyright © 1973 by The Human Factors Society, Inc., and reproduced by permission.)

6. *Specialization of tools and operator skills should be encouraged.* This principle outlines the classic division of labor philosophy. By dividing tasks into manageable sizes, special equipment and procedures can be developed that will permit the operator to become an expert at that job. Care should be exercised, though, to avoid the trap of overspecialization.

7. *Comfortable environmental conditions should be provided.* Lighting, ventilation, and temperature should be within the comfortable range. Sufficient light should be available.* Likewise, air flow and temperature must be kept within a reasonable range. When workers get too hot or too cold they cannot be as productive as they

---

*The Illuminating Engineering Society suggests standard levels for certain types of tasks. These levels may be found in many references, including the *Work Design* text by Konz (1979).

would be under better conditions. This work conditions guideline has to be, in today's energy-conscious society, a trade-off between what we would like, what we can afford, and what is safe.

8. *The workplace and work method should be designed so that the operator does not have to deal with any static loads.* The workplace design should not use the hands as holding devices.

9. *The workplace should be designed to minimize and/or eliminate the use of the eyes.* Eye motions should be few. Whenever the eyes enter the motion cycle, just about every other motion ceases. These unavoidable delays, while search and select occur with the eyes, are among the ineffective Therbligs.

Design of Tools and Equipment

These principles are concerned with the physical design of the tools the operator uses while performing a job. A significant amount of research in the human factors or ergonomics area has been done in identifying the specific designs that are most effective in production operations. These motion economy laws though, are once again some common sense applications of the more sophisticated theory.

1. *Whenever possible, multiple machining operations should occur simultaneously.* One of the best illustrations of the obvious productivity implications of this principle is found in the simple operation of slicing a loaf of bread. Freshly baked bread can be cut into slices either one slice at a time or by having the entire loaf sliced simultaneously by a set of blades. The multiple cuts, made with the same amount of effort as the single cut, are obviously a more productive way of performing the job. The byproduct of this productive way of performing this task is a better quality, i.e., more uniform, slice. Anyone who has ever cut bread a single slice at a time knows how much variation is present from slice to slice.

2. *Hand tools and control devices should be located within the normal work area, should be easily accessible to the operator, and should be operated by the muscle group that is best suited to that type of operation.* This rule involves the classic ergonomics maxim of designing the system to fit the person. Certain operations, such as the downward application of force, are best suited to a particular part of the anatomy, such as the leg or foot. When this type of operation is required, the tool should be easily reached by the foot and be in such a position that the foot and leg are free to apply the required downward force.

3. *Tools and materials should be prepositioned.* Not only should they be located within the normal work area, but they also need to be located or arranged in the sequence in which they will be used. Preposition is an effective Therblig.

4. *Handling of tools and materials by the operator should be kept to a minimum.* Whenever possible, tools should be designed so that the operator does not have to become directly involved in the tools' routine operation. Humans are generally slower and less reliable than tools, and hence, less productive.

5. *Jigs and fixtures should be used to hold parts in position.* These devices eliminate the need for the hands to function as holding devices and promote the performance of simultaneous motions by the hands.

6. *Tools should be designed for use by either hand.* Left-handed or right-handed tools can significantly delay the operator, thereby reducing productivity, when the wrong handed tool is used or provided for use.

7. *The surface where a tool is held by the operator should be slightly compressible, slip-free, and comfortable enough to be held for extended periods of time.* Tools that tire the hands when used for extended periods detract from productivity. Rest to overcome this fatigue results and is an unavoidable delay, an ineffective Therblig.

8. *Power assisted tools should be used whenever possible.* (No further explanation should be necessary.)

Most of these principles lead to the obvious conclusion that there are certain activities that people do better than machines and certain activities that machines do better than people. The following last formal principle lists some of the generally accepted distinctions between people and machines.

9. *Use people where they are most effective and use machines where they are most effective.*

MAN VS. MACHINE

Man is generally believed to be more capable of performing in the following:*

- Situations which require improvisation or adaptation
- Tasks which involve unusual or unexpected events
- Tasks in which there is a great deal of distracting information presented
- Tasks involving the use of new solutions or transferring similar solutions from other tasks
- Tasks using uncoded or raw information
- Situations requiring the use of subjective judgment based on experience
- Situations requiring anticipation of results
- Perceptions of complex stimuli in highly varied situations.

Machines are more capable, usually, in performing the following:

- Storing and retrieving coded information

*Reprinted with permission from *Human Behavior: A Systems Approach* by N. Heimstra and V. Ellingstad. Monterey, CA: Brooks/Cole, 1972, p. 440.

- Doing large quantities of work in short periods of time
- Reliably performing highly repetitive tasks
- Doing tasks where being distracted by surrounding extraneous factors detracts from productivity
- Sensing stimuli outside of the human's normal range
- Maintaining constant performance over long periods of time (with appropriate maintenance)
- Exerting great force precisely and consistently
- Deductive reasoning or categorization when established discussion rules are followed
- Simultaneously handling numerous inputs.

These machine limitations are subject to change as technology develops. The analyst is urged to remember that, regardless of capabilities and limitations, people and machines can and should complement each other. Their proper use will lead to increased productivity.

Example: Payroll Operations

Methods improvement is dependent upon the analyst understanding the task that is to be done. It can and should be practiced not only on existing jobs, but also on projected jobs. A complete description of the job, such as what is provided by standard charting procedures, is an excellent starting point. Likewise, a complete narrative can also be used effectively.

The payroll clerk in a certain organization performs the following operations each month in preparing salaried employees' paychecks:

1. Each employee's annual salary, claimed exemptions and insurance information are checked on the master list.
2. The clerk calculates the monthly gross pay amount for each salaried employee. This amount is determined by dividing the gross pay by 12.
3. The clerk enters this monthly gross pay on the backup ledger and on the computer data entry form.
4. The clerk checks Federal withholding tables to determine the amount of Federal tax to be withheld from the pay. This amount is entered on the backup ledger and on the computer data entry form.
5. The clerk checks State withholding tables to determine the amount of State tax to be withheld. This amount is entered on the backup ledger and on the computer data entry form.
6. The clerk calculates the Social Security deduction by multiplying the gross pay times the current percentage rate. This amount is then entered on the backup ledger and on the computer data entry form. Near the end of the year, the clerk checks the ledger to see if the employee has made the maximum contribution to Social Security

for the year. However, with the level of the maximum contribution and the normal pay scales at this organization, this part of the job rarely occurs. Those employees who might be involved are identified by a special "flag."

7. The clerk determines the employee's contribution to the retirement fund. This amount is calculated by multiplying the gross pay times the appropriate percentage. The amount is entered on the backup ledger and on the computer data entry form.

8. The clerk determines the employee's insurance contributions from the insurance table, based upon the coverage the employee has elected. This amount is entered on the backup ledger and on the computer data entry form.

9. The clerk determines if the employee has any miscellaneous deductions, such as credit union payments or the like. If miscellaneous deductions apply, these deductions are entered into the appropriate column on the backup ledger and on the computer data entry form.

10. All of these amounts—gross pay, Federal withholding, State withholding, social security, retirement, insurance, and miscellaneous deductions—are entered on the cumulative pay ledger. The net pay, the difference between the gross pay and all of the deductions, is determined. This month's amount is added to the yearly total. Social Security totals are double-checked with the maximum contribution limits. If maximum Social Security has been reached, the amount withheld is recalculated. The cumulative totals are then entered on the computer data entry forms.

11. After all the information has been entered on the appropriate ledgers, the data entry forms, one for each salaried employee, are delivered to the computer data entry clerk.

Can methods improvement procedures and/or motion economy and/or common sense be used to improve this process? The first question the analyst should ask is, "Is this operation necessary?" The question should be specifically asked about each step in the operation. The goal, where possible, is to eliminate unnecessary work.

For example, looking at Operation 2, the analyst could ask, "Is it necessary for the clerk to calculate the gross pay every month when the salary amount is fixed for 12 months?"

For Operation 3, the question might be, "Is it necessary for the clerk to enter the pay amount in two separate transactions, presuming that it is even necessary to re-enter the same amount every month?"

For Operation 4, the question might be, "Is it necessary to check the withholding? Won't it be the same every month?"

The analyst may also ask other types of questions, such as, "Can the operations be combined?" Or, more specifically, "Can one form be used for the backup ledger and data entry?" "Can photocopying provide the necessary duplicate record?"

The reader should be able to raise many other questions. Included among these should be the question, "Is a machine better equipped to perform the task than a person?" Also, upon examination of the workplace, the analyst should ascertain whether forms, references, and files are conveniently located.

Once a question has been answered to the analyst's satisfaction, a better or improved method should be proposed. A possible way to make the payroll operation more productive follows:

1. Each employee's file is checked to see if there have been any changes since the last check was prepared.
2. If no changes have been made, the same figures as last month are posted on a two-part form. The top form is the computer data entry form, the bottom is the backup ledger.
3. If there is a change, for example, in the number of deductions claimed, the new withholding rate is determined for both the Federal and State taxes, and entered on the two-part form.
4. The Social Security contribution is checked for the maximum contribution.
5. The new net pay is calculated and entered on the form.
6. The figures are posted to the cumulative ledger.
7. The data entry form is delivered to the computer data entry clerk.

An obvious improvement in productivity has been attained. The reader might be able to improve even more upon this improved method.

Example: Name Buttons

The Noname Manufacturing Company is in the business of manufacturing name buttons for conventions. It is a very small company that operates on a limited budget. Their capital is quite limited and the operating budget is even less. Their button is considered to be one of the best produced. Figure 4-7 shows a diagram of the name button. The part numbers refer to the following descriptions:

Material Description and Cost

1. Purchased from local art supply house; consists of $8\frac{1}{2}''$ by $11''$ multicolored cardboard $\frac{1}{32}$-inch thick and costs $3.00 per ream.
2. White mimeograph paper purchased in $8\frac{1}{2}$ by 11 reams from the local office supply company for $2.00 a ream.
3. Safety pin bought in lots of 1000 for $.50 a thousand from the same office supply house.
4. Laminated paper purchased in $4''$ by $5''$ sheets, six per package, at a cost of $1.00 per three packages.
5. Scotch Magic tape purchased in $\frac{1}{2}''$ by $500''$ rolls for $.89 per roll, six inches of tape is used on each badge.

Manufacturing Process   At each work station, the following procedure is followed. An example of a work station is shown in Figure 4-8.

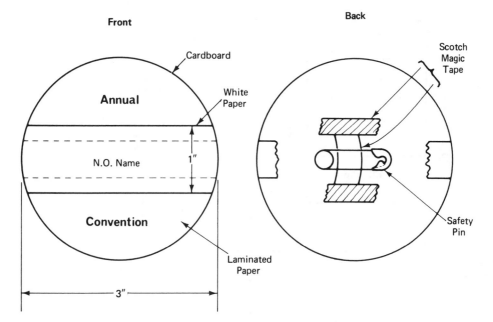

Figure 4-7   Diagram of Standard Name Button.

1. Assembler gets order from supervisor.
2. Assembler gets ream of cardboard and paper from stock, a distance of about 25 feet from the workplace.
3. Using wooden template, assembler cuts sufficient number of bases from the cardboard, with scissors, one sheet at a time.
   a. Additional cardboard is obtained as needed.
   b. Waste is disposed of as created.
4. Again, using a wooden template, the operator cuts, with the same scissors, one sheet at a time, part number 2.
5. Using a "kiddie type" removable type set, the assembler prints the convention details on the cardboard.

7. After printing, the white strips are taped to the cardboard and centered on the name badge to give the appearance of the badge as shown in Figure 4-7.
8. Laminated paper is cut to size and placed over the printing to protect it.
9. The safety pin is attached and fastened with the remaining tape.
10. Badges are placed in the carton.

The remainder of the process involves packing the badges in cartons and preparing the cartons for shipment. These tasks are done in the shipping room and do not affect the assembler.

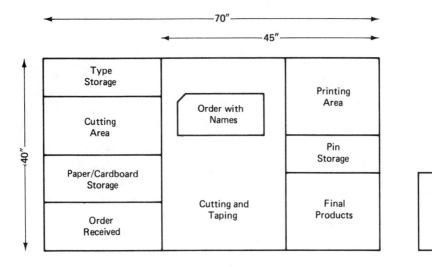

Figure 4-8   Workplace for Name Button Manufacture.

Following are the times for the assembly operations:*

1. 1.50 minutes per order.
2. 2.00 minutes per trip—two reams per trip.
3. 1.50 minutes per sheet.
4. 4.17 minutes per sheet.
5. .025 minutes per badge.
6. 1.67 minutes per badge.
7. .33 minutes per badge.
8. 2.00 minutes per badge.
9. 3.00 minutes per badge.
10. .017 minutes per badge.

A typical order consists of 500 badges. The assemblers, one per station, are paid $5.00 per hour. The cost of a name button consists of the material cost, direct labor cost, indirect labor cost, and the profit. Indirect labor is figured at the rate of $5.00 per hour for

*These are referred to as *historically set standards.* Chapter 5 contains a detailed discussion of standards and their development.

each worker. Profit, the difference between cost and selling price, is determined at the selling price of $.83 per badge in lots of 500 or more.

The company would like to use IE techniques to improve their productivity. Your assignment is to produce a proposal under the guidelines provided by your instructor, or specified in your company's normal operating procedures, for submission to the Noname Manufacturing Company.

Note: Although no correct answer is available to a problem of this type, almost everyone who attempts it has some ideas for a better way of performing the job. All improvements are steps in the right direction.

SELLING THE SOLUTION

It is part of the analyst's job to use methods improvement procedures to improve a job, to make it more productive and hence, more valuable to the company. It is equally important to convince the worker and supervisor that this improvement is indeed the better way of performing the job. Sometimes, selling the improvement can be more difficult than devising the improvement.

Resistance to Change

In many smaller organizations, a new IE is placed in a sink or swim situation. All too often, this analyst, after making the obvious improvements on paper, cannot implement the solution. The major reason for this road block often boils down to the way in which the changes were suggested to the people who must actually follow the new procedures. The novice analyst often studies the job himself. He then questions the method, uses common sense, and applies the principles of motion economy, all of which produce a significant improvement in work methods. He then pronounces that the new procedures will be used henceforth. The last portion of this process is usually rejection, if not downright sabotage by the worker.

Why does this situation happen? What caused this problem? (Sounds like a methods study of a methods study, doesn't it!) One of the major problems is that the analyst never consulted with the worker about the job. If the reader will give it the necessary thought, it should become obvious that the experienced operator usually knows his or her job far better than anyone else. If the analyst consults with the operator and supervisor, the selling is much easier.

Even if this most important part of the methods improvement process is followed, the analyst will still discover some resistance to the new method. This factor is found in many organizations—resistance to change.

Much has been written about this problem, some of the best of it by Krick (1962). People resist change for a number of reasons. To minimize, or eliminate, this resistance and sell the new procedure, these reasons for resistance must be understood and, perhaps more importantly, remembered when changes are to be made. Before any steps can be taken to overcome resistance, the reasons for it should be understood.

Inertia.   People become very content with the way things have always been done. When we drive to work we invariably select the same route just because it is our habit. When we read the newspaper, any newspaper, we will always start in the same place, perhaps the sports section, perhaps the front page. People often resist changes simply because they don't want to change.

Uncertainty.   Change brings with it some unknown consequences. Driving to work via our accustomed route, we may know we'll hit a big traffic jam and get mad about it, but we won't try a different route because there might be a bigger traffic jam if we follow that other route. (There also might not be, but we are hesitant to take a chance. Why look for new problems?)

Need.   More precisely, this reason should be called failure to see the need for making a change. Before people are willing to make a change, they must be convinced that something really needs changing. "If it ain't broke, don't fix it."

Understanding.   Failure to understand the change is a common reason for rejecting the change. If people don't understand what is happening they won't accept it. Most people have a tremendous fear of what they don't understand.

Obsolescence.   After a long period of time, individuals usually become skilled at a task. A change means they will have to start over again. These individuals become afraid they will not be able to regain their skill level.

Downgrading.   Changes in job methods often result in a job that is simpler to perform. When this simplification happens, the operator may be afraid that he will be replaced by a less talented worker or that, because he is now performing a job requiring less skill, he will lose status with his coworkers. Also, the supervisor may object because of the appearance that his workforce, because it is performing less skilled work, will be viewed as being less skilled. The supervisor may view this change as a loss of status for his position.

Support.   At times, supervisors will fight changes because they believe that this resistance will gain them favor with their employees and strengthen their own positions.

Personality.   The analyst and subject of the study or the analyst and supervisor may not get along. The same change suggested by a different analyst might readily be accepted. Some individuals just do not get along and, whatever one of the parties might say, the other will disagree just because of the source of the remark.

Resentment.   A supervisor may view the assistance of a methods analyst as interference in his affairs. This "help" may be viewed as an indication that the supervisor cannot do his job. This perception can cause the supervisor to reject all assistance, legitimate or not.

Participation.   The supervisor may resent and fight the change partly because he didn't think of it first. After all, it is his department and he should have discovered ways to make the jobs in his department more productive.

Tactlessness.   The method of presentation is important. Nothing will kill an idea more quickly than a ''know-it-all'' attitude. If the analyst embarrasses the supervisor or operator, resentment is almost sure to surface and change will be resisted.

Confidence.   New analysts must prove themselves. Wet-behind-the-ears analysts, often just out of college, often will not be believed just because ''someone that young, or inexperienced, cannot possibly know what to do.'' Sometimes, unfortunately, a few gray hairs are worth more in selling a change than the best idea in the world.

Timing.   The analyst may run into resistance just because the operator or supervisor is having a bad day. Many reasons, such as a fight with a spouse, a sick child, a bad day at the track, or the local team's loss, may cause rejection of a perfectly good idea.

Economics.   A change in job method might cause an operator to believe that there will be economic, i.e., pay, changes to follow. An improved job might require less skill and the job classification and pay rate might be lowered. The time required would possibly be less, or at least different, and operators generally fear that the new time standard would be tighter and require more work to maintain the same pay. The previously mentioned obsolescence is a real fear among many workers. They are concerned that they'll never become as proficient in the new method as they were in the old. As a result, they *know* their pay level will decline.

Social.   Alteration of work groups is a major source of resistance to change. As Roy (1960) pointed out, even the most boring jobs can be satisfying when the informal organization is pleasing to the workers. Redesign of a job or set of jobs can change work groups and lead, at least in the workers' minds, to perceived horrors. Change is often fought for this reason.

Overcoming Resistance to Change

There are many ways to botch the selling job. Although there is no magic formula to success, the following suggestions, again based on the work of Krick (1962), may make acceptance a little easier to come by.

1. *Explain the need for the change.* Don't overlook the worker in this respect. The change will directly affect the worker; if the worker is convinced that there is a good reason for making the change, then there is a better chance the change will be accepted.
2. *Explain the nature of the change.* Use straightforward, clear, well-organized language to insure that everyone understands the method or policy. Tailor your written and oral reports for the audience receiving it. For example, executives should be given a condensed description of the proposal, emphasizing the overall picture and making liberal use of charts, graphs, and other visual aids. Reports to persons who must administer the new procedure should include a thorough and easily understandable description of how the procedure is to function.

3. *Facilitate participation or the perception of participation in the formulation of the proposed method.* People are concerned about making their own ideas and recommendations succeed. The feeling of participation can be imparted in several ways.
   a. Consult operators, inspectors, supervisors, tool makers, maintenance men, managers, etc. to ask for information, opinions, and suggestions. Remember that the people who do the work on a regular basis know the procedures far better than any analyst. Show a real interest in what these people say. Seek advice even if you do not think that you will need it. Merely by being given the opportunity to express himself, the person will have a feeling of participation. And, who knows, you just might gain some valuable information or insights about the job!
   b. Whenever possible, suggestions should be incorporated into the final proposal with credit being given to the originator of the idea. Suggestions from these individuals need to be used on a regular basis. To solicit help and then ignore it kills any sense of participation. People will wonder why they should waste their time when "he never listens to me anyway." This acceptance of others' ideas may make the difference between acceptance and rejection of your proposal.

4. *Be tactful in introducing your proposal.* Watch your wording and mannerisms. Above all, avoid criticisms or anything that could even be construed as such. Just because a more productive procedure is being developed does not mean that the old way was necessarily bad. A good, better, best approach might be a way to interpret this point.

5. *Watch your timing.* In attempting to gain adoption of your proposal, avoid presenting it when the recipient is busy, upset, or otherwise distracted. Allow sufficient time for the concept to be thought about. Most new analysts want to change the world all at once and right now. Patience is indeed a virtue. If changes aren't rushed into, someday the recipient of the proposal may even make the proposal his own. Also, provide ample advance warning. Nobody likes to have new ideas, methods, or procedures sprung on them. Finally, changes should not be made during times of labor unrest. Any changes from the usual during such a time like would most likely add to the unrest.

6. *Introduce major changes in stages.* The size of some changes may frighten some people and cause resistance. Also, when people see how well the first stage of a proposed change works, they may be more receptive to later changes.

7. *Emphasize personal benefit.* In attempting to gain acceptance, capitalize on the features that provide the most personal benefit to the person or people you are trying to convince.

8. *Show a personal interest in the welfare of the person directly affected by the change.* Be aware of the social relationships and implications that changes will have on the relationships. If skilled workers will no longer need the skills formerly required, rather than underutilize talented people, try to find the person a job that will make the maximum use of his or her previously developed skills. If possible, try to guarantee continuing work at the previous developed skills. If possible, try to guarantee continuing work at the previous income level. This anxiety is a very real fear for anyone facing a job change. Make sure that the operators are thoroughly trained in

the new methods. Attempt to avoid drastic reduction in one important part of the job, such as responsibility or skill required; perhaps the job can be upgraded by letting the worker do his own setup or inspection.

9. *It is best to have the supervisor announce changes.* Most analysts serve in a staff or advisory capacity. Suggestions by the supervisor are much more likely to be accepted than those made by an "outsider."

Krick (1962) pointed out that, "The foregoing measures, concerning the minimization of resistance to a specific change are not substitutes for a long term conditioning for change. These measures should be supplemented by a long term effort to prepare personnel of the organization for, and harden them to, change in general" (p. 515).

A long-term process should include the following:

1. *Technical training.* Individuals should feel prepared to handle any change that might be made to improve productivity. Continual retraining will prepare employees to adapt to and accept new and different ideas.

2. *Education.* Individuals must understand the importance of and the need for change. What *was* good enough will not always be good enough.

3. *Communication.* Individuals need to be informed about what is happening and what is expected to happen regarding technology, competition, and the economy.

4. *Fairness.* How often this word enters the picture! Individuals must know they will be treated fairly at all times and especially when changes are made. Workers must know that matters of replacement, retraining, job content, and so forth will be dealt with fairly. If, when making the changes, the Golden Rule is observed, then the changes can be made with a minimum of disruption.

SUMMARY

Productivity improvement is important. Work must be performed smarter. We have to encourage all possible ways to increase productivity. This chapter will conclude with two additional illustrations of this necessary concept.

The first of these ideas was initially related by H. Myers in *Human Engineering* (1932). It is more a reflection on attitudes than on methods improvement.

A man about forty six years of age, giving the name Joshua Coppersmith, has been arrested in New York for attempting to extort funds from ignorant and superstitious people by exhibiting a device which he says will convey the human voice any distance over metallic wires, so that it will be heard by the listener at the other end. He calls the instrument a "telephone," which is obviously intended to imitate the word "telegraph" and win the confidence of those who know the success of the latter instrument without understanding the principles on which it is based.

Well informed people know that it is impossible to transmit the human voice over the wires as may be done with dots and dashes and signals of the Morse Code and that, were it possible to do so, the thing would be of no practical value. The authorities who apprehended this criminal are to be congratulated and it is to be hoped that his punishment will be prompt and fitting, that it may serve as an example to other conscienceless schemers who enrich themselves at the expense of their fellow creatures (p. 14).

Not only is resistance to change a tremendous factor in methods improvement, but the use of methods improvement in nontraditional applications is also an area that needs examination. The second of our illustrations represents this idea.

Productivity improvement analysis can be applied to many different situations. The author's four-year-old son recently started to help his father by setting the dinner table each night. His father, ever the astute analyst, observed the method the four-year-old used to set the table and believed that there was a better way for him to perform this task.

Figure 4-9 shows the layout of the kitchen where this activity took place. The operator performed the steps described below:

1. All of the knives were taken from the silverware drawer and placed on the table.
2. All of the forks were taken from the silverware drawer and placed on the table.
3. All of the spoons were taken from the silverware drawer and placed on the table.
4. One knife was placed at each place setting.
5. One fork was placed at each place setting.
6. One spoon was placed at each place setting.

Based on this description, the flow process chart in Table 4-1 was completed. The subject charted, remember, was the author's four-year-old son. Therbligs are shown on the chart as well as the standard flow process chart symbols.

The reader, after studying the method described and the flow process chart, should be able to suggest a method that will improve the productivity of this operation. An interesting problem to consider is how to implement the recommended changes under the analyst-operator relationship described in this situation. In other words, if the selling is

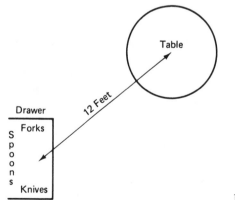

Figure 4-9 Table Setting Workplace.

TABLE 4-1 Table Setting Flowchart (Operator)

Activity	Symbol	Therblig
1. Select 3 knives from drawer	□	Reach, Search, Select
2. Grasp 3 knives from drawer	○	Grasp
3. Carry knives to table	⇨ ○	Move, Position
4. Place knives on table	○	Release
5. Return to silverware drawer	⇨ ○	Reach, Search
6. Select 3 spoons from drawer	○	Select
7. Grasp 3 spoons from drawer	○	Grasp
8. Carry spoons to table	⇨ ○	Move, Position
9. Place spoons on table	○	Release
10. Return to silverware drawer	⇨ ○	Reach, Search
11. Select 3 forks from drawer	○	Select
12. Grasp 3 forks from drawer	○	Grasp
13. Carry forks to table	⇨ ○	Move, Position, Release
14. Place forks on table	○	Release
15. Select knives and place one at each plate	□ ⇨	Select, Grasp, Move, Position, Release
16. Select forks and place one at each plate	□ ⇨	Select, Grasp, Move, Position, Release
17. Grasp spoons and place one at each plate	○ ⇨	Grasp, Move, Position, Release

not handled properly, the task may be done with the improved method, but it might be performed by the analyst rather than the intended operator!

This chapter illustrated the importance and the way to work at improving work methods. Just as important as making the improvements is being able to measure, in some way, the amount of work that is being performed. The next several chapters will address this issue.

REVIEW QUESTIONS

1. What is *methods improvement?*
2. What are the four major components of a methods improvement program?

3. How does methods improvement relate to productivity improvement?
4. What is a *Therblig?*
5. List and describe each Therblig.
6. What is an effective Therblig?
7. What are the *effective* Therbligs?
8. What is an *ineffective* Therblig?
9. What are the ineffective Therbligs?
10. What are the *principles of motion economy?*
11. What are the major classifications of the principles of motion economy?
12. What types of questions should the analyst ask when performing a methods improvement study?
13. Why do people resist change?
14. How can resistance to specific changes be overcome?
15. How can resistance to change be overcome in the long run?

PRACTICE EXERCISE

The Box Top Company manufactures the six cleat wooden packing crate shown in Figure 4-10. The crate is available to purchasers in the following sizes:

> *Width:* 22 to 36 inches (2-inch increments)
> *Depth:* 22 to 36 inches (2-inch increments)
> *Length:* 36 to 60 inches (4-inch increments)

The manufacturing process consists of five steps:

1. Obtain lumber from storage.
2. Cut lumber to size.
3. Take lumber to assembly area.
4. Assemble crates.
5. Take crates to storage.

Figure 4-11 shows the layout of the production area.

The following narrative describes exactly how the crates are currently manufactured:

Upon receipt of an order for a given size crate, the materials handling man pushes an empty truck to the materials storage area. There he loads the six-foot pieces of lumber onto the truck two at a time. He continues loading the truck until he has a sufficient amount of lumber for one complete crate. He then takes those boards to the cut-off operation. Boards for each size crate are stored in a specific location. The materials handling man then returns to lumber storage for more material until the order is filled.

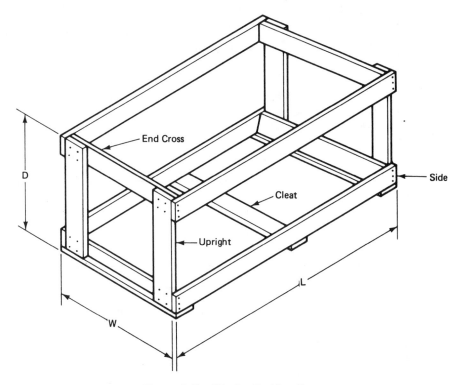

Figure 4-10   Wooden Packing Crate.

The cut-off man cuts boards to length, depending on the size of the order and number of boards required. First, he determines the number of boards required of each length for the particular order he is working on. He then cuts all the boards needed for each length, cutting all of one size at a time. If he should run short of boards, he goes to the storage area and gets the extra ones he needs. If he has extra boards, he returns these to the lumber storage area. Figure 4-12 shows this work area in more detail.

After the lumber is cut to size, it is placed on a truck and materials handling moves it to the nailing, or assembly, area. One assembler is responsible for each order. Each man builds each crate individually. There are four of these nailing stations at the assembly station. The actual method used in the assembly is:

1. End uprights nailed to end crosses at both ends.
2. Sides nailed to ends.
3. Cleats nailed to crate, six cleats per crate.

The work station has one vise for holding each part or one part of the crate while it is being assembled. Parts must be loaded into and removed from the vise several times

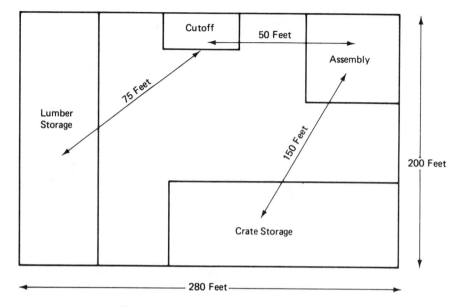

Figure 4-11   Packing Crate Production Area.

during assembly. All nails are stored in a common barrel with each assembler taking whatever nails he requires and laying them down near his work. Usually, each worker requires just a handful. The cut boards for each order are left on the truck on which they arrived from the sawing operation. Figure 4-13 shows this work area.

When the crates are completed they are placed on another truck and taken to the final storage area and again classified according to order. Depending on the size of the crates, six to eight are stacked on each truck. When a truck is loaded, a materials handling man is called and he takes the truckload to the final storage area. Then he unloads the crates and returns the truck to the assembly area.

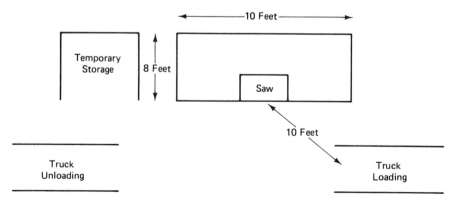

Figure 4-12   Detailed Work Area for Packing Crate Manufacture.

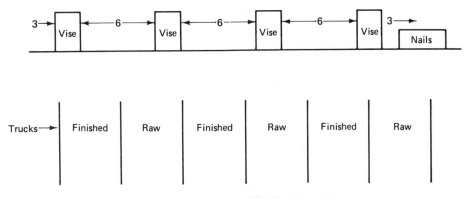

Figure 4-13   Crate Assembly Workplace Layout.

The personnel currently employed by the Box Top Company in the production area include: one supervisor, three materials handling men, one expert saw operator, and four assemblers. All belong to the local carpenter's union.

The production manager was satisfied with this method until just recently. The company received a very large increase in orders and must improve its manufacturing method if it hopes to meet the demand. Design an improved method for making these crates. Be sure to show that it indeed is a better method.

PART III

Measuring Productivity

Now that increased productivity has been established as an important goal, a plan of action to meet this goal must be developed. Before we can reach our destination, we must know where we have been. Before we can be sure that our actions have increased productivity, we have to be able to measure the current productivity level. A standard to compare productivity gains with is required.

Although historical data provide a rough estimate of past production, industrial engineers have traditionally been concerned with a more scientific measurement process. From the early work of Taylor and the Gilbreths to the present day, industrial engineers have been developing and refining the procedures used to measure work.

Standard methodologies, involving both direct and synthetic measurement procedures such as time study, work sampling, standard data, predetermined time systems, and physiological measures, are the measurement tools available to the current industrial engineer. The use of the newest computer technology and developments in electronics technology have changed and refined the traditional measurement devices, but the procedures remain the same. The newer systems allow the practitioner to develop and implement the best methods to measure all types of work.

Chapters 5 through 9 describe standard measurement procedures. The reader is cautioned that skill can be developed only through practice. These chapters illustrate how work measurement should be performed. The reader must develop his or her own skill.

CHAPTER 5

Time Study

The major objective of this chapter is to learn how to calculate a time standard based on stopwatch time study procedures. After completing this chapter, the reader should be able to

- Define a *standard time*.
- Define *performance rating*.
- Define *personal, fatigue,* and *delay allowances*.
- Calculate a standard from either continuous or snapback time study data.
- Calculate a standard when interference is present.
- Convert a standard to any of the standard formats.
- Understand the relationship between a standard and the compensation that workers receive.

    The reader will not, after completing this chapter, be able to conduct a time study with any degree of expertise. Time study is a skill that must be developed with practice.

INTRODUCTION

Work measurement is one of the oldest tools used by industrial engineers and, with the possible exception of the study of historical data, stopwatch time study is the oldest type of work measurement performed. According to Rice (1977), over 89 percent of the companies that perform work measurement use time study. This figure includes manufacturing companies making anything from food, clothing, farm tractors, electrical machinery, or computers, to service industries engaged in banking, health care, government services, and education.

Everyone is doing it. Time study is being done primarily to measure the present level of performance; that is, the productivity of those involved in creating goods and providing services in our economy. Not only are production workers subject to time studies, but so are materials handling operators, housekeeping and maintenance employees, clerical employees, first-line supervisors, and even engineers, technicians, and time study analysts.

In addition to determining current productivity levels, time study is also used to build wage incentive plans, estimate cost, schedule production jobs, and for other planning and control purposes.

PURPOSE OF WORK MEASUREMENT STUDIES

The immediate objective of all work measurement studies is the development of the standard. The time standard is an extremely important piece of management information. Regardless of the methodology used to determine standard time, the objective is the best estimate of the time required to perform a task. This time has many uses that are critical to the operation of any organization. The uses of the time standard include:

- Developing production schedules
- Determining wage payment plans
- Estimating manufacturing costs
- Providing a base for estimating productivity increases
- Identifying employee training needs
- Appraising employee performance
- Justifying the addition of production capacity.

Industrial engineers have traditionally defined the *standard,* or more correctly the *standard time*, as the time required by an average worker, working at a normal pace, to complete a specified task using a prescribed method. This definition seems to have several ambiguities within it. The first is the phrase, "average worker." To the student of statistics, the concept of average has several meanings. The first is the arithmetic mean. A second definition for *average* is the mode, or most frequently occurring value. The third expression for *average* found in statistics is the median, or middle value. Although none of these is entirely appropriate to the standard time use of *average*, a combination of the three definitions proves suitable. The average has also been described as the single best estimate of, or description of, a population. The "average worker" should be the typical worker, or a worker representative of all workers normally performing the work under study.

The average worker must be experienced in performing the job under investigation. The average worker is not the best nor the worst employee who normally performs the job. The supervisor should be able to identify the typical worker. If a union represents the employees, the local union official may also be able to help identify the average worker. Depending upon the exact nature of the collective bargaining agreement, the union may even have final approval of workers who are the subjects of a time study.

Identifying the average worker is the first of the apparent ambiguities in the definition of standard time. A second possible point of misunderstanding is the identification of "normal pace." A *normal pace* is anything you want it to be. It is arbitrary and very few organizations use the same normal pace. Some firms have established a normal pace as that pace required to walk six kilometers per hour on a level surface; others have selected shoveling three tons of coal per hour as normal, while yet others have determined that the average employee should be able to type 60 words a minute.

All of these designations have in common the arbitrariness and consistency within the particular organization that is necessary to make them effective bases for comparison. There is no one normal pace that is universal to all businesses. There are some very important concepts concerning *normal* that must be adhered to when determining the normal pace.

First of all, *normal* must be well-defined and understood by all parties affected. The normal pace used in evaluating one job within a company must be the normal pace used in evaluating all other jobs within the organization. Another word for this blanket definition is *consistency*.

Second, *normal* should be neither too fast nor too slow. The normal pace should be the pace that the average worker can work at for an entire day without undue physical strain. Third, the normal pace does not necessarily reflect the average pace or average productivity of a particular organization. The normal pace is a reflection of what should be performed, not what is being performed.

Finally, once established, the normal pace must not be changed. The pace should never be adjusted to raise or lower compensation levels. To develop meaningful standards, the normal pace must remain constant. Otherwise, any use of the standards becomes suspect.

The next possible point of difficulty in understanding the definition of standard time is describing the actual work done and the method by which it is completed. Tasks that are time studied must have a starting point and an ending point. Additionally, there must be an agreed upon way of performing the work. This method, or standard practice, must be followed every time the job is done. It should be acknowledged that the standard practice is the best way. It should be the result of a complete methods study, as described in Chapter 4. It is important that this method is considered best by the worker, time study analyst, supervisor, and, if applicable, the union representative. It is equally important that the job be performed according to this best method. Remember that determining and implementing the best method is a key part of the productivity improvement process.

To summarize, the standard time is that time required by the typical worker to complete a well-defined task at a normal pace using a defined procedure. Defining the standard time and creating a standard time through the use of time study are two different matters. Before we can discuss the measurement process, two important concepts must be understood: rating and allowances.

RATING JOB PERFORMANCE

The normal pace, regardless of the standard used, is an idealized pace. No individual is capable of consistently working at the normal pace for an entire work day. Some people work faster than normal and some work slower than normal. Some people change their pace as the work day progresses. Sometimes, the process of being time studied causes the subject to change the work pace to try to conform with what is believed to be normal or to conform to a peer group concept of normal. The industrial engineer must be aware of this tactic, whether it is consciously done or not, and compensate for it by rating the performance of the job being studied.

Rating is the process of comparing the actual work being performed with the analyst's concept of normal pace and evaluating the observed performance quantitatively. Rating generally uses the normal pace as a base of 100 percent. A worker performing at a pace 20 percent faster than the rater's concept of normal would be rated at 120 percent.

Similarly, a worker performing at a perceived pace of only three-fourths of normal would be rated at 75 percent. Rating factors are used to adjust the actual observed time when the standard for the job is calculated.

There is no magic to rating. To rate effectively, the analyst must be able to compare the observed activity to the predetermined concept of standard or normal pace. It is most important that every analyst have the same basic standard firmly entrenched in his or her mind. Consistency is of utmost importance. All jobs in an organization must be rated or compared to the same concept of normal performance. Companies use a variety of techniques to help achieve this consistency among the people responsible for performing the rating. Some purchase commercially available rating films or videotapes,* others produce their own, but the aim is the same: to provide a standard and to have the company's analysts practice on the same rating patterns. Generally, experienced time study analysts within one company will rate every job within five percent of every other analyst rating the same job. This skill is developed only through practice and experience. A new analyst often will work in conjunction with an experienced one to achieve the consistency that is necessary to set fair and effective time standards. Care must be taken to make sure that the job and not the worker is rated.**

Again, learning how to rate performance requires a significant amount of experience and practice. Many organizations train rating analysts by having them compare their perception of performance with that of an experienced or skilled analyst. Performance is typically influenced by speed and method. Sometimes performance rating is accomplished by evaluating the operator's performance on both the speed and method used.

PFD ALLOWANCES

In addition to acknowledging that workers can work at speeds other than the normal pace, it must be acknowledged that workers cannot work for an entire work day without some rest. Sometimes, rest is the worker's own doing, such as a visit to the water fountain or restroom; sometimes, it is beyond the control of the worker when, for example, the production line breaks down due to a broken piece of equipment. As the work day progresses, workers become tired and are less able to perform as they did early in the shift. The standard time must also be adjusted to reflect these personal, fatigue, and delay factors.

The PFD allowance, for personal, fatigue, and delay, as it is often called, is usually expressed as a percentage of the standard time and added to the time allowed to complete the particular task being studied. The amount of time provided for the PFD allowance is usually constant for all jobs within a given company, although it may vary due to particular working environments. For example, there may be a larger allowance permitted for a foundry than for other areas of a manufacturing company. Sometimes, the

*Two sources of commercially available rating films are Rath and Strong in Lexington, Massachusetts, and the Tampa Manufacturing Institute in Tampa, Florida.

**Workers sometimes believe that they are being evaluated rather than the job they are doing. As a result, they may try to confuse the novice analyst by varying their pace or their motion pattern to earn a higher rating.

allowance is developed after a careful study of all jobs in the organization; sometimes it is determined by experience, and other times it is negotiated and included as part of the collective bargaining agreement between the company and its union. The actual percentage allowed varies by company, from as low as five percent of the working day to as high as 30 percent.

CALCULATING THE STANDARD TIME

The standard time is the product of three factors: the actual observed time, the rating, and the PFD allowance. Once all the data are collected, the standard time is calculated according to the following relationship:

Standard Time = (Observed Time)(Rating Factor) +
(Observed Time)(Rating Factor)(PFD Allowance)
or
Standard Time = (Observed Time)(Rating Factor)(1 + PFD Allowance)

Example: Cashing Checks

To determine the number of tellers to provide in each of their branches, the Fourth National Bank of Marietta, Georgia, must know how long it takes a teller to perform each of the tasks normally handled at the bank's windows. One such operation that occurs with relatively high frequency is cashing checks for the bank's customers. This task includes the following steps:

The teller greets the customer, receives the check from the customer, examines the check for the amount and other pertinent information, verifies the account through a telephone call to the bank's computer, cancels the check, prepares a cash-out ticket, removes the money from the cash drawer, counts out the money to the customer, and wishes the customer a good day.

There are many distinct activities included in this rather routine task at the Fourth National Bank. The beginning and ending points of this job have been identified and the standard practice is well-established.

The industrial engineer for the bank has time studied the job and observed that it takes the teller .41 minutes to complete the process. The analyst also observed the pace at which the teller was working and, based upon experience and judgment, determined that this teller was working at 90 percent of the Fourth National Bank's concept of normal pace. The bank had long ago established a PFD allowance of 13 percent. The industrial engineer was then ready to calculate the standard time based on the time study performed.

The standard time (ST), or standard, for short, was calculated using the formula presented earlier in this section.

$$ST = (.41)(.9) + (.41)(.9)(.13)$$
$$= .37 + .05$$
$$= .42 \text{ minutes}$$

We have calculated the standard time to perform a specific job. This figure is the end result of a time study. Having identified our objective, we are now ready to examine the procedure used to reach this result.

CONDUCTING A TIME STUDY

The first requirement for making a time study is having a timing device. Traditionally, this tool has been a stopwatch; however, with the advent of new electronic technology, timing devices may be small electronic instruments that record time. Today, a wide variety of tools, ranging from electronic stopwatches to computerized timing devices that transmit the observed or directly measured times automatically to a processing unit for analysis, are available.* Figure 5-1 shows pictures of available timing devices, including some of the latest electronic and computer-linked timing devices on the market.

The discussion of the measurement process will assume that the analyst will make a modest start and begin making time studies with a standard decimal minute stopwatch like the one shown in Figure 5-2. The principles and procedures that will be explained on the following pages will be essentially the same regardless of the type of timing device used. By focusing on the application of the most basic tool, the reader should understand the method and be able to expand the principles to more sophisticated timing devices. The computerized devices will generally perform the analysis of the data much faster.

Timing Devices—The Stopwatch

The first task is to become familiar with the timing device, or in this case, the stopwatch. The watch winds at the stem, starts with the movement of the switch, and is set to zero by depressing the stem. Most stopwatches read in decimal minutes for ease in arithmetic when the standard is calculated. Hundredths of minutes are much easier to add and subtract than are seconds, which have a base 60 measuring system. Generally, one sweep of the large hand indicates that one minute has elapsed, and the total time elapsed since the watch was last reset is accumulated on the smaller dial located toward the top of the watch. The analyst should become proficient at reading the watch before starting to make a time study. Electronic timing devices have similar functions.

Usually, the stopwatch comes attached to a clipboard for ease in actually recording raw data at the time when the study is done. Figure 5-3 shows a typical stopwatch-clipboard combination. Once the analyst is thoroughly familiar with the timing device, it is time to conduct a study.

Conducting a Study

Obviously, the first step is to select a job to be studied. The selection of the task to be studied will give direction to the way the data will be collected. As the time study analyst

---

*By the time you read this text, the advances in electronics will make the descriptions of the latest available equipment obsolete.

Figure 5-1 Timing Devices. (Courtesy of Electro/General Corporation and the Meylan Corporation.)

Figure 5-2. Decimal Minute Stopwatch. (Courtesy of the Meylan Corporation.)

prepares to study a job, he should learn as much as possible about the job prior to actually performing the time study, including all available historical information, such as process instruction sheets, and talking with other engineers, technologists, and analysts who are familiar with other, similar jobs. Specifically, the analyst must identify and document the following for any job being time studied:

- Method
- Workplace layout
- Production equipment
- Machine speeds and feeds (if applicable)
- Inspection equipment (if applicable)
- Environmental conditions.

This information, along with any other pertinent information should be shown on the data collection form. Figure 5-4 shows a typical form with space for showing this information.

Although the industrial engineer has the responsibility of setting standards through time study, the first-line supervisor has direct authority in the production areas. After having identified the job that is to be studied, the analyst should ask the first-line supervisor for "permission" to conduct the study. In addition to allowing the study to take place, the supervisor can be a valuable resource. The foreman should be able to identify a typical worker. The supervisor can also provide some help with identifying the standard practice for the job. Finally, an introduction by the foreman of the time study analyst to the typical worker, or subject of the study, can be most helpful. The introduction by the supervisor will imply a sanction of the time study by the worker's boss.

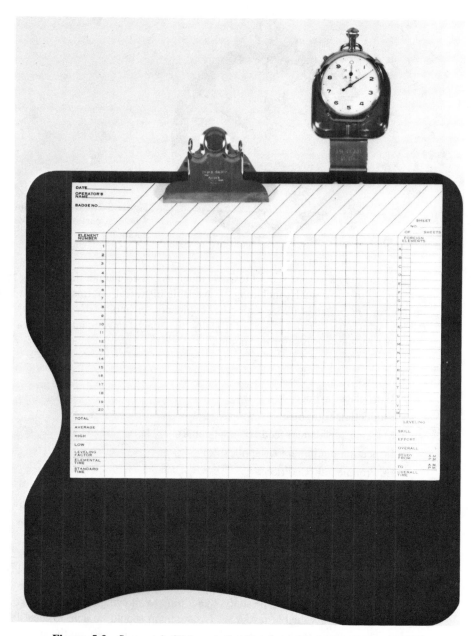

Figure 5-3 Stopwatch-Clipboard Combination. (Courtesy of the Meylan Corporation.)

OPERATION:

OP. NO:

PART NAME:

PART NO:

MACH. NAME:

MACH. NO:

OPERATOR NAME & NO:

MALE ☐
FEMALE ☐

EXPERIENCE ON JOB:

FOREMAN:

NO. MACHINES OPER'D:

MACH. SPEED:

DEPT. NO:

MATERIAL:

SKETCH OF WORKPLACE

SCALE:

DATE OF STUDY

OBSERVER

APPROVED

SUMMARY

NO.	ELEMENTS	NORMAL TIME	PAT'S PERS'L. ALLOW.	OTHER ALLOW.	STD. TIME

TOTAL STD. TIME PER CYCLE:

NO. PIECES PER CYCLE	STD. TIME PER PIECE

DRAWING OF PART:

Figure 5-4

Once introduced to the subject, the analyst must explain what will happen during the study. Depending on the company's history regarding the use of time study results, the explanation may be the most critical part of the entire study. If the company has been fair and consistent in its use of standards set in the past, then a simple explanation of what will happen, along with a request for an explanation of what happens in the operation, will suffice. The actual method used on the job may differ from the written instructions. The time study provides an opportunity to remove this inconsistency. As a byproduct of the time study, the written instructions should be reconciled with the actual practice, which should, of course, be the "best" method.

Unfortunately, not all organizations have used standards fairly. When this case applies, the industrial engineer has a much more difficult job. This situation is often manifested by hostility in the subject's behavior and the analyst has quite a selling job to perform before actually studying the job. One of the more common problems to overcome is the one already briefly mentioned, the worker's fear that, "*I'm* being evaluated." The job is to convince the worker that the reason the time study is being conducted is to set the standard time, not to judge the ability of the worker. Remember, the worker would not be the subject of the study unless his or her performance has not already been judged satisfactorily.

Sometimes, the subject will adopt a "me against you" attitude and try to sabotage the study by being less than cooperative with the analyst. The analyst must develop a rapport with the worker. He must display a confidence in his ability, yet not exhibit any brashness or behavior that would offend or upset the subject. He must make the subject feel as comfortable as possible in his presence, which is not always an easy task for the industrial engineer. Over time, the analyst will develop a personal style that will ease difficult situations. The ideal is to never allow an uncomfortable situation to arise. The key is always to remain consistent and fair in the development and use of the standards. However, one slip-up can do immeasurable harm. And, unfortunately, an error may have been committed long before the analyst joined the organization.

Once the job to be studied is selected, the following question concerning the nature of the standard time to be developed should be answered. "Why is the time standard being developed?" Some possible answers include:

- As a base from which to measure increases in productivity
- As a tool to develop standards for other jobs that are similar to the one being studied
- As a base for an incentive pay system
- As a tool for production planning
- As an audit to make sure that the standard is correct.

Depending upon the answer to the question, the time study may take two alternative paths.

If the reason the time study is being made is simply to see how long the task takes to complete, then the first set of instructions that follows regarding elemental analysis will be appropriate. If, however, more precise information is desired, then the second set of instructions about elemental analysis should be followed.

ELEMENTAL ANALYSIS

Elemental analysis is the process of dividing the job being studied into elements. Elements are components of a job or task that are logical divisions of the job, they have easily identifiable starting and stopping points, and repeat on a regular basis throughout the work day. Elements can be of fairly short or relatively long duration.

With a traditional stopwatch, it is virtually impossible to directly time an element of less than .03 of a minute duration. Even with electronic timing devices, it is difficult to directly time elements shorter than this division; .03 of a minute is the approximate physical limit on noting the starting and stopping point of an element and recording the required reading.

There are several characteristics that *must* describe every job that is divided into elements.

1. Every element must have an easily identifiable starting and stopping point.
2. There can be no discontinuities between elements. The ending point of one element must be the starting point of the next element. The instant one element ends and the next begins is known as the *breakpoint*.
3. The relative frequency of each element must be included at some point in the elemental description. This recording is conveniently done at the end of each elemental description, but there is no hard and fast requirement to do so.

Although some organizations do not use as detailed descriptions of elements as those provided in the proceeding examples, the starting and stopping points must be identifiable and there can be no discontinuities between elements.

The elements chosen depend on the use of the study. If the purpose of the study is to determine the standard time required to complete the job, several large elements may be selected. Referring to the bank example described earlier in the chapter, the Fourth National Bank's check cashing task may be divided into three rather large elements: (a) all activities involved in obtaining the check from the customer through examining the check, (b) all banking activities through preparing the cash-out ticket, and (c) the remaining actions of giving the customer the money and wishing him or her a good day.

On the other hand, the bank's IE department may want to use the data from this study to help set standards on similar jobs. In this case, the elements would probably be better defined and much more explicit. Each separate action, such as verifying the account balance, would be a separate element. If necessary, each action could be divided into several smaller elements. For example, the verification procedure, which includes calling the bank's computer on the telephone, could be subdivided to include dialing the phone number and keying in the customer's account number.

Example: Inserting Mailer in Envelope

One of the jobs in a mail house, an organization that prepares mass mailings for client companies, is "stuffing" pre-printed advertising in envelopes. This task, which is repeated

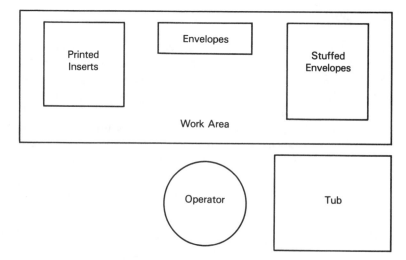

Figure 5-5 Envelope-stuffing Workplace.

many times during a typical work day, is an ideal job on which to set a time standard. The following paragraphs describe the work elements as they will be timed by a work measurement analyst. The workplace is shown in Figure 5-5.

Element 1: This element begins when the operator reaches, with the left hand, for a printed insert. It continues as the operator grasps the insert and brings the pre-printed material to the work area. The element ends when the right hand grasps the insert. This element occurs in every cycle.

Element 2: This element begins as the right hand grasps the insert. It continues as the mailer is turned 90 degrees, laid on the work surface, folded from the top to the one-third mark by the left hand, creased with the right hand, folded from the bottom to the one-third mark, and creased with the left hand. The element ends when the second crease is put in the insert with the left hand. This element occurs in every cycle.

Element 3: This element begins as the left hand completes the second crease in the mailer insert. It continues as the right hand picks up the insert and the left hand reaches for, grasps, and brings an envelope to the work area. The element ends with the right hand grasping the envelope. This element occurs in every cycle.

Element 4: This element begins with the right hand grasping the envelope. It continues as the left hand lifts the flap in preparation for the insertion of the mailer. The right hand inserts the mailer into the envelope while the left hand holds it. The element ends when the "stuffed" envelope is placed on the stack at the right of the work area by the right hand. This element occurs in every cycle.

Element 5: This element begins when the envelope is placed on the stack at the right of the work area. When 10 envelopes are stacked, the right hand picks the stack up and places the stack in the tub. The element ends when the envelopes are placed in the tubs and the right hand returns to the work area. This element occurs every 10th cycle.

Example: Mold Preforms

Rubber Products is an intermediate-size company that manufactures quality industrial rubber products. All of the end-products are molded on hot platen presses. The charge for each mold/press is prepared as a preform in the "material prep" department.

A majority of the preforms are made by performing exactly the same series of elements. These preforms are usually made in lots. The lots vary in size from 40 to 160 units. The foreman assigns the operator the lots he needs to produce to support the press room. The elements of a cycle are as follows:

Element 1-Set-up; Get Material; 1-80: The cycle and element begin with the operator assembling the correct die on the press platen. She clears the work area and assures that the proper tools and material are available (hot knife, plastic sheets, trays, and I.D. tags). She next obtains a four-wheel cart from the room and moves it 40 feet to the sheet storage unit. The appropriate number of $2' \times 3' \times \frac{1}{2}''$ sheets are removed from the storage unit and stacked on the cart. Study number 16 used 20 sheets. The element ends when she rolls the cart 20 feet and positions it in front of the press.

Element 2-Position Sheet to Press; 1-4: This element begins when the operator reaches for a $2' \times 3' \times \frac{1}{2}''$ rubber sheet. It is removed from the cart and positioned on the press platen three feet away. The element ends as the right hand reaches for the press striker button.

Element 3-Blank Forms (16); 1-4: The element begins as the right hand reaches for the press striker button. The left hand positions the press head and die over the material and the right hand presses the striker button. This element is repeated 16 times. The left hand pushes the press head and die aside. The element ends as the right and left hands reach for a cut form.

Element 4-Remove and Position Forms (16) to Table; 1-4: This element begins as the right and left hands reach for a cut form. Each hand grasps and removes a form from the platen. The operator makes a one-quarter turn and places the two forms on the work table to the right of the scale. She then makes a one-quarter turn to the right and reaches for another pair. This action is repeated eight times (16 forms). The element ends after the final turn to the right as the operator reaches for the scrap on the platen.

Element 5-Scrap Aside; 1-4: This element begins as the operator reaches for the scrap rubber on the press platen. The operator grasps the material with both hands, removes it from the platen, and places it in the tote pan about six feet away. This activity occurs once for every four parts. The operator turns to the work table. The element ends when the operator reaches with the right and left hands for two forms.

Element 6-Position 4 Forms to Scale; 1-1: This element begins as the operator reaches for the forms. She grasps a form in each hand and stacks them on the scale. This activity is repeated. The element ends as the hands release the forms on the scale.

Element 7-Read Scale, Remove One Form, and Trim to Weight; 1-1: The element begins as the form is released on the scale. The operator reads the digital readout on the scale and mentally compares the reading to the required weight. Usually the weight is more than is required. The operator reaches and grasps a form with the left hand and moves it to the table. With a simultaneous movement of the right hand, the operator secures the hot knife. She trims the form. The form is put back on the scale for a weight check. If the weight is correct, the hot knife is placed; if not, the process is repeated. The element ends when the operator grasps the stack of forms to remove them from the scale.

Element 8-Position to Tray; 1-1: The operation begins as the operator grasps the preform. It is removed from the scale and placed on the tray. Each layer of forms is leafed with plastic to prevent sticking. The element ends as the form is released.

Element 9-Position Tray to Shelves; 1-15: The element begins as the last (15th) form is released. The operator picks up the tray of forms and walks 15 feet to the shelves. The tray is placed in its labelled position. A "new" tray is secured and placed on the work table. This element ends the cycle.

Before looking at more examples, a brief discussion about recording observed times is needed. There are two standard ways to record time study data. The use of either method is a matter of analyst preference or company practice. These two methods are the *continuous method* and the *snapback method*. The adoption of new computer technology may eventually make the use of either of these obsolete.

After a job has been divided into elements, these elements must be listed on a data collection form. Figure 5-6(a) shows a typical continuous time study data form. Figure 5-6(b) shows a typical snapback time study data form.

The Continuous Method

The continuous method of data collection involves starting the stopwatch when the study begins and allowing it to run, uninterrupted, until the study is complete. At the conclusion of each element, the current reading, or the elapsed time, is recorded on the data sheet. The original data sheet becomes a cumulative record of elapsed time from the beginning of the study. Actual elemental times are calculated by subtraction when the timing is completed. For example, a job under study has eight elements called, respectively, elements 1, 2, 3, 4, 5, 6, 7, and 8. Figure 5-7 shows the raw data as they might appear immediately after being recorded. The observed elemental data are shown in Figure 5-8. Remember that the study began at time 0. These elemental times are the results of successive subtraction. As a matter of convenience, the decimals were omitted when the data were collected.

Problems may arise when using the continuous method because the analyst must take his eyes away from the operation momentarily while recording the time. When several elements of short duration occur sequentially, some errors or inaccuracies may occur in recording times. The analyst must also make a judgment about the location of the sweep hand at the conclusion of each element. Naturally, the use of one of the electronic digital watches eliminates the latter problem.

CONTINUOUS TIME STUDY OBSERVATION SHEET

ELEMENT DESCRIPTION & LEVELED MIN

T/S #

OPERATION:

PAGE #

DATE:

ENGR.: #

OPERATOR:

LIST PART CHARAC. & SKETCH ON BACK

REF #

PART #

DEPT:　　CC　　LC

DESCRIPTION	CD	READING	ELAPSED	%

Figure 5-6(a) Continuous Time Study Form. (Courtesy of Lockheed-Georgia Company.)

Figure 5-6(a) *(Continued).*

GA FORM GD3259 (FRONT)

146

THE WILLIAM CARTER COMPANY
TIME STUDY AND STANDARD

MILL NO. GARMENTS SIZE CODE NO.

OPERATION RATE MINS /

DOZENS	S. O. NO.	OPERATOR		DEPARTMENT		
MACHINE		R. P. M.	STITCHES	OBSERVER		DATE
START	FINISH	ADDITIONAL DATA		SPEED RATING		COMP. M HR
ELAPSED	QUANTITY PRODUCED			FATIGUE RATING		RATED M HR

ELEMENTS	Speed Rating	1	2	3	4	5	6	7	8	9	10	11	12

Figure 5-6(b) Snapback Form. (Courtesy of the William Carter Company.)

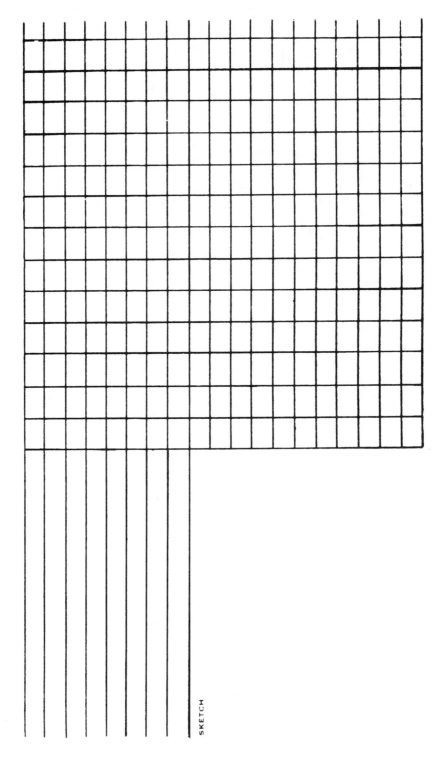

SKETCH

Figure 5-6(b) (*Continued*).

Element	Time
1	12
2	17
3	24
4	35
5	40
6	48
7	59
8	77

Figure 5-7 Raw Time Study Data—Continuous Method.

The Snapback Method

In the snapback method, the analyst resets the watch at the end of each element, thus eliminating any confusion that may result from reading the elapsed time. It also simplifies the analysis of the raw data. A drawback to using the snapback method is the time that

Element	Time	
1	12	12
2	17	5
3	24	7
4	35	11
5	40	5
6	48	8
7	59	11
8	77	18

Figure 5-8 Elemental Times for Sample Data—Continuous Method.

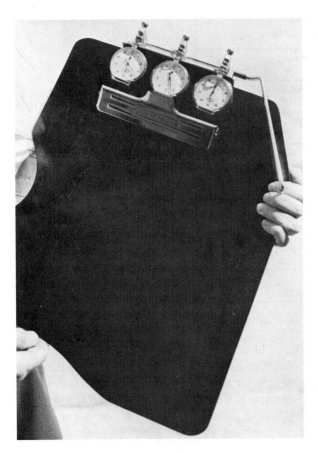

Figure 5-9   Three-Watch Board.

elapses while the sweep hand returns to zero. Although these omissions are small, some industrial engineers do not like working with data collected in this fashion because they know there are built-in errors.

A way to avoid this problem is to use a multiple watch board as shown in Figure 5-9. The electronic stopwatch shown in Figure 5-10 functions the same as the traditional three-watch board. By means of the electronics, the time for each element is temporarily frozen when that element concludes, while at the same instant, the timing for the next element is started. Otherwise, electronic timing devices automatically perform like the three-watch board.

In the future, the use of watches may be abandoned altogether. Instead, a device will be used that electronically allows time to elapse, and all that would be required of the analyst would be to depress a button every time an element is completed. At the termination of the study, the device would then be plugged into a computer and the results would be tabulated.

As was stated earlier, the data collection method used depends on personal preference and company policy. For our discussions, we will continue in the relatively

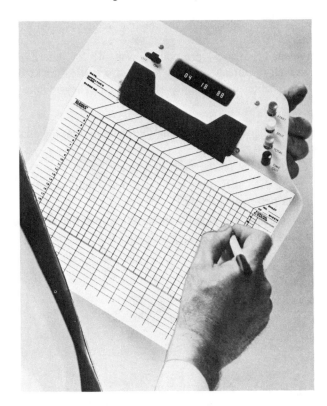

Figure 5-10 Electronic Watch. (Courtesy of the Meylan Corporation.)

modest vein, using a single stopwatch. We will also use the continuous method of collecting data.

   After the elemental analysis has been completed and before the job is timed, there are still two tasks to be performed. The number of cycles to time must be determined and data sheets for data collection must be prepared.

DETERMINING NUMBER OF CYCLES TO TIME

A *cycle* is the completion of the activities required to perform the job under study. Generally, all of the elements within a particular job occur within each cycle, but there are exceptions. These exceptions are what might be called regular nonrecurring elements, or noncyclical elements. Once again, referring to the teller at the Fourth National Bank, the bank procedures may state that account balances must be checked only when the value of the check to be cashed exceeds $50.00. Sometimes, this situation happens and at other times the individual cashing the check only wants, perhaps, $30.00, and the account balance is not checked. To determine the time required to perform this check cashing job, we must time enough cycles to feel confident that we have a fair representation of the time required for all the elements included in the job. Nonrecurring or noncyclical elemental times will then be prorated into the standard based on the relative frequency with which they occur.

There are many well-developed statistical procedures available for determining sample size. These methods are well-documented in any standard statistical text. They are summarized at the end of the chapter describing work sampling. They require the analyst to:

1. Specify a confidence level, that is, a level describing the certainty that the analyst would have in the results.
2. Estimate the percentage of time spent on the shortest element.
3. State a desired accuracy range, such as that the measured time should be within 10 percent of the true time spent performing the task.
4. Calculate a value for the sample size based on the information specified above and using the appropriate statistical formula.

Sample Size

The statistical formula used to estimate sample size is

$$n = [(p)(1 - p)(z^2)] / E^2$$

where

n is the sample size.
p is the proportion of time spent on the shortest element.
z is the standard normal curve ordinate corresponding to the desired confidence.
E is the accuracy desired in the final result.

Example: Sample Size Determination

A mailer-stuffing task has the five elements described earlier. The shortest of the elements is estimated to require 13 percent of the work day. Determine the appropriate number of cycles to time if the resulting standard is to be measured, with 95-percent confidence, to within plus or minus .01 minutes.

In the equation presented previously the following variables were identified:

$$p = .13$$
$$z = 1.96$$
$$E = .01$$

The sample size, n, is calculated to be 4344.8, or rounding up to the nearest whole number, 4345. Table 5-1 shows the sample sizes required for various confidence levels, proportions, and accuracy levels.

As a practical matter, if we remember that our objective is to develop a time standard that is fair and consistent we will end up timing enough cycles until we subjectively feel we have done our job regardless of what any statistical formula states. Experience plays a large role here. Sometimes, we will feel "sure" after 10 to 20 cycles

TABLE 5-1(a) SAMPLE SIZE FOR 90% CONFIDENCE WITH ABSOLUTE ACCURACY SHOWN

P	± .01	P	± .05	P	± .10	P	± .25
1	268	1	11	1	3	1	0
2	530	2	21	2	5	2	1
3	787	3	31	3	8	3	1
4	1039	4	42	4	10	4	2
5	1285	5	51	5	13	5	2
6	1526	6	61	6	15	6	2
7	1762	7	70	7	18	7	3
8	1992	8	80	8	20	8	3
9	2216	9	89	9	22	9	4
10	2435	10	97	10	24	10	4
11	2649	11	106	11	26	11	4
12	2858	12	114	12	29	12	5
13	3061	13	122	13	31	13	5
14	3258	14	130	14	33	14	5
15	3450	15	138	15	35	15	6
16	3637	16	145	16	36	16	6
17	3818	17	153	17	38	17	6
18	3994	18	160	18	40	18	6
19	4165	19	167	19	42	19	7
20	4330	20	173	20	43	20	7
21	4489	21	180	20	45	21	7
22	4644	22	186	22	46	22	7
23	4792	23	192	23	48	23	8
24	4936	24	197	24	49	24	8
25	5074	25	203	25	51	25	8
26	5206	26	208	26	52	26	8
27	5334	27	213	27	53	27	9
28	5455	28	218	28	55	28	9
29	5572	29	223	29	56	29	9
30	5683	30	227	30	57	30	9
31	5788	31	232	31	58	31	9
32	5888	32	236	32	59	32	9
33	5983	33	239	33	60	33	10
34	6072	34	243	34	61	34	10
35	6156	35	246	35	62	35	10
36	6235	36	249	36	62	36	10
37	6308	37	252	37	63	37	10
38	6375	38	255	38	64	38	10
39	6438	39	258	39	64	39	10
40	6494	40	260	40	65	40	10
41	6546	41	262	41	65	41	10
42	6592	42	264	42	66	42	11
43	6632	43	265	43	66	43	11
44	6668	44	267	44	67	44	11
45	6697	45	268	45	67	45	11
46	6722	46	269	46	67	46	11
47	6741	47	270	47	67	47	11
48	6754	48	270	48	68	48	11
49	6762	49	270	49	68	49	11
50	6765	50	271	50	68	50	11

TABLE 5-1(b) SAMPLE SIZE FOR 95% CONFIDENCE WITH ABSOLUTE ACCURACY SHOWN

P	± .01	P	± .05	P	± .10	P	± .25
1	380	1	15	1	4	1	1
2	753	2	30	2	8	2	1
3	1118	3	45	3	11	3	2
4	1475	4	59	4	15	4	2
5	1825	5	73	5	18	5	3
6	2167	6	87	6	22	6	3
7	2501	7	100	7	25	7	4
8	2827	8	113	8	28	8	5
9	3146	9	126	9	31	9	5
10	3457	10	138	10	35	10	6
11	3761	11	150	11	38	11	6
12	4057	12	162	12	41	12	6
13	4345	13	174	13	43	13	7
14	4625	14	185	14	46	14	7
15	4898	15	196	15	49	15	8
16	5163	16	207	16	52	16	8
17	5420	17	217	17	54	17	9
18	5670	18	227	18	57	18	9
19	5912	19	236	19	59	19	9
20	6147	20	246	20	61	20	10
21	6373	21	255	21	64	21	10
22	6592	22	264	22	66	22	11
23	6803	23	272	23	68	23	11
24	7007	24	280	24	70	24	11
25	7203	25	288	25	72	25	12
26	7391	26	296	26	74	26	12
27	7572	27	303	27	76	27	12
28	7745	28	310	28	77	28	12
29	7910	29	316	29	79	29	13
30	8067	30	323	30	81	30	13
31	8217	31	329	31	82	31	13
32	8359	32	334	32	84	32	13
33	8494	33	340	33	85	33	14
34	8621	34	345	34	86	34	14
35	8740	35	350	35	87	35	14
36	8851	36	354	36	89	36	14
37	8955	37	358	37	90	37	14
38	9051	38	362	38	91	38	14
39	9139	39	366	39	91	39	15
40	9220	40	369	40	92	40	15
41	9293	41	372	41	93	41	15
42	9358	42	374	42	94	42	15
43	9416	43	377	43	94	43	15
44	9466	44	379	44	95	44	15
45	9508	45	380	45	95	45	15
46	9543	46	382	46	95	46	15
47	9569	47	383	47	96	47	15
48	9589	48	384	48	96	48	15
49	9600	49	384	49	96	49	15
50	9604	50	384	50	96	50	15

TABLE 5-1(c) SAMPLE SIZE FOR 99% CONFIDENCE WITH ABSOLUTE ACCURACY SHOWN

P	± .01	P	± .05	P	± .10	P	± .25
1	657	1	26	1	7	1	1
2	1301	2	52	2	13	2	2
3	1931	3	77	3	19	3	3
4	2548	4	102	4	25	4	4
5	3152	5	126	5	32	5	5
6	3743	6	150	6	37	6	6
7	4320	7	173	7	43	7	7
8	4884	8	195	8	49	8	8
9	5435	9	217	9	54	9	9
10	5972	10	239	10	60	10	10
11	6496	11	260	11	65	11	10
12	7007	12	280	12	70	12	11
13	7505	13	300	13	75	13	12
14	7989	14	320	14	80	14	13
15	8461	15	338	15	85	15	14
16	8918	16	357	16	89	16	14
17	9363	17	375	17	94	17	15
18	9794	18	392	18	98	18	16
19	10212	19	408	29	102	19	16
20	10617	20	425	20	106	20	17
21	11009	21	440	21	110	21	18
22	11387	22	455	22	114	22	18
23	11752	23	470	23	118	23	19
24	12104	24	484	24	121	24	19
25	12442	25	498	25	124	25	20
26	12767	26	511	26	128	26	20
27	13079	27	523	27	131	27	21
28	13378	28	535	28	134	28	21
29	13663	29	547	29	137	29	22
30	13935	30	557	30	139	30	22
31	14194	31	568	31	142	31	23
32	14439	32	578	32	144	32	23
33	14672	33	587	33	147	33	23
34	14891	34	596	34	149	34	24
35	15096	35	604	35	151	35	24
36	15289	36	612	36	153	36	24
37	15468	37	619	37	155	37	25
38	15634	38	625	38	156	38	25
39	15787	39	631	39	158	39	25
40	15926	40	637	40	159	40	25
41	16052	41	642	41	161	41	26
42	16165	42	647	42	162	42	26
43	16264	43	651	43	163	43	26
44	16351	44	654	44	164	44	26
45	16424	45	657	45	164	45	26
46	16483	46	659	46	165	46	26
47	16530	47	661	47	165	47	26
48	16563	48	663	48	166	48	27
49	16583	49	663	49	166	49	27
50	16589	50	664	50	166	50	27

have been timed; at other times, hundreds of cycles will be timed before we feel confident. The number of cycles to be timed depends on the nature of the elements, the total cycle time, the number of nonrecurring elements, the cooperativeness of the subject, the skill and experience of the analyst, and, perhaps, the mood of the analyst when the study is being performed.

To summarize, the number of cycles to time can be stated as "enough." Time enough cycles until you, the analyst, feel that you will have "good" or consistent results.

Because the number of cycles to be timed will be unknown at the beginning of the study, enough data sheets should be prepared to cover any possible number of cycles that may be observed. Many companies have their own forms for collecting time study data; examples were shown in Figures 5-6(a) and (b) and (c). Although much information can be included on the form, the examples shown in Figures 5-6(a) and (b) and (c) show the minimum required. Additional information may be included, depending on the organization's needs.

POTENTIAL DATA RECORDING DIFFICULTIES

When the analyst records the time study data the following are some potential difficulties that might be encountered:

Foreign Elements are those events that occur while a job is being performed that are not part of the normal work assignment. A dropped tool, a fumble, or a sneeze is foreign to the work normally expected by the operator. Foreign elements are unnecessary, non-productive elements that are added to the job by the worker.

Missed Elements are those elements for which no time can be recorded. The analyst may have missed the breakpoint due to a variety of reasons, such as being distracted by a loud noise.

Out-of-sequence Elements occur when the operator varies the order of operations performed. Sometimes this deviation from the prescribed method is significant, while at other times it is merely a distraction. This difficulty may be caused by lack of training, lack of practice, or as an intentional effort by the operator to confuse the analyst.

Non-cyclical Elements are those necessary parts of the job that do not occur during every cycle. The ending breakpoints for these elements are recorded, in sequence, only during those cycles in which they occur. Their non-cyclical nature means that time is prorated based on its relative frequency of occurrence when the standard time is calculated. Set-up times are an extreme case of a non-cyclical element.

A Typical Study

If you will recall the hypothetical eight-element job mentioned earlier in this chapter, experience and judgment indicate to us that, barring anything unforeseen or too unusual, we should be able to complete this routine study by timing 10 cycles. Our next step is to fill out a data sheet, showing all the required background information on the top of the form

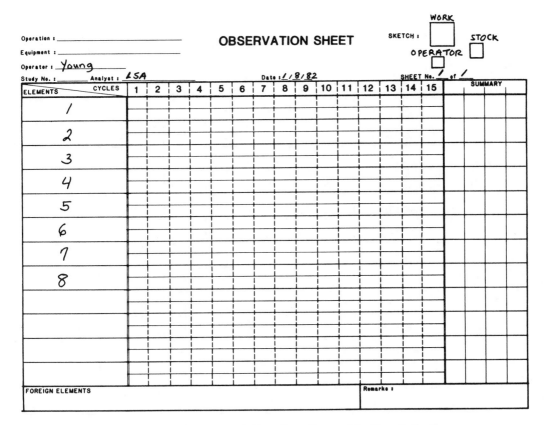

Figure 5-11 Sample Data Sheet Prepared for Data Collection.

and identifying the elements clearly and explicitly in the appropriate column. Figure 5-11 shows the sheet ready for data collection. The first two cycles were uneventful. Data were collected using the continuous method; Figure 5-12 shows these values. Beginning with the third cycle, however, the analyst missed the ending point of the first element. This oversight is not a catastrophe, *if* the analyst lets the watch continue running. Although the third cycle will not have a time for the first element, nor a time for the second element, the ending point of the second element will be recorded. This notation will permit calculation of the time for the third element. As a matter of fact, for future reference, the analyst should show some type of reason or code for the missed reading. In this particular case, the analyst sneezed when the observation should have been made. There are numerous reasons for a missed reading. As described previously, these unusual occurrences are usually called *foreign elements*. Figure 5-13 shows how these foreign elements might be coded on the time study data sheet. Our study proceeded with no further hitches until the sixth cycle. There, as shown in Figure 5-14, the fourth element shows what is obviously too much elapsed time. Something unusual must have happened. In this instance, the operator dropped a tool and had to pick it up. Again, the analyst should leave the watch

Operation :
Equipment :
Operator : _Young_

OBSERVATION SHEET

SKETCH : [WORK]   STOCK [ ]
OPERATOR [ ]

ELEMENTS	CYCLES	1	2	3	4	5	6	7	8	9	10	11	12	13	14	15	SUMMARY
1		16	179														
2		26	189														
3		60	222														
4		75	237														
5		98	261														
6		114	277														
7		140	303														
8		163	325														

Study No. : _____ Analyst : _LSA_   Date : _1/8/82_   SHEET No. _1_ of _1_

FOREIGN ELEMENTS   Remarks :

Figure 5-12 Initial Time Study Data (Times shown are in decimal minutes).

running and record the ending time. The letter "b" will note that something identifiable happened at this point. The foreign element, called "fumble," explains what the difficulty was. Although this particular time will be meaningless, unless the operator always develops fumble fingers right at that point, the rest of the times for the cycle will be valid.

As it happened, during this time study, when the analyst completed the sixth cycle and began the seventh, the watch passed the 10-minute mark. As a matter of convenience, the last three digits of the time study were the only ones recorded and the study proceeded. No further problems were experienced until the seventh element of the seventh cycle. Here, there was more than a three-minute delay while the operator replaced a broken tool bit. This type of delay occurs occasionally and it is one of the reasons for the PFD allowance. Again, as shown in Figure 5-15, this foreign element was listed and the analyst permitted the watch to continue running, resulting in only one lost time.

The industrial engineer, as the study progressed, evaluated each element in terms of how the operator's pace compared to normal. This rating factor was then recorded on the observation sheet at the conclusion of the study. Sometimes, an entire job is rated rather than the element-by-element rating described in this example.

a = Sneeze

Figure 5-13 Example of Missed Reading.

The remainder of the 10 cycles proceeded without a hitch. The entire data sheet is shown in Figure 5-16. The column headed ''R'' is the rating for each element.

DATA ANALYSIS

Once sufficient data are collected to satisfy the analyst that he will have sufficient confidence in his results, the actual analysis can be performed. This job is a clerical task, easily adaptable to computer analysis. It involves determining each elemental cycle time, calculating each elemental average, applying the rating factor to each elemental average, applying the rating factor to each elemental average, and adding the allowance. As with all routine tasks, this analysis is suitable for computerization. The elemental times for each cycle are obtained by subtraction. Because this operation is modest, it is noted that the analyst performed the entire analysis himself. The elemental times are shown in Figure 5-17. One interesting measurement appears for the fifth element during the third cycle. This reading, so significantly different from the others for that element, 13 as compared to 23 or 24, is quite suspect. Obviously, the analyst or the operator did something unusual at this point. Perhaps the operator left out a step or the analyst misread the watch. If proper investigation determines that this short elemental time is indeed a fluke, then it should be disregarded in the ensuing calculations. The time is circled for that reason.

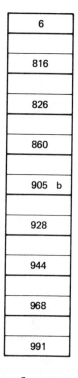

6
816
826
860
905   b
928
944
968
991

a = Sneeze   **Figure 5-14**   Example of Extended Cy-
b = Fumble   cle Time.

After all the subtractions have been performed, the elemental averages are calculated. These are shown in the appropriate column on the data sheet. The averages are
calculated using only the acceptable data. Values such as the 13 that was just investigated
are deleted from future use and the data sheet is annotated in the appropriate fashion (see
Table 5-2). Once this arithmetic is performed, the standard time for each element is calculated, using the rating for each element and the company standard of a 12 percent PFD

TABLE 5-2   ELEMENTAL AVERAGE CALCULATIONS

Element	Sum	Average	Decimal Minutes
1	145	16.1	.161
2	91	10.1	.101
3	338	33.8	.338
4	134	14.9	.149
5	209	23.2	.232
6	160	16.0	.160
7	220	24.4	.244
8	229	22.9	.229

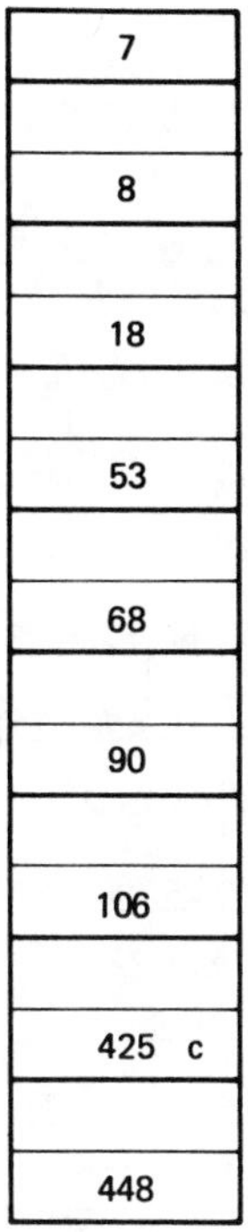

a = Sneeze
b = Fumble
c = Tool maintenance

Figure 5-15 Another Example of Extended Cycle Time.

allowance. These computations are shown in Table 5-3. All of the information should be shown on the original data sheet. This information, as shown in Figure 5-18, must be kept as part of the industrial engineering department's permanent record.

TABLE 5-3 STANDARD CALCULATION

Element	Average	Rating	Rated time	Allowance	Elemental standard
1	.161	1.05	.169	.020	.189
2	.101	1.05	.106	.013	.119
3	.338	1.00	.338	.041	.379
4	.149	.85	.127	.015	.142
5	.232	1.05	.244	.029	.273
6	.160	1.10	.176	.021	.197
7	.244	1.00	.244	.029	.273
8	.229	1.00	.229	.027	.256
					$\Sigma = 1.828$

Operation : _____

Equipment : _____

Operator : *Young*

Study No. : _____ Analyst : *LSA*

OBSERVATION SHEET

SKETCH :

WORK ☐ STOCK ☐

OPERATOR ☐

Date : *1/8/82*

SHEET No. *1* of *1*

ELEMENTS	CYCLES	1	2	3	4	5	6	7	8	9	10	11	12	13	14	15	R	SUMMARY
1		16	179	a	494	656	816	8	464	624	787						105	
2		26	189	352	504	666	826	18	474	635	797						105	
3		60	222	386	538	699	860	53	508	668	831						100	
4		75	237	400	553	714	905 b	68	523	683	846						85	
5		98	261	413	576	737	928	90	546	708	869						105	
6		114	277	430	592	752	944	106	562	724	885						110	
7		140	303	455	616	777	968 c	425	586	748	909						100	
8		163	325	478	639	800	991	448	609	771	932						100	

FOREIGN ELEMENTS

Remarks : a - *pneeze*
b - *fumble*
c - *tool maintenance*

Figure 5-16 Completed Raw Data Sheet.

INTERPRETATION OF RESULTS

The next step in making a complete time study is to present the results. There are several common ways of presenting these data. The first is simply to announce the total elemental standard time for the job in question. In the example we have been following, the standard was calculated to be 1.828 minutes, meaning that it will take the average worker, working at the normal pace under the defined work method, 1.828 minutes to complete the task. This time, of course, allows for the personal, fatigue, and delay allowances that normally occur during the work day.

Sometimes, a simple statement of the standard does not provide enough information for the person who will be using the standard. Rather than expressing the standard in minutes, a production scheduler might prefer the time expressed in hours. A simple conversion, based on the fact that there are 60 minutes in an hour, changes our standard of 1.828 minutes to

$$(1.828 \text{ minutes})(1 \text{ hour}/60 \text{ minutes}) = .030 \text{ hours}$$

	SKETCH :	WORK		STOCK

OBSERVATION SHEET

Operation : _____

Equipment : _____

Operator : _Young_

Study No. : _____ Analyst : _LSA_   Date : _1/8/82_   SHEET No. _1_ of _1_

ELEMENTS \ CYCLES	1	2	3	4	5	6	7	8	9	10	11	12	13	14	15	SUMMARY
1	16	16	/	16	17	16	17	16	15	16						
2	10	10	/	10	10	10	10	10	11	10						
3	34	33	34	34	33	34	35	34	33	34						
4	15	15	14	15	15	/	15	15	15	15						
5	23	24	(13)	23	23	23	22	23	25	23						
6	16	16	17	16	15	16	16	16	16	16						
7	24	26	25	24	25	24	/	24	24	24						
8	23	22	23	23	23	23	23	23	23	23						

FOREIGN ELEMENTS   Remarks :

Figure 5-17   Intermediate Calculations.

Another common way to express the results of a time study is in hours required to produce 1000 units, 100 units, or any other standard production quantity. Knowing how many hours one unit requires permits easy conversion to any standard quantity. If one unit of our product requires .030 hours, then 1000 units will require

$$(100)(.030 \text{ hours}) = 30 \text{ hours}$$

If the operator is paid on the basis of production with some sort of incentive plan, as described in Chapter 11, the paymaster will be more interested in how many units of production are required to earn a day's pay. This requirement might call for the expression of the standard in terms of the number of units expected to be produced in one hour or during an eight-hour work day. Using our sample standard, within one hour, the worker should be able to produce

$$(1 \text{ hour})/(.030 \text{ hours/unit}) = 33.33 \text{ units}$$

During a normal eight-hour work day, 266.67 units will be the expected production. If the worker produces more than this quantity, then the paymaster will know that a bonus or

OBSERVATION SHEET

Operation: _____
Equipment: _____
Operator: *Young*
Study No.: _____
Analyst: *LSA*
Date: *1/8/82*

SKETCH:
WORK ☐ STOCK ☐
OPERATOR ☐

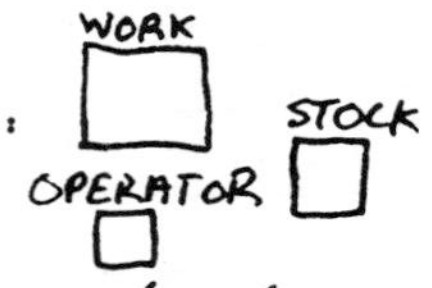

SHEET No. *1* of *1*

ELEMENTS	1	2	3	4	5	6	7	8	9	10	11	12	13	14	15	R	Σ	AVG	e.T.	ALL	S.T.
1	16	16	/a	16	17	16	17	16	15	16						105	145	.161	.169	.020	.189
	16	179		494	656	816	8	464	624	287											
2	10	10	/	10	10	10	10	10	11	10						105	91	.101	.106	.013	.119
	26	189	352	504	666	826	18	474	635	292											
3	34	33	34	34	33	34	35	34	33	34						100	338	.338	.338	.041	.379
	60	222	386	538	699	860	53	508	668	831											
4	15	15	14	15	15	16	15	15	15	15						85	134	.149	.127	.015	.142
	75	237	400	553	714	905	68	523	683	846											
5	23	24	13	23	23	23	22	23	25	23						105	209	.232	.244	.029	.273
	98	261	413	576	737	928	90	546	708	869											
6	16	16	17	16	15	16	16	16	16	16						110	160	.160	.176	.021	.197
	114	277	430	592	752	944	106	562	724	885											
7	24	26	25	24	25	24	c	24	24	24						100	220	.244	.244	.029	.273
	140	303	455	616	722	968	425	586	748	909											
8	23	22	23	23	23	23	23	23	23	23						100	229	.229	.229	.029	.258
	163	325	478	639	800	991	448	609	771	932											
																					Σ = 1.828

FOREIGN ELEMENTS

Remarks:
a - sneeze
b - fumble
c - tool maintenance

Figure 5-18 Completed Data Sheet.

incentive payment is deserved. Consistent production below this standard would indicate the need for retraining the worker or show the need for possible disciplinary action.

Expressing the standard in this last fashion also provides a suitable way to measure increases in productivity. Increases in production above the standard provide evidence that productivity has increased. As we have seen and will see in other sections of this text, there are numerous ways for industrial engineers to try to improve productivity. Standard time provides a tool to measure changes in productivity.

SPECIAL CASES

Unfortunately, setting time standards is not as straightforward as was just described. Not all production tasks are entirely operator-controlled. Many jobs have certain peculiarities that might make standard development more difficult than described. Because these special cases are deviations from the norm, there can be no all-encompassing solution presented. Instead, this section will present some of the special problems that the industrial engineer may encounter as time standards are set. Each organization will provide its own solution to these problems.

Machine Interference Time

Not all production jobs are entirely controlled by the operator. Sometimes, the operator must wait for the machine to complete its portion of the job before proceeding. No matter how hard the worker may be capable of working, there will be times when the operator stands idle while waiting for the machine. The nature of the process may interfere with the worker while he is performing his job and it may prohibit him from performing any other task while he is waiting for the machine to finish its work. The process may need monitoring because it may be positioned too far away from other equipment and the time the machine interferes with the operator's actions may force the operator to spend some seemingly unproductive time just waiting. Any standards set on this type of job should reflect this machine interference time.

A common way to "handle" machine interference is to simply measure it and count it as a separate and distinct element. The worker is rated at 100 percent during this time. If the worker is subject to an incentive pay system, a straight hourly or day rate would be paid to the worker for the time the machine controls activity.

Multiple Machine Responsibility

Often an operator may have more than one machine to tend. When this situation occurs, the machines may not be sequenced so that the operator is always busy. There may be times when the machines are busy and the operator is idle; at other times, the machines may be awaiting operator attention. Sometimes, even three or more machines may be operated by one person. The difficulties in setting standards compound as more machines are idle. There is, unfortunately, no easy solution to setting standards when the operator has multiple machine responsibility. The interference problem in this case is very real and significant.

Irregular Elements

Many jobs that require standards set as a base for measuring productivity increases have so many exceptions to standard operating procedures that the whole standard-setting procedure becomes very complicated. Take, for example, the seemingly straightforward task of preparing meals for patients in a hospital. The preparation of a simple meal such as toast and eggs can have so many combinations that a standard-setter might end up needing the hospital just to recover from the complications! Differences due to medical reasons in areas such as seasoning (salt, no salt, special salt), type of bread (white, wheat, bran, low-cal, high fiber), type of spread (butter, margarine, jam, nothing, salted, unsalted, saturated, polyunsaturated), and who knows what else can create some difficulties for the analyst who must set the standard on what at first seems like a very simple and straightforward production task. It is easier for restaurants with a standardized product to set comparable standards, especially the fast food chains that have a relatively limited menu.

We have already mentioned regular elements that don't occur on a cyclical basis. For example, the keypunch operator who must periodically gather new work and pass along processed work of differing quantities. This example poses some special problems in setting rates. However, by simply prorating the relative frequency with which the noncyclical elements occur into the final standard, this problem is virtually eliminated.

Set-up time is a special type of noncyclical element. It is usually handled as a separate component of the time standard.

Volume Irregularities

There are also situations where the volume of work varies and is not under the control of the operator or maybe even the company. One day a bank teller may be swamped with customers wanting to deposit checks and the next day have very few customers, each of whom only wants to handle company payroll transactions. This variation represents the nonstandardized procedure. It is compounded by the unpredictability, especially in the short run, of the type of task to be performed.

This list could be virtually endless. These examples have been cited primarily to make the reader aware of some of the problems that might be waiting when time standards are set. There are solutions to all of these problems; some are rather unique and others require no more than common sense. These problems are what add the challenge to this portion of the analyst's job.

The following pages include two examples of setting time standards with stopwatch time study. The first example uses a detailed statement of elements. The second example illustrates the use of larger elements.

Example: Collating Tests

In reaction to Proposition 13 and other tax-cutting moods in the nation, the State University decided to establish a productivity measurement and improvement program. The consulting firm of Industrial Engineers, Inc., was engaged by the university system to start the program. The initial approach of I.E., Inc., was to set up a trial productivity monitoring program for the clerical and secretarial positions within the university.

One clerical task that seemed to occupy significant amounts of the departmental secretaries' time during the day was test preparation. There are several types of activities performed during the preparation of exams: typing, reproducing, and collating. The collating task, as it is performed at State U., is an entirely manual and operator-controlled production operation. Although the methods study that preceded the time study questioned the need for this operation to be done manually, the administration of the school held firm and insisted that this method was the only way the job could be done. There was absolutely no money, the administration insisted, for luxuries such as automatic collating machines.

The manual collating task is performed with the aid of an expandable wire rack that holds up to 300 copies of 26 different pages. Each page of a test is placed in a separate division of the holder. Figure 5-19 shows a typical collating rack. Test pages have traditionally been placed in the rack in numerical order. The clerk then removes pages, beginning in reverse numerical order, with the right hand and places each page in the left hand. (The process is just the opposite for left-handed clerks.) After all the pages have been removed in this fashion, both hands are used to "square" the tests by tapping the stack, usually four times, against the table. The test is then moved into the stapler. The left hand places the finished product on the stack and the clerk is ready to begin again. After several tests are collated, the clerk moistens his fingers with beeswax so that the pages will be easier to grasp.

Industrial Engineers, Inc., believes that this operation is a candidate for a significant productivity improvement project, even within the limitations placed by the administration. However, before this improvement can be made and measured, I.E., Inc., must determine the current level of performance. They need a base from which to show how much they have helped. This base will be provided by means of a stopwatch time study. A study of work logs indicates that a five-page test is not only the average test length, but also the most frequently occurring test length. Therefore, the five-page test is considered

Figure 5-19   Collating Rack.

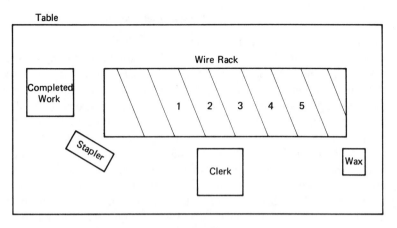

Figure 5-20   Collating Work Area.

to be most representative of the tests collated at State U. Careful development of the elements should allow for setting standards for tests of other lengths.

The analyst assigned to this study has prepared the following elements to be used in setting the standard for the test collating operation. She noted that the tests arrive from the duplicating process and are placed in the expandable wire rack before the collating process begins.

Figure 5-20 shows a sketch of the work area. Following is a complete description of the operations performed by the test collator:

ELEMENT NUMBER	DESCRIPTION
1	Place page 5 in left hand. Element starts when page 5 is grasped and concludes when page 5 is released by the right hand.
2	Place page 4 in left hand. Element starts when page 5 is released by the right hand, continues through removing page 4 from the wire rack, and concludes when page 4 is released by the right hand.
3	Place page 3 in left hand. Element starts when page 4 is released by the right hand and concludes when page 3 is released by the right hand after being placed in the left hand.
4	Place page 2 in left hand. Element starts when page 3 is released by the right hand and concludes when page 2 is released by the right hand after being placed in the left hand.
5	Place page 1 in left hand. Element starts when page 2 is released by right hand and concludes when page 1 is released by the right hand.

ELEMENT NUMBER	DESCRIPTION
6	Firm stock. Element starts when page 1 is released by the right hand, continues through grasping stack of pages on ends and tapping four times on table. Element concludes at the end of the last tap.
7	Staple test. Element starts at the conclusion of the test tapping, continues with the movement of the test into the stapler, and concludes after the right hand releases the stapler following the insertion of the staple into the test.
8	Stack tests. Element starts after test is stapled, includes placing finished test on stack, and concludes when simultaneously the left hand releases the test and the right hand touches page 5 in the wire rack. This element normally completes a cycle.
9	Periodically, the clerk applies beeswax to the fingers of his right hand. This regular, but noncyclical element begins after the test is stapled and includes reaching to the open wax dish and rubbing the fingertips through the wax, and concludes when the right hand touches page 5 in the wire rack.

For the purposes of this study, the analyst determined that 15 cycles would give satisfactory results. After explaining the study to the clerk and receiving agreement that the method described was indeed standard practice, the analyst was ready to collect the time study data. The completed data sheet is shown in Figure 5-21(a) and (b).

Industrial Engineers, Inc., suggested, and the dean of State University agreed, that a PFD allowance of 15 percent was reasonable for clerical employees. The analyst then calculated the elemental standard times. Table 5-4 shows the intermediate calculations.

The standard time for this particular task is the sum of the times for the first eight elements plus the inclusion of a portion of the ninth. The analyst observed that, on the average, element 9 occurred on one-third (5/15) of the timed cycles. Based on this observation, she calculated the standard time to collate one five-page exam as the sum of the elemental standard time for elements 1 through 8 plus ⅓ of the elemental standard time for element 9:

$$.082 + .154 + .135 + .137 + .133 + .102 + .252 + .066 + .207/3 = 1.13$$

The analyst concluded that the typical clerk should be able to collate a five-page test in 1.129 minutes. (Although these numbers seem long, which they probably are, they were selected to illustrate the procedure for making a time study.) Expressed in alternate formats, the standard becomes:

53 per hour
18.8 hours per thousand

Operation : _____

Equipment : _____

Operator : _____

Study No. : _____ Analyst : _____

OBSERVATION SHEET   SKETCH :

Date : __/__/__   SHEET No. ___ of ___

ELEMENTS / CYCLES	1	2	3	4	5	6	7	8	9	10	11	12	13	14	15	Σ	SUMMARY AVG.	R%
1. Page 5 to left hand	7	7	8	7	7	8	7	7	7	7	7	7	7	7	7	107	.071	100
	7	106	222	329	428	630	729	828	938	57	157	274	374	470	577			
2. Page 4 to left hand	15	14	15	16	/	15	15	15	15	15	15	14	15	15		209	.149	90
	22	120	237	345	b	645	744	843	953	72	172	289	388	485	592			
3. Page 3 to left hand	13	13	13	13	/	13	13	a	(22)	13	13	13	13	13	13	156	.130	90
	35	133	250	358	541	658	757	865	975	85	185	302	401	498	605			
4. Page 2 to left hand	13	13	13	13	13	13	13	15	15	13	13	13	12	13	13	198	.132	90
	48	146	263	371	554	671	770	880	990	98	198	315	413	511	618			
5. Page 1 to left hand	13	13	a	13	13	13	13	13	13	13	13	13	12	(3)	13	168	.129	90
	61	159	284	384	567	684	783	893	3	111	211	328	425	514	631			
6. Firm tests	9	9	9	8	8	9	9	9	9	9	9	9	9	9	9	133	.089	100
	70	168	293	392	575	693	792	902	12	120	220	337	434	523	640			
7. Staple tests	22	21	22	22	22	22	22	22	(13)	22	22	23	22	22	21	307	.219	100
	92	189	315	414	597	715	814	924	2	142	242	360	456	545	661			
8. Stack tests	7	7	7	7	7	7	7	7	7	8	7	7	7	7	7	106	.071	80
	99	196	322	421	604	722	821	931	32	150	249	367	463	552	668			
9. Apply beeswax	18			18			18			18			18			90	.180	100
	214			622			50			267			570					
a - discard blank sheets																		
b - dropped reading by analyst																		

FOREIGN ELEMENTS   Remarks :

Figure 5-21(a)   Collating Time Study Sheet.

Industrial Engineers, Inc., is now ready to study and suggest some procedures for improving the productivity of this particular operation. The standard that has just been developed provides a convenient basis for comparison.

For example, one suggestion that the analyst immediately had was to replace the manual stapler with an electric stapler. Trying this device in one department reduced the elemental standard for stapling, element 7, from .252 minutes to .086 minutes per test, thus reducing the standard time required by .166 minutes, which represented an increase in productivity of

$$.166/1.13 = .147 \text{ or } 14.7 \text{ percent}$$

Other similar changes should be obvious to the reader. The data on productivity improvement can then be used not only to demonstrate that State U. is getting more for its money, but also to justify further capital expenditures in future efforts to increase productivity. Perhaps the relatively modest investment in the electric stapler and the almost 15-percent improvement that resulted will permit the analyst to sell the administration on purchasing an automatic collating device. Such a device would eliminate the need

OPERATION: Collating

PART NAME:	Test	OP. NO: . N/A
MACH. NAME:	Collator	PART NO: N/A
OPERATOR NAME & NO:	McGuire 1426	MACH. NO: N/A
EXPERIENCE ON JOB:	3 Years	MALE ☐ FEMALE ☒
		FOREMAN: Rezak
NO. MACHINES OPER'D:	1 MACH. SPEED: N/A	DEPT. NO: N/A
MATERIAL:	Paper	

SKETCH OF WORKPLACE   SCALE:

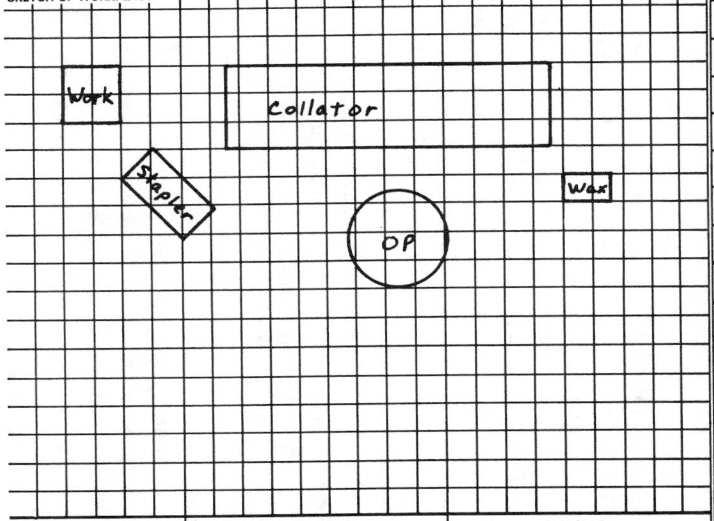

Work

Stapler

Collator

OP

Wax

DATE OF STUDY 7/6   OBSERVER LSA   APPROVED

SUMMARY

NO.	ELEMENTS	NORMAL TIME	PAT'S PERS'L ALLOW.	OTHER ALLOW.	STD. TIME
1)	5 to LH				
2)	4 to LH		(SEE		
3)	3 to LH			TABLE 5-3)	
4)	2 to LH				
5)	1 to LH				
6)	Firm				
7)	Staple				
8)	Stack				
9)	Wax				

TOTAL STD. TIME PER CYCLE: .0188 Hrs.

NO. PIECES PER CYCLE: 1   STD. TIME PER PIECE: .0188 Hrs.

DRAWING OF PART:

Figure 5-21(b).

TABLE 5-4   INTERMEDIATE CALCULATIONS

Element	Average	Rating	Rated average	Allowance	Elemental standard
1	.071	1.0	.071	.011	.082
2	.149	.9	.134	.020	.154
3	.130	.9	.117	.018	.135
4	.132	.9	.119	.018	.137
5	.129	.9	.116	.017	.133
6	.089	1.0	.089	.013	.102
7	.219	1.0	.219	.033	.252
8	.071	.8	.057	.008	.065
9	.180	1.0	.180	.027	.207

for elements 1, 2, 3, 4, 5, 6, and 9. The additional increase in productivity would be significant.*

Example: Notebook Lamination

The Amalgamated Paper Products Union has a provision in its contract with Noteworthy Notebook Company that permits grievances to be filed on time standards, or rates, that union members feel are unfair. These rates can be unfair in either of two ways. First, the rates can be "loose," meaning that the standard permits more time to complete the operation than is really required. An employee who is paid based on production output can easily earn a lot of money when working a job with a loose rate. Workers may feel that the rate is unfair because all the other rates are not as loose. Rarely will a grievance be filed by a union member based on a loose rate, but it poses a problem for management. It can also create a problem for the union leadership because payments will not be consistent for all employees. Different amounts of work might result in the same pay and this situation is not healthy for management or for union leadership.

Rates that are too "tight" are also unfair. With a tight rate, the standard permits less time to complete the job than is usually required. A worker on an incentive pay plan faced with a tight rate would have difficulty earning incentive pay. Members of Amalgamated do not hesitate to file a grievance about rates that were viewed as being tight. In situations where the standards were not correct, the company management and union leadership were very interested in correcting the real or perceived inequities. If workers think they are not getting the same pay for doing the same work as someone else, problems could result.

Naturally, not all rates that union members view as being unfair are really unfair. However, grievances can be filed any time a union member thinks the rate is unfair. The grievance procedure, as spelled out in the contract between Amalgamated and Noteworthy, provides for the following steps to be taken when a complaint is received:

*After following the calculations found in this example, the reader has probably concluded that this portion of time study can be very tedious. Appendix A of this book includes a computer program written in BASIC that handles the arithmetic involved in time standard calculation. This Appendix also includes other useful BASIC computer programs.

1. File a written grievance to the immediate supervisor.

2. The industrial engineering department of Noteworthy will (a) make sure the standard was set with the current standard practice or method in effect and (b) audit (restudy) the standard to check its validity.

3. If the grievance has not been resolved then the union may, at its expense, hire an outside expert to audit the rate. The expert must be a qualified industrial engineer. The company retains the right of approval for the industrial engineer hired.

4. If the union's IE sets a standard that differs by more than five percent from the company's standard, the grievance will be presented to an arbitrator selected from a list provided by the Federal Mediation and Conciliation Service, according to standard methods, for arbitration.

A member of the Amalgamated Paper Products Union has filed a grievance about the rate set by Noteworthy for the operation officially known as "Hand Laminate Cover and Back to Sewn Notebook." This procedure is usually referred to as "laminating" in the language of the company and union. As would be expected, the employee claims the standard is too tight. She has been unable to earn any incentive pay when working this operation, despite the fact that her coworkers average a 20 percent bonus on incentive pay.

The laminating operation involves gluing a hard cardboard cover to presewn composition books. The 60-page books are produced in sets of three, so that once the covers are secured and the glue has dried, the product is cut twice to produce three separate notebooks. Figure 5-22 shows a typical notebook produced in this fashion. Figure 5-23 shows the workplace layout for the lamination operation.

Noteworthy's Chief IE divided the job into many elements when she reset the standard. These elements, which occur in a number of different styles of notebooks, reflect the detailed division of the job into the fine elements required by Noteworthy. These elements included elemental descriptions such as:

1. Operator places aside the finished strip of three laminated composition books with right hand.

2. Operator reaches to stack to grasp next cover with the left hand.

3. Right hand reaches to grasp next strip of pages from stack.

4. Right hand positions pages in fixture.

5. Left hand of operator transfers cover to both hands.

These elements continue until the books are placed aside and the cycle begins again. The Chief IE's reset standard produced the same standard time of 1.273 hours per thousand notebooks that led to the initial grievance.

Because the employee was still unhappy with the rate, the union chose to proceed to the next step of the grievance procedure. A local consulting IE was hired by the union to audit the standard for the laminating operation. After carefully studying the job, the consultant determined that because his objective was to measure the length of time the average worker should take under standard working conditions, working at a normal

Figure 5-22   Notebook.

pace, to complete the laminating process, that the job could be studied effectively by dividing the procedure into three large elements. In his report to the union, he described these broad divisions as follows:

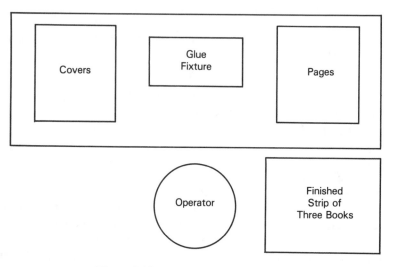

Figure 5-23   Lamination Room Layout.

TABLE 5-5 SUMMARY OF TIME STUDY RESULTS

Element	Sum of valid readings	Average	Rating	Elemental average
A	18.46 min.	.072	1.15	.083
B	26.97	.105	1.00	.105
C	8.85	.034	.90	.031

• **Element A:** Preparation. This element involves placing the notebook components in the proper position for laminating. The element starts when the operator reaches for the covers and strips of paper and ends when both hands grasp the covers.

• **Element B:** Assembly. This element involves fastening the covers to the strips. The element starts when the operator, while grasping the covers with both hands, moves toward the glue. The element ends when the covers are attached to the strips of paper.

• **Element C:** Removal. This element involves moving the laminated set of three notebooks to the stack at the side of the work area. This element starts after the covers are attached to the strips of paper and ends after the operator has placed the completed product on the stack and prepares to reach for the covers and strips required for the next set of three notebooks.

Although these elements are not as detailed as those described by Noteworthy's Chief IE, the results of the study will yield the same type of information—the standard time or rate.

Because the consultant was not an expert in the assembly of paper notebooks, he decided that timing 275 cycles would give the confidence he desired to ensure that the standard he set was an accurate and fair representation of the work being performed. Due to fumbles, missed readings, defective covers, lumpy glue, and other foreign elements, the consultant was actually able to time 257 cycles. His results are summarized in Table 5-5.

Noteworthy Notebooks and the union agreed to an 8.4 percent PFD allowance. The standard time for the laminating operation was determined by the consultant to be

$$(.083 + .105 + .031) + (.084)(.083 + .105 + .031) = .237 \text{ minutes}$$

The standard set by the consultant was .237 minutes per strip. The company-set standard was 1.273 hours per thousand books. Because three notebooks are produced from each strip that is laminated, the standard time per book is

$$.237/3 = .079 \text{ minutes per book}$$

During one minute, an operator should be able to produce

$$(1 \text{ minute}/.079 \text{ minutes/book}) = 12.66 \text{ books}$$

In one hour, the operator would produce

$$(60 \text{ minutes/hour})(12.66 \text{ books/minute}) = 759.6 \text{ books}$$

The time required to produce 1,000 books is

$$(1000 \text{ books})/(759.6 \text{ books/hour}) = 1.316 \text{ hours}$$

The consultant has set a standard of 1.316 hours per thousand for the union. This standard is looser than the company standard by a factor of

$$[(1.316 - 1.273)/(1.273)] \times 100 = 3.38 \text{ percent}$$

Because this difference is less than five percent, according to the collective bargaining agreement, the standard set by the company must be allowed to stand. The grievance was settled in favor of the company.

In this example, a valid standard was set using rather large elements. When the main purpose is to determine how long a specific job takes to complete, this practice is acceptable. The standard must reflect the work done and whether it involves measurement of very fine divisions or large segments of the task. The methodology is the same in both cases. The type of elemental division naturally reflects the end use of the data as well as the policy of the organization performing the study.

SUMMARY

The time study is the traditional method IEs have used to set standards. Until very recently, standards have been set primarily on production or "blue collar" jobs. Other methods, including historical data, predetermined time systems, standard data systems, work sampling, and physiological measures of human work performed are all legitimate paths to the same goal. These tools are being used more frequently in some nontraditional areas to measure productivity. Care must be taken by the analyst to use time study and all of the other measurement tools as tools and not as solutions to productivity problems.

Standards are a common denominator in measuring productivity. Before action can be taken to help improve productivity, a base of reference is necessary. Standards provide this foundation. To be useful however, standards must be fair and consistent. When using time study to set standards, the reader is cautioned that valid standards can only be developed by experienced time study analysts. Factors such as watch operation and rating, while not complicated, require experience and practice to develop the skill and consistency that is required for standards to be valid representations of the time required to complete the particular task in the desired way and with the desired quality.

REVIEW QUESTIONS

1. What is a *standard?* Why is it important?
2. Why do standards have to be consistent?
3. What is an *average worker?*
4. What is *rating?* How is it performed?
5. What is an *allowance?* How is it determined?
6. Describe the continuous method of recording time study data.
7. Describe the snapback method of recording time study data.

OBSERVATION SHEET

Operation :							SKETCH :	
Equipment :								
Operator :								
Study No. : ___ Analyst : ___			Date : _/_/_				SHEET No. ___ of ___	

ELEMENTS	CYCLES	1	2	3	4	5	6	7	8	9	10	11	12	13	14	15	R	SUMMARY
A		6	66	155	238	358	426	501									110	
B		15	ᵃ86	165	299	367	440	510									110	
C		28	99	178	312	380	455	518									110	
D		35	106	⊘	319	387	462	525									110	
E		46	115	246	328	396	471	535									110	
F		60	140	273	352	420	495	560									110	

FOREIGN ELEMENTS	Remarks :

Figure 5-24

8. Summarize the procedure that should be used when a stopwatch time study is performed.
9. What is a *loose rate?*
10. What is a *tight rate?*
11. What is *machine interference time?* What problems does it present to setting a time standard?
12. What is an *element?*
13. What is a *noncyclical element?*
14. How does *set-up time* relate to time standards?

PRACTICE EXERCISES

1. Calculate the standard time for the data in Figure 5-24. Use an allowance of nine percent.
2. The Ace Hexnut Company sells hexnut and bolt assemblies. Their rather primitive assembly method is to have the operator place the bolt, head down, into the holding fixture. The operator then places the hexnut on the threaded end and manually tightens the nut. The completed assembly is then removed from the fixture and placed in a shipping barrel. Each operator is provided with two fixtures, two stockpiles of raw material, and two shipping barrels. These

Operation : _____

Equipment : _____

Operator : _____

Study No. : _____ Analyst : _____   Date :__/__/__   SHEET No.____ of____

OBSERVATION SHEET   SKETCH :

ELEMENTS \ CYCLES	1	2	3	4	5	6	7	8	9	10	11	12	13	14	15	R	SUMMARY
1	8	8	8	a	7	5	8	8	11	/							
	8	68	132	199	260	320	393	458	487	610						100	
2	4	4	4	4	4	5	4	4	4	4							
	12	72	136	203	264	325	397	462	491	614						120	
3	10	10	10	11	10	10	10	10	9	10							
	22	82	146	214	274	335	407	472	500	624						100	
4	5	7	6	6	6	5	6	6	5	6							
	27	89	152	220	280	340	413	478	505	630						110	
5	25	27	26	25	27	37	29	28	b	26							
	52	116	178	245	307	377	442	468	583	656						80	
6	8	8	8	8	8	8	8	8	c	8							
	60	124	166	253	315	385	450	476	/	664						100	

a - sneeze

b - dropped hexnut

c - missed reading

FOREIGN ELEMENTS   Remarks :

Figure 5-25

materials permit the operator to simultaneously assemble two products at one time. The following elements have been developed for performing a time study:

(1) Reach for and grasp bolt.

(2) Place bolt in fixture.

(3) Release bolt; reach for and grasp hexnut.

(4) Place hexnut on end of bolt.

(5) Fasten nut to bolt.

(6) Place bolt in shipping barrel.

Ten cycles were timed and each element was rated. These results appear in Figure 5-25. Use a PFD allowance of 12 percent.

(a) Calculate the standard time to complete one assembly.

(b) How many hours per thousand are required?

(c) How many assemblies should the worker be able to complete in one hour?

(d) The worker normally is expected to earn $40 a day for this task. If the worker is paid on a straight incentive system, how many assemblies would the worker be expected to complete in a normal eight-hour work day? How much is each assembly worth?

(e) If the worker produced 1700 parts, how much would her daily earnings be?

(f) How would the standard be affected if element 5 were performed by an air wrench instead of manually?

Operation : _27_

Equipment : _____

Operator : _____

Study No. : _____ Analyst : _SRA_   Date : _9/13/82_   SHEET No. _1_ of _2_

OBSERVATION SHEET   SKETCH :

ELEMENTS	1	2	3	4	5	6	7	8	9	10	11	12	13	14	15	SUMMARY
1	20	185	295	470	600	765	870	10	160	300	445	600	750	890	70	
2	80	285	370	530	690	800	940	80	220	375	500	670	825	00	150	
3	95	240	385	545	705	815	955	95	235	390	515	685	840	15	165	
4	120	265	410	570	730	840	980	120	260	415	540	710	860	40	195	
5	130	275	420	580	740	850	990	130	270	425	550	725	870	50	205	
6			440													
7	150										575					

FOREIGN ELEMENTS   Remarks :

Figure 5-26

(g) Are you satisfied that a sufficient number of cycles were timed? Why?

3. The junior pharmacist at Major Mercy Hospital spends several hours each day preparing medications to be administered to each of the patients requiring them. Certain medications such as routine pain killers, antibiotics, and sleeping pills occur with such frequency that the management engineer for the hospital believes that standards can be set for filling many of the prescriptions. The steps involved in filling any prescription are:

(1) Identify drug to be administered.

(2) Locate drug in storage.

(3) Place proper dosage in paper cup.

(4) Identify paper cup as to drug, dose, and patient.

(5) Place drug in shipping area so orderly can claim.

(6) Refill label supply occasionally, as needed.

(7) Refill cup supply occasionally, as needed.

The management engineer is planning to determine the average time required to fill a prescription. He feels that by using each of the seven steps described above as an element he will have sufficient data to develop the standards required. Because of the nature of the work, the hospital uses a 20 percent PFD allowance when setting standards on professional tasks. The time study data collected are shown in Figure 5-26.

Operation : 27

Equipment : _____

Operator : _____

Study No. : _____   Analyst : SRA   Date : 9/13/82   SHEET No. 2 of 2

OBSERVATION SHEET   SKETCH :

ELEMENTS / CYCLES	16	17	18	19	20	21	22								R	SUMMARY
1	275	375	530	900	820	970	140								75	
2	300	460	590	750	900	45	215								120	
3	315	475	605	765	915	60	240								100	
4	340	500	630	735	940	85	275								100	
5	350	510	640	860	950	95	285								100	
6		670													100	
7					120										100	

FOREIGN ELEMENTS   Remarks :

Figure 5-26   (Continued.)

(a) Determine the average time required to fill a prescription.
(b) Why is there so much variability in the times recorded for element 2?
(c) Why do you think the rating on element 1 is below 100 percent?
(d) Would it be appropriate to place this task on an incentive system? Why?
(e) To what uses might the hospital management engineer put the results of his study?
(f) What are some problems of attempting to improve productivity on a job like this?

CHAPTER 6

Standard Data
Systems

OBJECTIVES

This chapter introduces the use of standard data systems for setting time standards. To complete this chapter, the reader should be able to

- Describe the general concept of the standard data system in setting time standards.
- Understand the process used to develop a standard data system.
- Calculate a time standard from a standard data system.
- Develop and use a relatively simple standard data system.
- Understand the advantages and limitations of using standard data systems.

INTRODUCTION

There are times when it is not practical to set time standards with a stopwatch or even with any direct measurement procedure. Many situations are impractical or impossible. For example, if there are many different but similar jobs there may not be enough time study analysts available to set all the standards. A large aircraft manufacturing facility has an active list of over 30,000 different part numbers. Each component part requires a number of machining and forming operations. It is not cost effective, nor is it a productive use of resources, to try to set a standard on each job simply through the use of direct observation.

An apparel firm contracts most of its production to a large department store chain. Because of the rapid changes in style, this company does a significant amount of low-volume work. It is constantly changing the product it manufactures and therefore, the method it uses to manufacture its products. There are many times when the production run is complete before it would be even physically possible to use time study to set a rate.

In both of these cases, it is still desirable to have a standard available for the job. An appropriate way to set standards in these cases is through the use of macroscopic standard data or elemental standard data systems.

THE PURPOSE OF STANDARD DATA SYSTEMS

The major purpose of a standard data system is to set time standards. As was discussed in the last chapter, standards are used for measuring increases in productivity, determining wage payment levels, and as tools for production planning and cost estimating. Standard data systems are excellent tools for setting time standards. Standard data systems do not use direct study of every job that requires a standard. Rather, as Mundel has said, standard data systems organize standard times from a number of related jobs into a database from which the standard times for related jobs may be constructed or synthesized (Abruzzi, 1952).

Although a company may produce a tremendous number of different products, many parts are produced in similar ways. Certain manufacturing processes have identifiable characteristics that occur no matter what the shape of the product. Machine tools can make metal chips at certain rates in certain materials regardless of the end use of the product. The time required depends only upon certain identifiable variables such as length of cut, depth of cut, material being cut, material feed, machine speed, and so on.

A reliable set of time standards, developed by skilled analysts using direct measurement, in which appropriate variables are identified, provides the database necessary for a standard data system. These standards are rated to the company's concept of normal performance. The PFD allowance is usually added after the standards are synthesized with the standard data system.

CONSTRUCTING STANDARD DATA SYSTEMS

Rather than explain the procedure for developing a system in general terms, it is more productive to explain the process via an example.

Example: Manufacturing Washers

Traditionally, the Zero Washer Company has manufactured a wide variety of different washers. They currently market eight different washers. All of these washers have the same outside diameter, the same thickness, and are made of the same material. The only differece between these different washers is the sizes of their inside diameters. The washers are standardized products and, as a result, Zero Washer has developed a set of standards showing the time required to produce 1000 washers of each different inside diameter. These characteristics are shown in Table 6-1.

A cursory examination shows that, as the inside diameter gets bigger, it takes longer to complete 1000 washers. Logically this relationship makes sense, the more material that is removed, the longer it should take.

Although Zero had a reasonably profitable business manufacturing and selling this standardized product line, the marketing director knew there was a market for more sizes of washers than jut the eight currently in production. As a matter of fact, she knew she could sell any size washer with an inside diameter of up to one inch as long as it had the standard outer diameter, was of standard thickness, and was made out of the standard material.

The price of a new model washer was almost entirely dependent on the labor required to manufacture it. The labor cost was dependent on the time required to

TABLE 6-1 TIME STUDY DATA FOR THE ZERO WASHER COMPANY

Model	ID	Hours/1000
A-1	1/16	.60
A-2	1/8	.65
A-3	1/4	.70
A-4	3/8	.76
A-5	1/2	.82
A-6	3/4	.97
A-7	5/8	.90
A-8	7/8	1.03

manufacture the washer. The cost, obviously, had to be ascertained before the washers could be priced and sold. Thus, the time required had to be determined before the washers could be manufactured and before a time study could be performed.

This situation is a textbook example of an ideal situation for the development of a standard data system. Because preliminary analysis seemed to show that manufacturing time increased as the inside diameter increased, it was logical and appropriate to conclude that time was a function of inside diameter. In common symbols this relation was written as

$$T = f(\text{ID})$$

Development of a standard data system will determine exactly what the functional relationship is. Although there are many analytical tools available for this process, it is best to keep the analysis as simple and straightforward as possible.

One of the simplest and most direct methods of analysis is the graph. In this example, because time depends on or is a function of inside diameter (ID), it is logical to graph observed time versus inside diameter (Figure 6-1). A simple algebraic analysis of these data, using the slope-intercept method, gives a slope of .52 and an intercept of .57, meaning that the time in hours required to produce 1000 washers is equal to .52 times the inside diameter plus .57. We can predict time standards simply by substituting the washer's inside diameter into the equation

$$T = .52(\text{ID}) + .57$$

For washers with an ID of .6 of an inch, the required time would be .88 hours for manufacture.

A question should immediately be raised in our minds: How good is this standard? After all, we never made a single measurement. We only used existing data to predict the standard. Our standard data set time standard is virtually worthless unless we can show how closely our predicted standards compare with existing and accepted standards.

A relatively simple way to show this comparison is through the use of a table that compares standards predicted by the standard data system with observed standards set through direct observation. Table 6-2 shows actual standards, predicted standards, the ab-

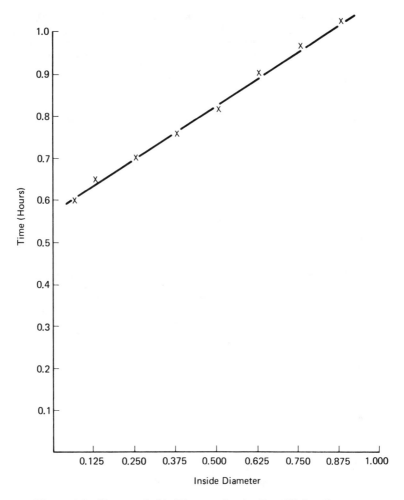

Figure 6-1 Time vs. Inside Diameter for the Zero Washer Company.

solute difference, percentage difference, and the average difference. As can be seen, the average differences are quite small. Additionally, individual differences are never more than 2.3 percent from measured standards. Table 6-3 shows a typical set of ± 10 percent limits. It must be pointed out and emphasized that the standard data relationship is valid only for washers with an inside diameter up to one inch. It is extremely dangerous to extrapolate beyond the one-inch ID.

Although the Zero Washer Company developed a standard data system, it dealt with a very simple situation. The actual time required to produce a given product is usually dependent on more than one variable. Also, the products most companies make generally are not so similar that only one characteristic, such as inside diameter, varies from product to product. The relationship between time and important variables may not be a

TABLE 6-2 ACTUAL TIME DIFFERENCE AND PERCENTAGE TIME DIFFERENCES FOR
THE ZERO WASHER COMPANY'S STANDARD DATA SYSTEM

Model	Predicted time	Time study time	Actual difference	Percent difference
A-1	.603	.600	−.003	−.5
A-2	.635	.650	+.015	+2.3
A-3	.701	.700	−.000	0
A-4	.765	.760	−.005	−.7
A-5	.832	.820	−.010	−1.2
A-6	.963	.970	+.010	+1.0
A-7	.895	.900	+.005	+0.6
A-8	1.025	1.030	+.005	+0.5

simple linear relationship. Determination of the relationship may require the use of some advanced analytical tools.

TABLE 6-3 10% CHECK CURVE FOR THE ZERO WASHER COMPANY

Model	Predicted time	+10% (of Predicted)	Actual time	− 10% (of Predicted)
A-1	.603	.663	.600	.543
A-2	.635	.698	.650	.571
A-3	.700	.770	.700	.630
A-4	.765	.842	.760	.689
A-5	.830	.913	.820	.747
A-6	.960	1.056	.970	.864
A-7	.895	.985	.900	.806
A-8	1.025	1.128	1.030	.923

The two examples that follow illustrate the development of standard data systems for situations more complicated than our introductory example.

Example: Zero Washer Company

Actually, the Zero Washer Company has a wider and more diversified product line than was described in the previous example. Not only does the inside diameter vary, as the "A" series of washers showed, but washer thickness varies as well. Luckily for Zero, there is a good database of time standards for the current models in production. Table 6-4 shows the characteristics for the various washers produced by Zero as well as the direct observation stopwatch time study values for the time required to produce 1000 units of each model.

Because three variables are involved, it is impractical, if not impossible, to plot time as a function of the variables, although it does seem that time is indeed a function of the inside diameter, the outside diameter, and the thickness of the washer. However, it is possible and desirable to plot time as a function of each of the variables. Figure 6-2

TABLE 6-4   TIME STUDY DATA FOR ZERO WASHER COMAPNY'S DATABASE

Model	ID	OD	Thickness	Hours/1000
A-1	1/16	1.5	1/16	.60
A-2	1/8	1.5	1/16	.65
A-3	1/4	1.5	1/16	.70
A-4	3/8	1.5	1/16	.76
A-5	1/2	1.5	1/16	.82
A-6	3/4	1.5	1/16	.97
A-7	5/8	1.5	1/16	.90
A-8	7/8	1.5	1/16	1.03
B-1	1/16	2.0	1/8	.64
B-2	1/8	2.0	1/8	.78
B-3	1/4	2.0	1/8	.96
B-4	3/8	2.0	1/8	1.09
C-1	1/16	1.0	3/32	.81
C-2	1/8	1.0	3/32	.91
C-3	1/4	1.0	3/32	1.04
C-4	3/8	1.0	3/32	1.16
C-5	1/2	1.0	3/32	1.22
E-1	1/16	1.25	1/32	.67
E-2	1/8	1.25	1/16	.73
E-3	1/4	1.25	3/32	.84
E-4	3/8	1.25	1/8	1.00
E-5	1/2	1.25	1/8	1.11
F-1	1/16	1.75	1/16	.92
F-2	1/2	2.5	1/8	1.16
F-3	7/8	3.0	3/16	1.28

shows time plotted versus ID for all the washers. Figure 6-3 is a graph of time and outside diameter and Figure 6-4 shows the relationship between time and washer thickness.

An examination of each graph reveals the following preliminary information:

- *Time vs. ID*—In general, the larger the ID, the more time required to produce the parts.
- *Time vs. OD*—No significant or discernible relationship between time and outside diameter exists.
- *Time vs. Thickness*—There is perhaps a relationship between time and thickness of each washer. Although not obvious, it appears that the time seems to increase as the thickness of the washer increases.

It is appropriate now to use the statistical technique of correlation analysis to test our hypotheses about these relationships.* The reader is reminded that a coefficient of correlation approaching $+1.00$ means that a strong positive relationship between variables

---

*For those who have forgotten or who have never been exposed to statistical tools such as correlation, there is a special appendix at the end of this chapter that briefly reviews the methods of calculation for these and other statistics.

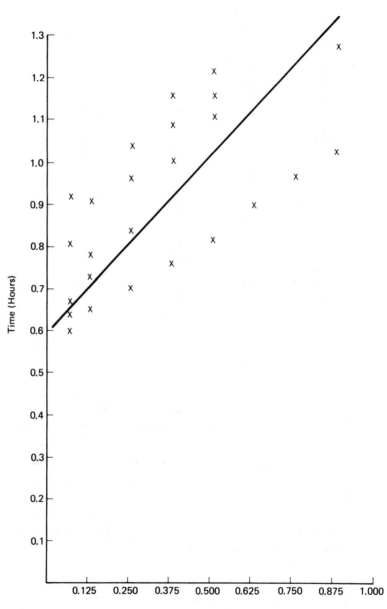

Figure 6-2   Time vs. Inside Diameter for the Zero Washer Company.

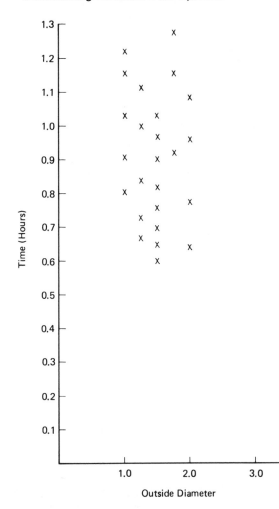

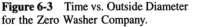

Figure 6-3 Time vs. Outside Diameter for the Zero Washer Company.

exists or, in this case, an increase in a physical characteristic relates to an increase in the time required to produce that piece. A coefficient of correlation approaching 0.00 means that no significant relationship between variables exists. A coefficient of -1.00 means a strong negative relationship or, in this case, a decrease in a physical characteristic means an increase in the time required for processing.

Using the standard statistical formula for the correlation coefficient r we can determine the strength of the relationship between time and ID, between time and OD, and between time and the thickness of the material.

Time and ID: $r = .67$
Time and OD: $r = .21$
Time and Thickness: $r = .57$

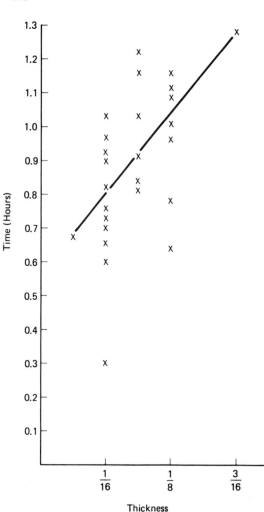

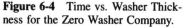

Figure 6-4 Time vs. Washer Thickness for the Zero Washer Company.

Statistical analysis, specifically the test of hypotheses, as illustrated in the appendix to this chapter, indicates that there are significant positive relationships between time and the inside diameter and between time and the thickness of the material. It is appropriate to say that time is a function of the ID and the thickness, or

$$T = f(\text{ID, Thickness})$$

By examining the graphs, it can be reasonably assumed that there is a linear relationship between time and the independent variables of inside diameter and washer thickness. Because two variables are involved, multivariate linear regression is used to fit the general equation

TABLE 6-5 COMPARISON OF STANDARD DATA RELATIONSHIP WITH OBSERVED TIME
STANDARDS FOR ZERO WASHER COMPANY

Model	Predicted time	Observed time	Actual difference	Percentage difference
A-1	.72	.60	−.12	−20.00
A-2	.74	.65	−.09	−13.85
A-3	.80	.70	−.10	−14.29
A-4	.85	.76	−.09	−11.84
A-5	.91	.82	−.09	−10.98
A-6	1.02	.97	−.05	−5.15
A-7	.96	.90	−.06	−6.67
A-8	1.07	1.03	−.04	−3.88
B-1	.87	.64	−.23	−35.94
B-2	.90	.78	−.12	−15.38
B-3	.95	.96	.01	1.04
B-4	1.01	1.09	.08	7.33
C-1	.80	.81	.01	1.23
C-2	.82	.91	.09	9.89
C-3	.88	1.04	.16	15.38
C-4	.93	1.16	.23	19.83
C-5	.98	1.22	.24	19.67
E-1	.64	.67	.03	4.47
E-2	.75	.73	−.02	−2.74
E-3	.88	.84	−.04	−4.76
E-4	1.01	1.00	−.01	−1.00
E-5	1.06	1.11	.05	4.50
F-1	.72	.92	.20	21.73
F-2	1.06	1.16	.10	8.62
F-3	1.37	1.28	−.09	−7.03
		SUM		−39.82
		AVERAGE		−1.59%

$$T = b_0 + b_1(\text{inside diameter}) + b_2(\text{thickness})$$

(*Reminder:* The appendix to this chapter illustrates typical regression analysis procedures.)

Regression indicates that time can be synthesized or predicted by using the following equation:

$$T = .54 + (.43)(\text{ID}) + (2.44)(\text{Thickness})$$

Table 6-5 shows, for the database, the time standards of the standard data relationships compared with the actual observed standards. As can be seen, the standard data relationship does a reasonably good job of predicting time standards for the existing product. This same relationship can be used to predict how long it will take to produce a similar washer with different ID and thickness. For example, the time required to produce a washer with an ID of 3/32 and a thickness of 1/16 would be predicted by the following standard data equation:

TABLE 6-6 ELEMENTAL TIME STUDY DATA FOR THE ABD COMPANY

	Model						
Element	119	130	220	310	311	322	329
10	.24	.22	.23	.23	.24	.22	.23
20	.38	.35	.35	.37	.36	.36	.37
30	12.06	10.44	8.71	6.58	10.83	6.34	7.25
40	3.66	4.81	—	2.79	5.84	4.55	4.10
50	—	1.63	1.91	1.69	1.80	1.45	—
60	.12	.12	.13	.11	.14	.14	.13

$$T = (.43)(3/32) + (2.44)(1/16) + .54$$
$$= .73 \text{ hours}/1000$$

Not all products are as similar as washers. Sometimes, when the products a company produces are similar and have many of the same operations, they are excellent products to use in developing and applying standard data systems. When products have many elements in common, standards set by standard data can be most productive.

Example: The ABD Company

The ABD Company produces seven different products. These products are produced with a maximum of six operations. Some of the products require all six operations. Table 6-6 summarizes time standards for each of the elements for each of the products. Upon visual examination of the data, one fact seems to leap out at us. The time for some of the elements seems to be constant from model to model. Elements 10, 20, and 60 require almost exactly the same time regardless of the model being produced. If we were to check the elemental descriptions, we would discover that they are almost exactly the same. Thus, virtually constant elemental times would be expected.

The remaining elements (30, 40, and 50) show noticeable variation from model to model, indicating that the time required to perform these elements must be a function of, or be dependent upon, some variable or variables. Some possible sources of variation might be machine-controlled elements such as feed, speed, or length of cut; differences in material if, for example, each model was made with a different alloy steel; or a difference such as dimensions between the various models produced.

An examination of the manufacturing specifications for different models of this product rules out machine-controlled differences, material differences, and finishing differences. Dimensional differences appear to be the source of the variability in time standards.

Further examination of the dimensional characteristics shows that two dimensions differ in every model. These differences are really the only items that differentiate one product from the next.

In basic algebraic terms, the time required for elements 30, 40, and 50 is a function of two dimensions:

$$T(30, 40, 50) = f(D_1 \text{ and } D_2)$$

Once this relationship is established, we can develop our standard data system. In this type of a situation the standard data system development involves two analytical tasks. First is the calculation of the standard time for the constant elements—i.e., those elements that require the same time in every similar job.

For example, for element 10,

$$(.24 + .22 + .23 + .23 + .24 + .22 + .23)/7$$
$$= .23 \text{ minutes}$$

for element 20,

$$(.38 + .35 + .35 + .37 + .36 + .36 + .37)/7$$
$$= .36 \text{ minutes}$$

and for element 60,

$$(.12 + .12 + .13 + .11 + .14 + .14 + .13)/7$$
$$= .13 \text{ minutes}$$

The average base time is the sum of these average constant times:

$$.23 + .36 + .13 = .72 \text{ minutes}$$

This constant of .72 minutes will be added to each unit time when the standard data system is developed.

The second analytical task is determining the relationship between the normal cycle times attributable to elements 30, 40, and 50, and the varying dimensions D_1 and D_2. Table 6-7 shows the normal time, for each model, attributable to the three elements under study and the dimensions D_1 and D_2 for each model.

Visual examination of these data provides little information. The data are graphed in Figure 6-5. In this graph, time is plotted as a function of D_1, the upper line. Time as a function of D_2 is plotted on the lower line. As can be seen, straight lines fit these data fairly well.

It might not always be possible to draw a nice curve. Not all data necessarily fit a nice straight line or even a "nice" curvilinear relationship. If this situation should occur one of the following reasons may apply:

TABLE 6-7 VARIABLE TIME DATA FOR THE ABD CCOMPANY

Model	Normal time (30, 40, 50)	D_1	D_2
119	15.72	18.00	27.25
130	16.88	23.50	33.50
220	10.62	2.50	1.50
310	11.06	3.00	7.25
311	18.47	25.00	44.00
322	12.34	12.00	18.25
329	11.35	3.00	8.00

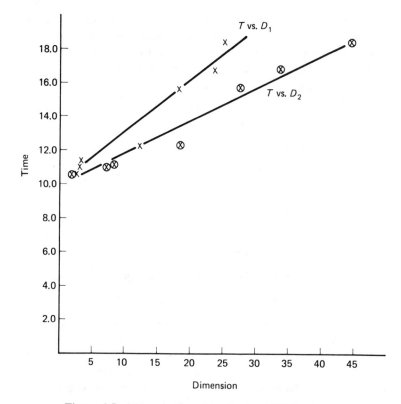

Figure 6-5   Time vs. D_1 and D_2 for the ABD Company.

1. The original data, from time studies, were not valid.
2. Variables other than those plotted are influencing the lines.
3. The work method and the real way the work is performed may not agree.

It also might be possible for the data to have formed two separate and disjointed straight lines in one or both of the curves. If this situation occurs, a recommended procedure would be to establish a special relationship between time and the respective independent variable that is valid only in specific ranges.

Because our graphical analysis shows straight lines, there is obviously a linear relationship between time and the dimensions. Unfortunately for us, the rate of change is much more rapid for the time and D_1 than for the time and D_2.

Using a version of regression analysis, multivariate linear regression, the general expression for time can be developed to fit the equation as

$$T = b_0 + b_1 D_1 + b_2 D_2$$

As explained in the appendix to this chapter, the actual regression and analysis calculations yield values for b_0, b_1, and b_2 of 9.77, .100, and .138, respectively, thus making the equation become

$$T = 9.77 + .100D_1 + .138D_2$$

The standard data system is almost, but not quite, completely developed. The regression study only accounts for predicting the times of the variable elements. The average base times, representing the constant elemental times, must be added to the equation. The average base time of .72 minutes, when added to the variable portion, gives the following standard data equation:

$$T = 10.49 + .100D_1 + .138D_2$$

Before we can say that the development of the system is complete, we must once again answer the question of just how good the system is. An examination of the output of the relationship shows that, indeed, it is a good predicting relationship. Table 6-8 shows the relationship between actual observed times and standard data predicted times for each of the models.

Although this relationship appears to be good, we need a consistent way to answer the question, "How good is the relationship?" We might use the concept of statistical confidence interval. This way is probably the best, but it presumes that the people who have to accept the use of the system are knowledgeable or at least acquainted with this statistical methodology. Perhaps a better or at least a visually more impressive way of showing how well the system predicts is through the use of the "check curve." The check curve is actually a table of values. It uses a percentage relationship between the system times for each model and the observed times. The table shows, for a particular percentage such as 10 percent, the predicted times, the upper 10-percent limit, the observed time, and the lower 10-percent limit. The fact that all of the observed values fall within the 10-percent limits around the predicted value tells the user of the system that he or she can be fairly confident that the results obtained from the standard time system accurately predict the time required for the job. Table 6-9 shows a 10-percent check curve for this example.

TABLE 6-8 OBSERVED VS. PREDICTED TIME FOR THE ABD COMPANY'S STANDARD DATA SYSTEM

Model	Observed time	Predicted time	Actual difference	Percentage difference
119	16.46	16.05	+.4095	+2.49
130	17.57	17.46	+.1070	+.61
220	11.33	10.95	+.3830	+3.38
310	11.77	11.79	−.0205	−.17
311	19.21	19.06	+.1480	+.77
322	13.06	14.21	−1.1485	−8.79
329	12.08	11.89	+.1860	+1.54

TABLE 6-9   10% CHECK CURVE FOR STANDARD DATA SYSTEM FOR THE ABD COMPANY

Model	Predicted time	+ 10 Percent	Actual time	− 10 Percent
119	16.05	17.66	16.46	14.45
130	17.46	19.21	17.57	15.71
220	10.95	12.05	11.33	9.86
310	11.79	12.97	11.77	10.61
311	19.06	20.97	19.21	17.15
322	14.21	15.63	13.06	12.79
329	11.89	13.08	12.08	10.70

The standard data system can be used to estimate the time standard for similar jobs as long as the values for D_1 and D_2 are known and as long as D_1 and D_2 fall within the actual maximum and minimum values for D_1 and D_2 that were used to determine the variable portion of the relationship. Extrapolation beyond the observed data can be dangerous. There is no reason to believe, or disbelieve, that the same relationship would exist outside the known limits.

Formal presentation of this system might look like the following:

Family: Product Line Q
Relationship: $T = 10.49 + .100D_1 + .138D_2$
Restrictions:
1. $2.5 \leq D_1 \leq 25.0$
2. $1.5 \leq D_2 \leq 44.0$

If the ABD Company planned a new product or family of products in this particular line, a standard could be estimated as soon as the values for D_1 and D_2 were known. Naturally, once the system was developed, using it would be a very productive way to set standards.

METHODOLOGY FOR DEVELOPING STANDARD DATA SYSTEMS

Before looking at more examples, it would be appropriate to summarize the methodology used to develop standard data systems. Development of a general procedure will help with future system development.

Developing a Database

The first requirement of any standard data system is the development of a database. Unless a sufficient number of elemental times are available, it would be impossible to proceed any further.

Once the data are collected, they must be sorted through and summarized. Like elements, that is elements that describe similar operations, they must be identified and summarized. These data, like elements, will be used to determine the predicting relationship. Furthermore, once these similar elements are identified, they must be further subdivided as being either constant or variable. A constant element occurs in almost every comparable job and its time does not change from job to job or from product to product. Variable elements, as the name indicates, depend upon certain characteristics. As the characteristics change, the time to complete the operation changes.

Calculating Average Base Time

The next task in developing a standard data system is calculating the average base time. This time is the sum of the average times allowed for constant elements. A much more difficult analytical task follows. The predictive relationships between variable elements and the time required for each variable element must be determined. Although there is no single way to do this prediction, techniques such as linear regression, multivariate regression, and curvilinear regression are helpful. If these aids do not give a suitable result, more sophisticated mathematical techniques such as factor analysis might be worth using to develop the relationship. Sometimes, several different relationships must be developed for different portions of the variables' ranges. If an initial attempt fails to find a "good" relationship between time and the variables, it would be logical to double-check the variables to make sure that they have been properly identified. It is very easy to include variables that don't belong and even easier to exclude variables that belong. A careful reexamination of the data would be very helpful if a predicting relationship cannot be developed initially.

Once the constant and variable segments are identified and calculated, these components are combined in one standard data equation (or set of equations). If there are limitations on any variable values, these limits must be specified.

Preparing the Check Curve

After the standard data system is developed, the check curve is prepared. This curve shows how well the standard data relationship fits the existing data. It should build the confidence of anybody using or being affected by the standards set by the standard data system.

Preparing Documentation

Finally, as in the development of any system, proper documentation must be prepared. The basic time study data, the graphs showing the relationship of the variables to time, and whatever else in the way of calculations must be preserved. If there is ever a question as to the validity of a standard, this information is absolutely necessary to substantiate the system.

Example: Painting Estimates

The Polka Dot Painting (PDP) Company is a house-painting firm. It is constantly called upon to give estimates for house-painting jobs. Based on experience, the company has

TABLE 6-10 PAINTING TIMES FOR THE PDP COMPANY

Time	Area	Perimeter	Paint grade	Priming*
56	4200	1000	3	1
59	4500	1100	3	0
64	4800	1200	3	1
66	5100	1300	2	0
71	5400	1400	2	1
76	5700	1800	1	0
88	6000	2600	1	1
85	6500	1700	2	0
100	7000	2800	1	1

*1 = yes; 0 = no.

Note: Time is in worker hours to complete the job; Area is in square feet to be painted; and Perimeter is in feet.

determined a number of variables that influence the time required to paint a house. These variables include whether the house needs a primer coat, the square footage to be painted, the perimeter of the area to be painted, and the grade of paint to be used. Paint used by PDP can be one of three grades. Houses either need priming or they don't. Based on past experience, PDP has assembled the database shown in Table 6-10.

Initial examination of the data seems to indicate that there are no constant elements. All of the variables seem to take on different values for each situation. A correlation analysis shows the following values for the relationship between time and each of the variables. A test of significance, as explained in the appendix at the end of this chapter, identifies the relationships that are significant:

Time and Area: $r = .977$ (significant)
Time and Perimeter: $r = .947$ (significant)
Time and Paint Grade: $r = -.837$ (significant)
Time and Priming Coat of Paint: $r = .154$ (not significant)

Time is a function of the area painted, the perimeter, and grade or quality of the paint used. In general terms, we would expect a multiple regression relationship, either linear or curvilinear. Figures 6-6 , 6-7, and 6-8 examine the relationships between time and each variable. Upon visual examination, all of the graphs appear to be linear. Multiple linear regression analysis would lead us to expect a general relationship of

$$T = b_0 + b_1V_1 + b_2V_2 + b_3V_3$$

where V_1 is the area to be painted, V_2 is the perimeter of the area to be painted, and V_3 is the grade of paint to be used. A complete analysis of data gives the following equation:

$$T = -3.059 + .01V_1 + .01V_2 + 1.66V_3$$

Table 6-11 shows a five percent check curve. As can be seen, all predicted values are within five percent of the observed values, indicating a fairly high degree of confidence

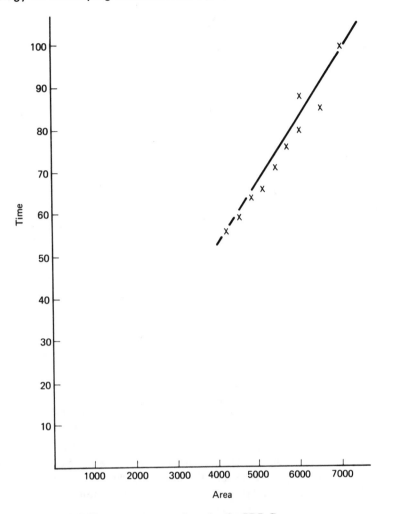

Figure 6-6 Time vs. Area for the PDP Company.

about the results we might obtain from using this relationship. We must qualify the use of this standard data system by limiting the potential values of the variables V_1, V_2, and V_3. The formal presentation of the system would be as follows:

$$T = -3.059 + .01V_1 + .01V_2 + 1.66V_3$$
subject to
$$4200 \leq V_1 \leq 7000$$
$$1000 \leq V_2 \leq 2800$$
$$1 \leq V_3 \leq 3$$

If PDP wants to prepare a bid on a new job, one with an area within the 4200 to 7000 range, with a perimeter of 1000 to 2800 feet, and using paints of grades 1, 2, or 3, they

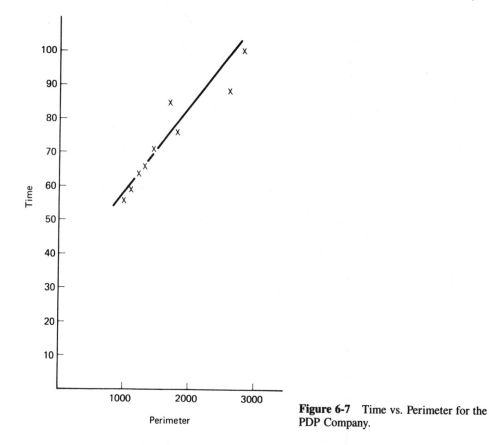

Figure 6-7 Time vs. Perimeter for the PDP Company.

will be able to predict the time required by using the standard data relationship without previously having done the job. For example, PDP can estimate a job with an area of 5500, a perimeter of 2100, and a paint grade of 2 as follows:

$$T = -3.059 + (.01)(5500) + (.01)(2100) + 1.66(2)$$
$$= -3.059 + 55 + 21 + 3.32$$
$$= 76.261 \text{ hours}$$

The next example illustrates how a company uses standard data and a computer to set time standards.

Example: Computerized Database

The Gigantic and Large Manufacturing Company produces a wide variety of different products, each assembled with a large number of component parts. All of the component parts undergo one or more of the company's 150 different manufacturing operations. The parts vary from each other in many respects, but they all have certain common characteristics. It is almost as if there were a buffet of characteristics, such as length, hole lo-

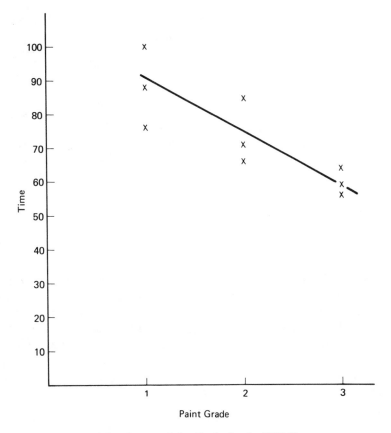

Figure 6-8   Time vs. Paint Grade for the PDP Company.

cation, material, shape, and so on, that is used to determine the unique characteristic of each part. The company is also always adding new parts and dropping old parts.

This situation is obviously ideal for the use of standard data. It would not be economical to perform complete time studies of each new or changed part as it enters production. Once the company has developed a sufficiently large database of time standards for the various characteristics and processes, a computer can search the database of time standards and locate all the appropriate constant and variable elements necessary for producing a part. All that is required by G and L is identification of the appropriate characteristics and specification of the dimensions.

For example, the manufacture of a particular product, say a cylindrical pin, might be accomplished with the following steps:

1. Material from stock to cutoff saw.
   (L = _____ Dia = _____ Material Code No. _____)
2. Cut required quantity: _____

TABLE 6-11   5-% CHECK CURVE FOR THE PDP COMPANY

Predicted time	+5 Percent	Actual	−5 Percent
53.92	56.61	56	51.22
57.92	60.82	59	55.02
61.92	65.02	64	58.82
64.26	67.47	66	61.04
68.26	71.67	71	64.85
73.60	77.28	76	69.92
84.60	88.83	88	80.37
82.26	86.37	85	78.15
96.60	101.43	100	91.77

3. Deburr
4. Inspect
5. To stock

The specifications of length, diameter, quantity, and raw material code, along with the identification of "cutoff saw," provide G and L's computerized database with enough information to compute the time standard using a variety of standard data relationships.

The preceding example merely illustrates the possibilities available when a standard data system is used to its fullest advantage. Not every organization has the resources to develop such a sophisticated database and computerized retrieval system. However, some of the country's largest companies, such as Lockheed Aircraft Corporation and General Motors, presently operate such systems.

Example: Form Tubing Ends

Some companies rely on standard data to set most of their production standards. One such organization is the Lockheed-Georgia Company.* Lockheed-Georgia Company has developed a procedure to follow whenever a standard must be developed. This example will illustrate the method used by the company to develop time standards by tracing the development of the standard data system for the operation called "form tubing ends." This operation is performed on metal tubing. The results of this particular operation are shown in Figure 6-9.

The first step in the standard data development process is the preliminary survey, or research step. This step involves gathering information, such as prints, manufacturing instructions, load center** information, and so on. Once all of the background data are compiled, the analyst making the study divides the job into the elements necessary for its completion. For the forming tubing ends operation, the following elements have been defined:

*The information contained in this example is reproduced with permission of Lockheed-Georgia Company.

**In Lockheed terminology, a *load center* is a description of the actual manufacturing equipment that is to perform the operation.

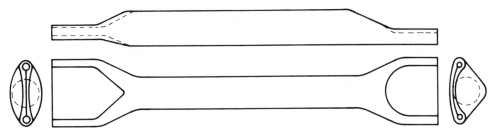

Figure 6-9   Aluminum Form Tubing End.

Setup

1. Select new order
2. Get and read blueprint
3. Get parts from icebox
4. Unwrap parts
5. Position parts
6. Get oil
7. Set up die/inserts/mandrel
8. Check length of part
9. Adjust set up
10. Stamp shop order
11. Make tag for order
12. Fasten tag to order
13. Aside order to outgoing
14. Return oil
15. Setup to handform

Run

16. Get fillers
17. Prepare one filler
18. Assemble one filler
19. Assemble two fillers
20. Flatten one end
21. Inspect part
22. Put parts into oil can
23. Handform

After the elements have been identified and described to company standards, the analyst performs sufficient time studies to obtain enough data to use in developing the system. Studies are made on a sufficient number of different models of pipes to ensure that a

TABLE 6-12   OBSERVED TIME AND VARIABLE LISTING

OD	E/T	F	STDHRS/100
.750	2.0	1.0	2.400
.375	2.0	1.0	2.180
.750	2.0	1.0	2.840
1.000	2.0	1.0	3.130
.750	2.0	1.0	2.660
.750	2.0	1.0	2.350
1.000	2.0	1.0	2.890
1.000	2.0	1.0	2.610
.750	2.0	1.0	2.360
.750	2.0	1.0	1.990
.500	2.0	1.0	2.440
.750	2.0	1.0	2.390
1.000	2.0	1.0	3.240
1.000	2.0	0.0	1.980
2.075	2.0	0.0	3.670
.625	1.0	0.0	.800

OD is outside diameter in inches; E/T is the number of ends formed on a tube; F is filler—1.0 indicating a filler was used and 0.0 indicating no filler.

significant sampling of different characteristics is made. Lockheed knows they won't time study every possibility, but they try to study as many different elements as possible.

Once the time studies are complete, a spreadsheet is prepared. This spreadsheet shows the time, elements, and variables, as identified by the analyst, that might affect the times required for the completion of the task. These variables are plotted against their times on a scatter diagram in a preliminary attempt to identify those variables that are most likely to have a direct bearing on the time required for the operation. Lockheed is insistent that the analysts perform this graphing exercise. The industrial engineering supervisor is convinced that the results are worth the investment in time.

The analysis of this operation shows that time is probably related to outside diameter, the number of ends of the tube formed, and whether or not a filler is used in the tube. Table 6-12 shows the standard hours per hundred observed for different combinations of these variables. Figure 6-10 shows time plotted as a function of each of these variables.

A computer analysis indicates that there is a significant relationship or correlation between these variables and the standard hours allowed. After these variables have been identified, the computer is used to determine the relationship between time and the independent variables.*

The 16 different samples used to generate the data are shown as a check curve, Table 6-13. Because Lockheed-Georgia is an Air Force contractor, they must comply with Military Standard 1567** which calls for a ±25 percent accuracy on standards. Standard statistical

*The computer analysis used some sophisticated regression analysis. The program currently used was developed by the IE department at Lockheed-Georgia.

**Mil-Std 1567, which pertains to work measurement programs, is reproduced as an appendix to this text.

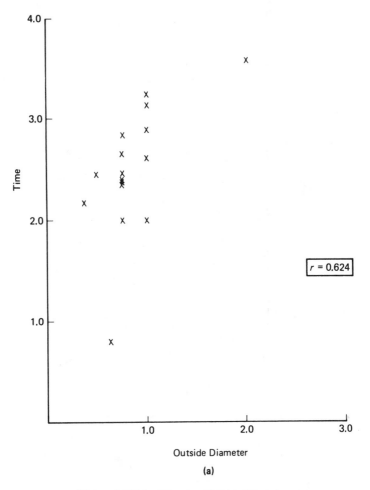

Figure 6-10(a)   Time vs. Outside Diameter.

analysis indicates that Lockheed-Georgia can be 97 percent confident that the formula developed is within ±25 percent of the true time required to perform this particular job.

However, after developing the formula, Lockheed-Georgia is not finished with its standard data development. Their IE department prepares a chart or nomograph for each operation. This chart permits rate clerks to set a standard for the job by following a simple set of instructions or decision rules. Figure 6-11 shows the standard data formula table for the form tubing ends operation. The remainder of this example will illustrate how the rate clerk would use such a table. After all, the purpose of the analysis performed is to prepare information that will facilitate, or make more productive, the actual standards-setting procedure.

As Figure 6-11 shows, there is a constant time of .33 hours required every time the job runs. This time is required for the setup. Although there may be a varying number of

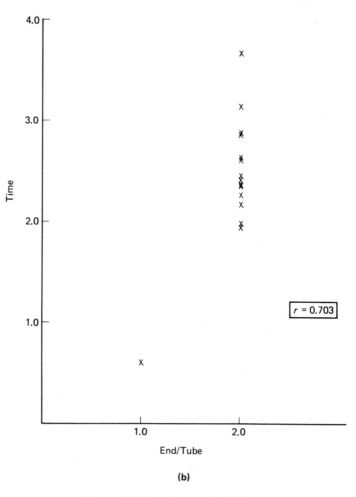

(b)

Figure 6-10(b)   Time vs. End/Tube.

different tube ends formed, the setup occurs only once per run. The analyst identifies this set-up time as the constant portion of the standard. The analyst then examines the specifications to see if there is a filler used with the tubing. If there is no filler, the top line of Figure 6-12 is followed; if there is a filler, the bottom line is followed. An examination of the print will indicate whether one or two ends of the tube are formed. If no filler is used and one end of the tube is formed, the first column of Figure 6-13 is followed; likewise, the other possibilities: no fill and two ends formed; filler and one end formed; and filler and two ends formed, correspond to the next set of columns of this figure.

If no filler is required and there is only one formed end, the outside diameter is used as the final indicator of allowed time. For outside diameters of up to .208 inches, a standard of .35 hours per hundred is allowed for the run or the variable part of this job. If the outside diameter is greater than 1.467 inches but less than or equal to 1.803 inches, 2.50

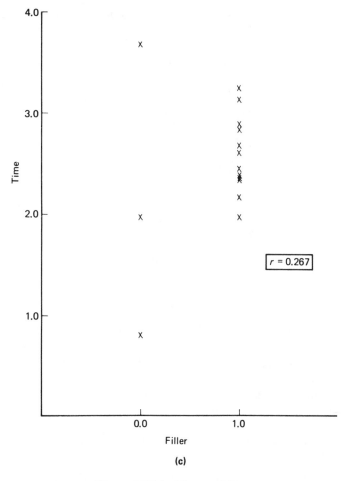

Figure 6-10(c)   Time vs. Filler.

hours per 100 is the variable time allowed. Similarly, if filler is used, two ends are formed, and the outside diameter is between 1.367 and 1.702 inches, the variable part of the standard is 3.00 hours per hundred. In each case, the standard time for the job must be combined with the set-up time. Although this chart handles almost all cases, a formula is provided for the exceptions, those cases where the ODs are outside the bounds of the table. Figure 6-13 shows the formula as part of the entire standard data system for this operation.

It must be pointed out that, although Lockheed-Georgia performs the bulk of its data analysis on the computer, the same analysis can be done manually. More importantly, this analysis shows how a standard data analysis can be used to develop the standards for parts without the need for additional direct observation. In a company with a variety of products, such a system is a necessity if standards are to be set for the products fabricated.

TABLE 6-13 CHECK CURVE FOR DATA: MIL SPEC TEST
ON EQUATION

Equation	T/S	Diff.	Percent dev.
2.532	2.400	.13	5.52
1.974	2.180	−.21	−9.44
2.532	2.840	−.31	−10.83
2.905	3.130	−.23	−7.20
2.532	2.660	−.13	−4.80
2.532	2.350	.18	7.76
2.905	2.890	.01	.50
2.905	2.610	.29	11.29
2.532	2.360	.17	7.31
2.532	1.990	.54	27.26
2.160	2.440	−.28	−11.46
2.532	2.390	.14	5.96
2.905	3.240	−.34	−10.35
2.062	1.980	.08	4.15
3.588	3.670	−.08	−2.24
.800	.800	−.00	−.00

Without such a system, there is little doubt that the IE staff would not be able to keep up with the demand for time standards. Every company, regardless of the quantity or variety of parts produced, can use the advantages of the principles underlying standard data development.

After Lockheed-Georgia has developed its charts, the development of the system is completed with the following:

- *Cost center effect*—The impact of the standard on the particular department is determined.
- *Photographs*—For future reference, a pictorial record of the method used is established.
- *Training application*—Data and charts are prepared to train the clerical personnel in the application of the standard. The tables, charts, formulas, and instructions must be clearly understood by those who will have to use them on a regular basis. (Many parts are, naturally, much more complicated than this tube forming example.)

As a final step in the development process, each standard data application is subject to the scrutiny and analysis not only of the supervisor and the department head, but also of all the members of the IE group charged with standard data development. A regular peer evaluation serves as a rigorous backup to the check curve to assure the validity of each set of standards that will be set by the system.

Lockheed-Georgia Company
Form Tubing Ends
Load Center 528/534/847

FORM TUBING ENDS

SET UP:	Form Tubing Ends 0.33 Std. Hrs./Occ.

FORM TUBING ENDS

Ends/Tube

| Filler | Yes | 1 | 2 | | | Std. Hrs. |
	No			1	2	per 100
		OUTSIDE DIAMETER (Up to)				
		0.208				0.35
		0.258				0.40
		0.325				0.50
		0.392				0.60
		0.460				0.70
		0.560				0.80
		0.695	0.258			1.00
		0.829	0.392			1.20
		0.997	0.560	0.359		1.40
		1.199	0.762	0.560	0.292	1.70
		1.467	1.031	0.829	0.560	2.00
		1.803	1.367	1.165	0.896	2.50
		2.139	1.702	1.501	1.232	3.00
		2.475	2.038	1.837	1.568	3.50
				2.341	2.072	4.00

| Without Filler: | Std. Hrs./100 | $= (OD \times 1.49 + Ends \times 0.7) - 0.83$ | LIMIT: | $E/T = 2$ |
| With Filler: | Std. Hrs./100 | $= (OD \times 1.49 + Ends \times 0.7) + 0.01$ | | $OD = 3$ |

Figure 6-11 Standard Data Formula Presentation.

ADVANTAGES OF STANDARD DATA SYSTEMS

Having examined, in some detail, the procedure and method used for developing macroscopic standard data systems, it is worthwhile to review some of the advantages that have been suggested for using standard data systems to set time standards.

First among these benefits is *consistency*. Standards set with standard data systems are more consistent than any one specific time study because the standards come from a larger database. They are developed from multiple time studies, performed by different

Lockheed-Georgia Company
Form Tubing Ends
Load Center 528/534/847

FORM TUBING ENDS

SET UP: Form Tubing Ends 0.33 Std. Hrs./Occ.

FORM TUBING ENDS

Ends/Tube

Filler	Yes	1	2			Std. Hrs.
	No			1	2	per 100

Figure 6-12 Standard Data Formula Presentation.

analysts, on many different operations. Random errors, due to rating inconsistencies, are minimized. It is much more important in setting time standards to be consistent than to be accurate. The author's intent is not to downplay accuracy, but the value to the organization of the standards is much greater if they are consistent from job to job. Ideally, the standards should be correct and consistent. Having standards that agree with each other is most important.

Second, when standard data systems are used, standards can be *synthesized*. Time standards can be set before production on a new product begins. Using a stopwatch to set standards naturally requires a job to be performed by an experienced operator before it can be timed. In addition to the operator, there must be tools, equipment, and material. Standard data can provide standards ahead of time for cost estimates methods analysis and production planning.

The *cost* of setting standards is lower when standard data are used. Once a system is in operation, the cost to set each standard is significantly lower.

The *time* required to set each standard is reduced when standard data are used. Again, once the elemental standard data system is developed and in place, the time required by the IE to set each standard is much less than with stopwatch time study.

The use of standard data systems requires careful study and the standardization of work *methods*. Similar products and similar processes will be more likely to be produced according to the same procedure.

Finally, employees and management can have *confidence* in the fairness of the standards set. The elemental standard data system permits easy explanation of the allowed time for each operation or product. The standards, because they have been developed from a large database, will be accepted as being unbiased. Additionally, the use of the system eliminates direct contact between specific analysts and specific operators, eliminating opportunities for charges of personal bias.

Lockheed-Georgia Company
Form Tubing Ends
Load Center 528/534/847

FORM TUBING ENDS

SET UP: Form Tubing Ends 0.33 Std. Hrs./Occ.

FORM TUBING ENDS

Ends/Tube

| Filler | Yes | 1 | 2 | | | Std. Hrs. |
	No			1	2	per 100
		OUTSIDE DIAMETER (Up to)				
		0.208				0.35
		0.258				0.40
		0.325				0.50
		0.392				0.60
		0.460				0.70
		0.560				0.80
		0.695	0.258			1.00
		0.829	0.392			1.20
		0.997	0.560	0.359		1.40
		1.199	0.762	0.560	0.292	1.70
		1.467	1.031	0.829	0.560	2.00
		1.803	1.367	1.165	0.896	2.50
		2.139	1.702	1.501	1.232	3.00
		2.475	2.038	1.837	1.568	3.50
				2.341	2.072	4.00

Figure 6-13   Standard Data Formula Presentation.

LIMITATIONS OF STANDARD DATA SYSTEMS

Besides having advantages, there are a number of disadvantages or limitations that must be recognized when developing and using standard data systems. The relative *complexity* involved in deriving the relationship between time and the independent variables that affect the time may be excessive. Not all standard data systems fit the nice and neat linear multivariate regression model that was used in the first example problem, and not every company has access to a computer routine as sophisticated as Lockheed-Georgia's. Sometimes, the statistical analysis may be too difficult for the available resources and, sometimes, there just may be no observable relationship between time and any of the variables.

A relatively simple change in a work *method* may cause an entire relationship to be scrapped and a new one to be developed. The new method then must be integrated into the database, taking even more time.

A major source of error in the development of any standard data system can be attributed to the *validity,* or lack of validity, of the original data. A standard data system is only as good as the original data that were used to develop the relationships between the work elements and the allowed time.

A final limitation is *cost.* The development cost, when a standard data is system is started, is fairly large. A significant volume of standards must be set using the system to justify the start-up costs.

SUMMARY

This chapter has summarized the development and use of macroscopic standard data, also called elemental standard data, to set time standards for productivity measurement. Every possible relationship has not been shown. Instead, a methodology has been illustrated through several examples whereby standard data systems have been developed using this procedure. When properly used, the system can measure and improve not only the productivity of the jobs the standards directly measure, but also the productivity of the analyst setting the standard.

APPENDIX—CORRELATION ANALYSIS

The following section describes the calculations required to determine the coefficient of correlation between two variables and shows a quick test to determine whether or not the coefficient of correlation is significant. This section in no way purports to be a complete statistical treatment of the topic, nor does it presume a prior knowledge of statistics on the part of the reader.

The coefficient of correlation r measures the strength of the relationship between two variables. A high correlation only indicates that there is a strong relationship. Correlation does not justify any assumption of a cause-effect relationship between the variables. For example, it has been determined that there is a high correlation between the number of sick people and the number of patients visiting physicians. Just because there is a strong relationship does *not necessarily* mean that visiting a physician's office causes illness.

Specifically, a coefficient of correlation of $r = 1.0$ indicates a perfect, positive relationship. As the value of one variable increases, the other variable increases by a similar amount. If $r = -1.0$, a perfect negative correlation exists, meaning that as one variable increases, the other decreases by a like amount. A coefficient of correlation of $r = 0.0$ means there is no relationship between the two variables. As one variable increases, the second variable either increases or decreases according to no demonstrable pattern. The coefficient of correlation is calculated, for variables X_1 and X_2 using the following formula, where n is the number of pairs of data for the two variables:

$$r = \frac{n\Sigma X_1 X_2 - (\Sigma X_1)(\Sigma X_2)}{\sqrt{[n(\Sigma X_1^2 - (\Sigma X_1)^2][n(\Sigma X_2^2) - (\Sigma X_2)^2]}}$$

Example

Find the correlation between the following variables X_1 and X_2:

X_1	X_2
14	5
15	4
16	3
17	2
18	4
14	5
16	4
17	2
18	4
16	3
14	5
15	3
15	4
16	4

The first step is to find the arithmetic totals necessary to perform the required calculations:

$$\Sigma X_1 = 14 + 15 + 16 + \ldots + 16 = 221$$
$$\Sigma X_1^2 = 14^2 + 15^2 + 16^2 + \ldots + 16^2 = 3513$$
$$\Sigma X_2 = 5 + 4 + 3 + \ldots + 4 = 52$$
$$\Sigma X_2^2 = 5^2 + 4^2 + 3^2 + \ldots + 4^2 = 206$$
$$\Sigma X_1 X_2 = (14)(5) + (15)(4) + (16)(13) + \ldots + (16)(4) = 811$$

$n = 14$

$$r = \sqrt{\frac{(14)(811) - (221)(52)}{[(14)(3513) - (221)^2][(14)(206) - (52)^2]}}$$

$$r = \sqrt{\frac{-138}{(341)(180)}}$$

$$r = \frac{-138}{247.75}$$

$$r = -.56$$

To determine whether this value of r is significant, another calculation must be performed. This calculation will test the hypothesis that the value of r is not significantly different from zero at a given level of confidence. Confidence is an indication of how certain we are that our statement is true. Common confidence levels are 90 percent, 95 percent, and 99 percent.

To test our hypothesis that r is not significantly different from zero, that is, that there is no relationship between the variables X_1 and X_2, the following equation is used:

$$z = \left| \frac{\sqrt{n-3}}{2} \ln \frac{1+r}{1-r} \right|$$

For our example,

$$z = \left| \frac{\sqrt{14-3}}{2} \ln \frac{.44}{1.56} \right|$$

$$z = \left| (1.66) \ln (.28) \right|$$

$$z = 2.10$$

This test value of z is compared to one of the following table values, depending on how sure, or how confident, we want to be of our results.

CONFIDENCE	COMPARISON VALUE
90%	1.646
95%	1.960
99%	2.576

If the calculated value of z is less than or equal to the table value we have selected, we accept our hypothesis that r is not significantly different from zero and therefore, there is no relationship between the variables X_1 and X_2. If the calculated value of z is greater than the table value we have selected, then we are confident that there is a significant relationship, either positive or negative, depending on the arithmetic sign of the calculated value of r.

In our example, if we want to be 95 percent confident that our conclusion is accurate, based on our sample data, we compare our calculated z value with 1.960. Our value of 2.10 is greater than 1.960; therefore, we can say, with a 95 percent chance of being correct, that there is a significant relationship between X_1 and X_2. Because the actual value of r was $-.56$, we are 95 percent confident that there is a negative relationship between X_1 and X_2. Based on an examination of the original data, we can say that as X_1 increases, X_2 decreases.

Regression

Having established that X_1 and X_2 are related, we might want to predict the relationship. This prediction can be accomplished, when there are two variables, by performing a bivariate regression analysis. If an examination of the data, via a graph, makes us believe that there is a linear relationship between the variables, the equation relating these variables will be of the following general format where X_1 is the independent variable and X_2 the dependent variable:

$$X_2 = b_0 + b_1X_1$$

The values of b_0 and b_1 correspond to the intercept and slope of the line, respectively. We can determine the values for b_0 and b_1 by solving a set of simultaneous equations called *normal equations*, derived from sample data. These equations take the form:

$$\Sigma X_2 = nb_0 + b_1\Sigma X_1$$
$$\Sigma X_1 X_2 = b_0\Sigma X_1 + b_1\Sigma X_1{}^2$$

Consider the calculations for r. If we arbitrarily call our X_1 the independent variable and our X_2 the dependent variable, we obtain the following equations when we perform the indicated summations on the original data:

$$52 = (14)b_0 + (221)b_1$$
$$881 = (221)b_1 + (3513)b_1$$

Solving these equations simultaneously, we determine that,

$$b_0 = 10.03$$
$$b_1 = -.40$$

Thus, the regression equation becomes

$$X_2 = 10.03 - .4X_1$$

When the dependent variable is significantly correlated to two or more variables, it may be advantageous to use multiple linear regression. The methodology is the same as for bivariate linear regression, except that each additional independent variable X_1 requires an additional "normal" equation to solve for an additional value of b_1. For two independent variables and a dependent variable, y,

$$y = b_0 + b_1X_1 + b_2X_2$$

The normal equations, to be solved simultaneously for values of b_0, b_1, and b_2, are

$$\Sigma y = nb_0 + b_1\Sigma X_1 + b_2\Sigma X_2$$
$$\Sigma X_1 y = b_0\Sigma X_1 + b_1\Sigma X_1{}^2 + b_2\Sigma X_1 X_2$$
$$\Sigma X^2 y = b_0\Sigma X_2 + b_1\Sigma X_1 X_2 + b_2\Sigma X_2{}^2$$

For three independent variables, the general form of the regression equation is

$$y = b_0 + b_1X_1 + b_2X_2 + b_3X_3$$

With normal equations

$$\Sigma y = nb_0 + b_1\Sigma X_1 + b_2\Sigma X_2 + b_3\Sigma X_3$$
$$\Sigma X_1 y = b_0\Sigma X_1 + b_1\Sigma X_1{}^2 + b_2\Sigma X_1 X_2 + b_3\Sigma X_1 X_3$$
$$\Sigma X_2 y = b_0\Sigma X_2 + b_1\Sigma X_1 X_2 + b_2\Sigma X_2{}^2 + b_3\Sigma X_2 X_3$$
$$\Sigma X_3 y = b_0\Sigma X_3 + b_1\Sigma X_1 X_3 + b_2\Sigma X_2 X_3 + b_3\Sigma X_3{}^2$$

In each case the actual values of y, X_1, X_2, and X_3 and the summations are substituted into the normal equations, which are solved simultaneously for the values of b_0, b_1, b_2, and b_3.

If it appears that there are more than three independent variables and if these variables are linearly related to the dependent variable, then this model can be expanded using the same format shown. Computer programs in BASIC are provided in this book's Appendix that will facilitate the development of the correlation coefficient and multiple regression relationship.

REVIEW QUESTIONS

1. When is it appropriate to use standard data to set time standards?
2. What are the two major elements found in most standard data systems?
3. What statistical procedure helps to determine whether or not there is a significant relationship between two variables?
4. What statistical procedure determines the nature of the relationship between two variables?
5. What is meant by the statement that time standards set using standard data are consistent?
6. How are time standards synthesized?
7. What relationship does standard data have with the cost of setting time standards?
8. How can standard data help with methods improvement?
9. What is the major drawback to developing standard data systems for all jobs within an organization?

PRACTICE EXERCISES

1. The Second National Bank believes that the time required for its keypunch operators to punch a card depends on several variables. A correlation study has eliminated all of the potential variables except for the number of characters to be punched on each card. A number of time studies were performed and the following data were collected:

NO. CHARACTERS	STD. TIME CARD (HOURS)
5	.008
10	.014
15	.018
20	.021
30	.027
40	.032
50	.039
60	.045
70	.051
80	.055

Develop the standard data relationship to determine how long it should take an operator to keypunch a card with 44 characters. How good is the estimate?

2. Prototype, Inc., is a small laboratory that performs a variety of tests on metal specimens submitted by clients. To assure that the test equipment consistently gives the same results when performing the same test on the same type of material, Prototype is constantly running verification tests on standard test specimens. Because Prototype limits itself to testing carbon alloy steels, it has a limited number of different standards. Prototype purchases ¾-inch diameter steel rods in the following alloy configurations:

C1020	C1144
C1035	C1115
C1045	C1095
C1141	

(*Note:* The last three digits indicate the percentage of carbon in the alloy, e.g., the 1020 carbon steel has .2 percent carbon.) Properties such as hardness, tensile strength, and so on under varying heat treatment conditions are well-documented. By comparing their test results in these known quantities with the standards readily available in handbooks. Prototype can determine whether or not their equipment is testing accurately. This process calls for a large number of test specimens.

Test specimens are manufactured according to the following procedure, as shown in time study elemental descriptions:

- *Load steel rod into cutoff saw.* The element begins when the operator grasps the rods and concludes when the rods are set in the saw and the first test specimen is in position to be cut. This task occurs every tenth cycle.
- *Cut specimen to length.* The element begins when the preceding test specimen is in position to be cut. It concludes when the specimen is cut.
- *Prepare next specimen for cutting.* The element begins when the preceding specimen is cut and concludes when the next specimen is in position to be cut. Occurs in 9 out of every 10 cycles.

Time study data for the different models are tabulated below. Every time a rod is cut, 10 specimens are prepared.

	ELEMENT		
MATERIAL	**1**	**2**	**3**
C1020	1.68	.40	.50
C1035	1.68	.60	.51
C1045	1.70	.75	.49
C1141	1.67	1.18	.48
C1144	1.69	1.23	.51
C1115	1.68	.92	.49
C1095	1.67	.87	.50

a. Develop a standard data system that will predict times to produce specimens for any alloy steel Prototype might need.
b. Show how good the system is at predicting time standards.

3. A paper manufacturing company produces pads of paper in varying sizes. Depending on trends or fads, they must be able to adapt to market demand on very short notice. To price their pads at levels that are representative of the actual costs involved in producing the pads, they need labor standards for model changes literally before production begins.

   An examination of the standards and characteristics of the existing product line gives the following data:

MODEL	LENGTH	WIDTH	NO. PAGES	STANDARD TIME
A1	7	5	25	.104
A2	7	5	50	.116
A3	7	5	75	.119
D4	8½	5	50	.108
F6	8½	7	50	.111
L10	8½	8½	25	.117
S20	11	8½	25	.114
S22	11	11	25	.110
S24	14	8½	50	.121
S30	14	8½	25	.116

   a. Determine the standard data system that will predict standard times for pads of paper.
   b. Demonstrate how good the system is in predicting time standards.
   c. How long will it take to produce a pad of paper having 50 sheets that are 8½" by 11"?
   d. How long will it take to produce a pad of paper having 50 sheets that are 9" by 12"?

CHAPTER 7

Predetermined Time Systems

OBJECTIVES

The purpose of this chapter is to introduce the reader to the concept of using predetermined time systems, or microscopic standard data, for setting time standards. After completing this chapter, the reader should be able to

- Describe the general concept of a predetermined time system.
- Understand the difference between predetermined time systems (microscopic standard data) and other standard data systems (macroscopic standard data).
- Understand the procedure for calculating a time standard with a predetermined time standard.
- Know the advantages and disadvantages of using a predetermined time system for setting time standards.
- Know the significant characteristics of the most frequently used of all standard time, or predetermined time systems.

INTRODUCTION

There are times when it is neither possible nor practical to set time standards using stop-watch time study. There are also times when it might not be possible to use a company-developed database to synthesize time standards. When this situation applies, and when the company desires to develop measures of productivity, the use of predetermined time standards might be appropriate.

Predetermined time standards are standard data systems that are designed to be used in a wide variety of product and process applications. These systems of times are standard data systems that have been generalized to fit a wide variety of applications rather than just one particular company's needs. Predetermined time systems have been designed either as general purpose systems, with the intention of being applicable to all types of jobs, or as special purpose data, with applications to just one type of work such as sewing. There are literally thousands of such systems available.

THE PURPOSE OF PREDETERMINED TIME STANDARDS

The use of predetermined time systems in productivity measurement and work improvement represents some of the most sophisticated applications of industrial engineering/work measurement in this area. These systems are designed as management tools with the specific objectives of ascertaining labor costs and ways to reduce them. The systems find the best way to do the work, standardize this method, and determine the normal time required to perform the task. Another major use of predetermined time systems is to set time standards. Standards set with these systems represent the best estimate and measurement the analyst can develop for a particular job based on the careful analysis of the motions required to complete the job. These motions are usually very small, very fine, or microscopic divisions of the work performed.

Generally, predetermined time systems, such as Methods Time Measurement (MTM), Basic Motion Timestudy (BMT), Work Factor (WF), or the hypothetical Always Fair Times (AFTWAYS) are presented as systems that can be used to set time standards for workers working at normal pace. The normal pace used often varies from system to system, but almost all systems present their data as representing normal performance. Regardless of the rating system used, however, the standards set with each system are consistent as long as they are set by a qualified analyst. PFD allowances must be added to the standards set with these microscopic standard data systems.

USE OF PREDETERMINED TIME SYSTEMS

Regardless of the system used, the general procedure for applying predetermined time systems is similar. The actual use of each system is unique to the particular system.

Whichever system is used, the following procedure is generally followed. The job for which the standard is to be set is analyzed in terms of work elements to be performed. These may be as microscopic as reaches, grasps, and insertions or they may be more broadly defined as gets or places. The analysis of the job is recorded on a left-hand/right-hand chart showing all of the motions each hand performs. These motions are expressed in the terminology of the particular system being used.

By carefully describing all of the motions required to perform a particular job, the analyst will have to carefully study the method being used to perform the job. At least theoretically, this study should ensure that the most productive method is being used. Once all of the motions have been described, these descriptions are compared with the standard motions of the predetermined time system.

When the motions required to complete the work have been identified, the standard can be set. In predetermined time systems, each motion that is described and coded has a specific time allowed for its completion. These times have been developed by the system's organization based on direct observations of a large number of significant time studies on similar operations. Each motion that is defined has a standard time specified by the system. By completely identifying all of the motions required, the entire time for a sequence of motions, or for an entire operation, can be synthesized. Once the allowance* is applied, a reasonably accurate time standard can be prepared. This procedure, of course, is based on the assumption that the correct motions have been identified before the times are assigned.

Predetermined time standards permit organizations to synthesize time standards without developing their own databases. These synthesized standards are as good or as biased as the analyst's ability to analyze the motions being studied. If the analyst knows

*The PFD allowance was described in Chapter 5.

how to use the system, the results should be consistent. Some of the larger and widely used predetermined time systems, such as MTM, offer a certificate program for people using their system. The MTM "blue card" indicates that an individual is competent to apply the MTM system of predetermined times. No system should ever be applied without adequate training in the rules and definitions of that system.

APPLICATION OF A HYPOTHETICAL PREDETERMINED TIME SYSTEM—AFTWAYS*

For any analyst to use any predetermined time system, a complete understanding of the system is required. For example, the AFTWAYS (Always Fair Time) predetermined time system is defined as follows:

> In AFTWAYS there are several basic motions. The time allowed for each of these basic motions is expressed in terms of AFT Time of Measurement Units, or ATOMS. These times are subject to a number of variables.

The Carry Motion

AFTWAYS motions are defined in the following way: The first of the AFTWAYS motions to be defined is Carry. *Carry* is defined as the process or action of transporting the hand(s) from one location to another. Carry begins when the hand has secured control of the object it is to transport and concludes when the object it is to transport has reached its destination or a specific location. The time required for the carry motion in AFTWAYS depends upon the weight of the object carried and the distance the object is carried. Table 7-1 shows the times, in ATOMS, that AFTWAYS allows for each of these possibilities. In AFTWAYS, the Carry motion is designated through the use of the symbol *C* followed by a dash and the distance carried, in inches, followed by another dash and the weight of the object carried. For example, carrying a 12-pound object 26 inches would be designated as C-26-12. The carrying distance is measured as the actual distance traveled. The weight of the object carried is determined by a standard scale. Using Table 7-1, we can determine that the time for a C-26-12 is 44.46 ATOMS.

The Grab Motion

The second of the AFTWAYS motions is Grab. *Grab* is defined as the process of gaining complete control of an object. In AFTWAYS, an object must be grabbed before it can be carried. There are three possible cases of grab: GR-1 is grabbing a small object, GR-2 is

---

*AFTWAYS is a predetermined time system used to illustrate the application of this type of microscopic standard data system. While it would be preferable to illustrate the use of a real system, such as MTM, it is not the intent of this section to create licensed or certified users of these systems. The scope of this example is limited to demonstrating the use of these systems.

TABLE 7-1 AFTWAYS CARRY TIMES

Distance (in inches)	Time (in ATOMS)	Weight factors	
		Weight up to (pounds)	Factor
2	1.0	5	1.00
4	1.4	10	1.60
6	1.8	15	1.95
8	2.2	20	2.30
10	2.6	30	2.75
12	3.0	Over 30	3.50
14	3.4		
16	3.8		
18	4.2		
20	4.6		
22	15.0		
24	18.4		
26	22.8		
28	26.2		
30	26.6		
40	39.0		
50	52.4		
60	78.1		
70	97.6		

used when a medium-sized object is grabbed, and GR-3 is specified when a large item is grabbed. The times allowed for each case of Grab are shown in Table 7-2.

TABLE 7-2 AFTWAYS GRAB TIMES

Case	Time (in ATOMS)
GR-1	11.1
GR-2	8.6
GR-3	9.7

The times are read directly from the chart, depending on whether, in the analyst's judgment, the grabbed item is small, medium, or large.

The Load Motion

The next motion defined in the AFTWAYS system is Load. *Load* is defined as the act of placing a part in a machine for the part to be operated on in some fashion. The time allowed for Load depends on the class of fit of the object when it is placed in the machine. An L-1 case indicates that there is no resistance. The L-2 case is called a close fit. In a close fit there is some resistance, but not much. In other words, there is no significant binding. An L-3 indicates a tight fit. Some extra effort is required for this case of Load. The fourth and final case of Load, the L-4, indicates a snug fit. A snug fit is characterized by extreme difficulty in loading the part in the machine. Table 7-3 shows the times for each case of Load. As with Grab, the times for loads are read directly from the table. The analyst must decide which case is appropriate.

TABLE 7-3   AFTWAYS LOAD TIMES

Case	Time (in ATOMS)
L-1	4.4
L-2	9.2
L-3	14.4

The Unload Motion

Unload is the opposite of Load. It is defined as the act of removing a part from a machine after it has been acted upon in some fashion. As was the case with Load, the time required for Unload depends on the case. Just as the definition of Unload was the mirror image of Load, so too the four cases of Unload are mirror images of the four cases of Load. UL-1 is a loosely binding unload. In the UL-1 case, parts are removed from a loosely fitting machine. In this case there is no resistance. A UL-1 must, as is the rule with all unloads, follow the same case of Load, namely the L-1 case. The UL-2 is a close unload. In the UL-2 case, there is some resistance involved in removing the part from the operation, but it is minor or slight resistance. The UL-3 case is called the tight unload. In this case, there is a significant resistance to removal of the part. Again, the rules of AFTWAYS state that a UL-3 may only follow an L-3. The final case of Unload, the UL-4, is the case used when there is a very tight fit and a very large amount of resistance is encountered. Table 7-4 shows the times allowed for each case:

TABLE 7-4   AFTWAYS UNLOAD TIMES

Case	Time (in ATOMS)
UL-1	2.6
UL-2	5.2
UL-3	10.4

The Letgo Motion

The AFTWAYS motion *Letgo* is defined as the severance of operator control of an object. This motion, similar to the MTM release motion, has one case. LG-1 is a standard breaking of control that is marked by a distinct release of the object. The LG-1, which is the only way an object may be released by the operator, may occur only after a Grab has preceded it. The time for LG-1 is shown in Table 7-5.

TABLE 7-5   AFTWAYS LETGO TIME

Case	Time (in ATOMS)
LG-1	1.0

The Place Motion

The next motion in the AFTWAYS system, Place, always precedes Load or another, yet to be defined motion, Put Together. *Place* is defined as the process of locating a part so that it can be loaded or put together with another part. The motion Place has several cases. Pl-A is used when the object is to be loaded or put together with another object with which it is symmetrical. Two objects that are symmetrical have an infinite number of ways in which they can be placed together. Pl-B is a semi-symmetrical placement. Semi-symmetrical placement of two objects means there is a finite number of ways the two objects can be placed together. Symmetrical placement might be illustrated by placing a round peg in a round hole; semi-symmetrical placement would be analogous to placing a square peg in a square hole. Pl-C is a non-symmetrical placement. When a Pl-C is identified, there is only one way the parts may be placed for loading a part together. Placing a typical door key in a position to be inserted in a lock to unlock a door would involve the use of the Pl-C case.

The Place motion in AFTWAYS is not fully described until object size is determined as well. AFTWAYS defines objects as being either large, L, or small, S. Large objects are those that are easily grabbed. Small objects take some degree of muscular control. Placing a large symmetrical object in preparation for loading it into a processing operation would be designated as PL-A-L. The AFTWAYS ATOMS for place are shown in Table 7-6.

TABLE 7-6   AFTWAYS PLACE TIMES

Case	Time (in ATOMS)
PL-A-S	4.4
PL-A-L	4.8
PL-B-S	5.9
PL-B-L	5.6
PL-C-S	7.7
PL-C-L	8.2

The Put Together Motion

The AFTWAYS motion *Put Together* is defined as the process of joining two objects so that they become one. PT-S is the case of put together used when no difficulty is encountered in joining the two objects. PT-A is used when there is some difficulty, but not extreme difficulty, in assembling the two objects.

PT-D is the Put Together case used when it is extremely difficult to put two objects together. All of the cases Put Together must be preceded by the AFTWAYS motion Place. The times for this motion are shown in Table 7-7.

TABLE 7-7   AFTWAYS PUT TOGETHER TIMES

Case	Time (in ATOMS)
PT-S	10.5
PT-A	14.0
PT-D	20.5

The Take Apart Motion

Take Apart is the opposite of Put Together. The various cases of Take Apart, designated by TA and the case, are the same as PT except, in this instance, two objects are being disengaged instead of being put together. TA-S is used for easy-to-separate objects, TA-A is for objects that require some, but not excessive force to disengage, and TA-D is used when extreme difficulty is encountered in taking two objects apart. The Grab motion must be performed before Take Apart can occur. The times for the cases of Take Apart are shown in Table 7-8.

TABLE 7-8   AFTWAYS TAKE APART TIMES

Case	Time (in ATOMS)
TA-S	3.0
TA-A	4.0
TA-D	6.0

The Push-Pull Motion

Push/Pull is the AFTWAYS motion employed to change the location of an object on a flat surface such as an assembly bench. Push/Pull must be preceded by a Grab and must eventually be followed by a Letgo. The time required for Push/Pull depends on the distance the object is to be moved and the size of the object. Objects are classified with regard to Push/Pull as being either large, L or small, S. The times for the basic motions PP-S and

TABLE 7-9 AFTWAYS PUSH/PULL TIMES

Distance object pushed or pulled	Small objects	Large objects
Up to 1 inch	3.5	4.9
2	4.5	6.3
3	5.5	7.7
4	6.5	9.1
5	7.5	10.5
6	8.5	11.9
9	14.0	19.6
12	18.0	25.2
15	22.0	18.6
18	26.0	33.8
21	30.0	39.0
24	35.0	45.5
30	48.0	62.4
36	60.0	78.0

Weight factor of Push/Pull	
Weight up to	Factor
5	1.00
10	1.25
15	1.40
20	1.75
30	2.00
Over 30	3.00

PP-L are shown in Table 7-9. The distance is adjusted similarly to the way distance is applied to Carry.

Body Motion

The final major AFTWAYS motion is the general category of Body Motion. Body Motions are any motions not performed by the hands. The cases of Body Motion indicate the part of the body performing the motion. A BM-1 is the body motion in which the upper body is turned or moved. A BM-2 involves the use of the operator's legs. A BM-2 is applied each time the legs take a step or operate a foot control. For example, an operator who steps one step forward and then steps back one step has performed two BM-2s. The BM-3 is motion performed by the head. Any distinct motion made by the head requires a BM-3 designation. Table 7-10 shows the times for body motions:

Applications Rules

AFTWAYS, as is the case with most predetermined time systems, has a number of applications rules in addition to the motion definitions. For accurate and consistent time

TABLE 7-10 AFTWAYS BODY MOTION TIMES

Case	Time (in ATOMS)
BM-1	12.0
BM-2	17.0
BM-3	24.0

standards to be set, it is absolutely necessary for all rules of the system to be followed at all times.

Some of the rules of AFTWAYS, in addition to the ones already stated, include the following:

- No two AFTWAYS motions may be performed simultaneously unless they are the same motions.
- Distances involved in all AFTWAYS motions are actual distances moved. (The user of AFTWAYS is reminded that rarely does a body motion follow a straight path. Rather, natural arcs are followed for all movements.)
- Weights that are used in some of the AFTWAYS motions are actual weights of the objects being transported. These weights should be either carefully estimated or measured on a reliable scale.

Review of AFTWAYS Motions

Before looking at some examples illustrating the use of AFTWAYS, it would be helpful to the applicator of this hypothetical system of predetermined times to review the definitions of each basic AFTWAYS motion.

CARRY is the process of transporting the hands from one location to another. The motion begins when the hand has secured control of the object and concludes when the object is transported to its destination.

GRAB is the process of gaining complete control of an object. There are three cases of Grab. These cases are dependent on the size of the object being grabbed.

LOAD is the act of placing a part in a machine for the part to be operated on in some fashion. There are four cases of Load. The case is determined by the class of fit.

UNLOAD is the act of removing a part from a machine after it has been operated upon. There are also four cases of Unload, each defined by the class of fit.

LETGO is the severance of operator control of an object. There is only one case of Letgo. For an object to be let go, it must first have been grabbed.

PLACE is the process of locating a part so that it can be loaded or put together with another part. Cases of Place depend on the symmetry of fit as well as the size of the objects being fit together.

PUT TOGETHER is the act of joining two objects so that they are one. The three cases of Put Together reflect the ease or difficulty of putting the objects together.

TAKE APART is defined as separating two objects that have been joined. The cases of Take Apart depend on the amount of force required to disengage the objects.

PUSH/PULL is employed when the intent is to change the location of an object on a flat surface. The time allowed for Push/Pull depends on the distance the object is pushed/pulled.

BODY MOTIONS are any motions not performed by the hands. The three general classifications of body motions are upper body, lower body, and head motions.

Use of AFTWAYS

The previous sections have described the AFTWAYS predetermined time system. Once the user has mastered the definitions of the motions and rules of application, it is appropriate to use this hypothetical system to set standards. After a job has been selected as needing a standard set, the analyst must prepare a left-hand/right-hand chart of the operation. For example purposes, let us examine some common, every-day tasks. The first of these is the simple process of removing a ballpoint pen from a desk drawer and placing it on a desk. This operation, found in many office jobs, begins when the operator reaches for the desk drawer and continues as the drawer is opened and the pen is selected. Once the pen is selected, it is moved to the top of the desk in the general vicinity it will be needed when the operator is ready to write. Figure 7-1 shows the detailed motions of the two hands on a left-hand/right-hand chart designed specially for use in AFTWAYS. As is consistent with the applications rules of AFTWAYS, no two motions are performed simultaneously unless they are identical.

Notice that each motion of the analysis chart has been numbered. Following is a discussion of the reasoning used to assign the symbol and calculate the allowed number of ATOMS for each of the basic AFTWAYS motions performed and shown in Figure 7-1. The numbered paragraphs correspond to the numbered motions in Figure 7-1. After the motions and times have been identified, they are entered onto the left-hand/right-hand chart to complete the analysis (Figure 7-2).

1. The left hand initially reaches to the desk drawer. This motion is a Carry. The hand has been transported from the initial position to the drawer. The hand moved an actual distance of 14 inches; hence, the AFTWAYS symbol specified is C-14. Because the empty hand carries no weight, a C-14 requires 3.4 ATOMS. Had the hand been transporting an object of a given weight exceeding five pounds the appropriate weight factor would have been applied. An item carried 14 inches weighing between 10 and 15 pounds would have had its ATOM time calculated as

$$T_A = (3.4)(1.95) = 6.63$$

PRE-DETERMINED TIME ANALYSIS

PROCESS: Pen to Desk   METHOD: (PRESENT / PROPOSED), STUDY Nº 1
OPERATOR: JWB   ANALYST: CAW
DATE: 5, 18, 91   SHEET Nº 1 of 1

LEFT-HAND ACTIVITY	SYMBOL	ALLOW. TIME	SYMBOL	RIGHT-HAND ACTIVITY
1 Reach to desk drawer				
2 Grasp desk drawer				
3 Open desk drawer				
				4 Reach to pen
				5 Grasp pen
				6 Move pen to center
				top of desk
7 Close desk drawer				
8 Let go of drawer				
9 Reach to pen				
10 Grasp pen top				
11 Remove pen top				
12 Place top on pen				
				13 Place pen on desk
14 Release pen top				
				15 Release pen

Figure 7-1 Left-hand/Right-hand Analysis for Pen Removal.

2. The left hand, after being carried to the desk drawer, grabs the drawer in anticipation of opening it. A desk drawer cannot be classified as either large or small, so, primarily by default, it is called medium in size. A GR-2 is specified, with the time of 8.6 ATOMS read directly from the Grab table.

3. Opening the desk drawer is viewed as removing or taking apart the desk. In a normally operating desk, there is no reason to believe that this activity would be any-

PRE-DETERMINED TIME ANALYSIS

PROCESS Pen to Desk   ,   METHOD (PRESENT)/ PROPOSED), STUDY Nº __1__
OPERATOR: JWB   ,   ANALYST: CAW
DATE: _5_ / _18_ / _91_   SHEET Nº __I__ of __1__

LEFT-HAND ACTIVITY	SYMBOL	ALLOW. TIME	SYMBOL	RIGHT-HAND ACTIVITY
1 Reach to desk drawer	C-14	3.4		
2 Grasp desk drawer	GR-2	8.6		
3 Open desk drawer	TA-S	3.0		
		3.6	C-15	4 Reach to pen
			GR-2	5 Grasp pen
			C-20	6 Move pen to center
		4.6		top of desk
7 Close desk drawer	PT-S	10.5		
8 Let go of drawer	LG-1	1.0		
9 Reach to pen	C-20	4.6		
10 Grasp pen top	GR-2	8.6		
11 Remove pen top	TA-S	3.0		
12 Place top on pen	C-7	2.0		
		1.6	C-5	13 Place pen on desk
14 Release pen top	LG-1	1.0		
		1.0	LG-1	15 Release pen

Figure 7-2   Complete AFTWAYS Analysis for Pen Removal.

thing but a simple task. Therefore, the motion TA-S and the required ATOMS of 3.0 are indicated.

4. After the drawer is opened, the right hand reaches for the pen in the drawer. This motion is a Carry in the AFTWAYS system and, as was the case in the first element, there is no weight factor involved. The actual distance the hand was carried was 15 inches. Because the Carry time table only lists the ATOMS for C-14 and

C-16, the time for C-15 must be determined by interpolating between the two table values. In this instance, the time for C-15 is 3.6 ATOMS.

5. Once the hand has reached the pen in the open drawer, the pen must be grabbed. Relatively speaking, a pen is an average-sized item, thus, a GR-2 would be required.

6. The right hand moves the pen to the center top of the desk. In this motion, the pen is carried by the right hand to the work area, a distance of 20 inches. The AFTWAYS motion involved is a C-20.

7. After the pen has been removed from the desk drawer, the left hand closes the desk drawer. Because opening the drawer was classified as a Take Apart, closing the drawer must be viewed as a Put Together. The rules of AFTWAYS specify that when motions appear in pairs within any motion analysis, they always require the same case of take apart as put together. Therefore, a motion of PT-S is specified.

8. Before the left hand can do anything else, it must let go of the desk drawer. The only case of Letgo is LG-1.

9. The left hand reaches for the pen. In AFTWAYS terminology, the left hand is carried 20 inches to the pen in the center of the desk.

10. Upon arriving in the vicinity of the pen, the left hand grabs the pen top. As established earlier in the analysis, the pen is a medium-sized object and the motion is a GR-2.

11. The next action for the left hand is the removal of the pen top. This activity fits the Take Apart AFTWAYS motion. In terms of the effort required, this motion requires relatively small effort. The symbol for the motion is TA-S.

12. After the pen top is removed, it is placed on the desk. This Carry requires traveling a distance of seven inches; the C-7 motion requires 2.0 ATOMS.

13. The right hand immediately places the pen in the desk. The pen, because it was closer to the table after the cap was taken apart from it, requires a carry of only five inches. The C-5 requires 1.6 ATOMS.

14. As this job is defined, the last motion the left hand performs is the release of the pen top, an LG-1.

15. Similarly, the right hand lets go of the pen. The motions required for this job are not complete.

The analysis concludes with the calculation of the total number of ATOMS required. An appropriate PFD allowance is added and the time, in ATOMS, required for a typical operator working at a normal pace, to complete the job, is established.

Example: Packaging Replacement Sewing Machine Needles

The following example illustrates the use of our hypothetical predetermined time system (AFTWAYS) to synthesize a standard for a job at the Winslow Needle Company. As with the preceding example, a complete rationale, as would be required in any such system, is presented for the assignment of the AFTWAYS motions and times.

A new packing job is being contemplated in the shipping room of the Winslow Needle Company. Currently, all needles manufactured are shipped in bulk containers to the Hollar Sewing Machine Factory where they are obviously used in sewing machines. However, Winslow's management believes that there is a market for replacement needles, packaged five to an envelope, and shipped 100 packets to the box.

Because management knows how much it costs to manufacture each needle, they must know how much it will cost to package the needles. Envelope costs and box costs are readily available but, because this job is brand new, they need to be able to predict the labor cost without actually performing the task. This use is ideal for any predetermined time system.

As the analyst envisions this job, it begins when the packer picks up the 40-pound bulk container of needles. (A typical bulk container holds 4000 needles.) The container is then carried 10 steps to the packing workbench. Envelopes are filled with the required five needles and the filled envelopes (with self-sticking flaps) are placed 100 to each box. When a box is filled, the cover is placed on the box and it is set on a shipping pallet to the left of the packer. When the pallet is loaded, the materials handling staff will carry the loaded pallet away and bring an empty one. After 800 envelopes have been filled, the empty storage container, now weighing 25 pounds, is returned the 10 steps where the packer originally picked it up. The packer then picks up another loaded bulk storage container and the job repeats. The analyst wants to determine the time required to fill each box with 100 packets of needles, each containing five needles. A 15 percent PFD allowance will be specified.

A sketch of the workplace is prepared; it is shown in Figure 7-3. Figure 7-4 shows the complete left-hand/right-hand analysis. Two important points should be noted. First, body motions in AFTWAYS are always shown as right-hand activities. Second, when

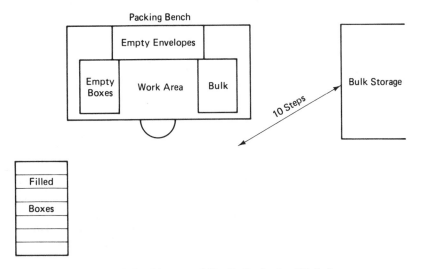

Figure 7-3   Diagram of Needle Packaging Workplace.

PRE-DETERMINED TIME ANALYSIS

PROCESS <u>Needle Packaging</u>   ,   METHOD (PRESENT/ PROPOSED), STUDY Nº <u>1</u>
OPERATOR: <u>C. Babbage</u>   ,   ANALYST: <u>LSA</u>
DATE: <u>6</u> / <u>14</u> / <u>91</u>   SHEET Nº <u>1</u> of <u>2</u>

LEFT-HAND ACTIVITY	SYMBOL	ALLOW TIME	SYMBOL	RIGHT-HAND ACTIVITY
1 Reach to bulk needles				1 Reach to bulk needles
2 Grab container				2 Grab container
3 Pick up container				3 Pick up container
				4 Walk 10 steps to
				workbench
5 Place container on				5 Place container on
bench				bench
6 Release container				6 Release container
7 Reach for empty				
envelope				
8 Grasp envelope				
9 Place envelope in				
work area				
				10 Reach for needle
				11 Pick up needle
				12 Place needle in
				envelope
				12A Release needle
13 Place envelope in box				
14 Release envelope				
15 Reach for cover				
16 Grab cover				

Figure 7-4 Left-hand/Right-hand Chart for Needle Packaging.

both hands perform exactly the same motion, although the motion is shown actively for each hand, the time allowed is the time required of just one hand. Figure 7-5 shows the AFTWAYS motions and allowed ATOMS for each motion. The interpretations of the AFTWAYS rules for all of the motions appear in the following paragraphs. The numbers refer to the numbers of the lines on the left-hand/right-hand chart.

PRE-DETERMINED TIME ANALYSIS CONTINUED

STUDY N⁰ _1_   SHEET N⁰ _2_ of _2_

LEFT—HAND ACTIVITY	SYMBOL		SYMBOL	RIGHT—HAND ACTIVITY
17 Carry cover to box				
18 Place cover on box				
				19 Reach to box
				20 Grab box
21 Carry box to pallet				21 Carry box to pallet
22 Release box				22 Release box
23 Return to work area				23 Return to work area
24 Reach to bulk				24 Reach to bulk
container				container
25 Grab empty container				25 Grab empty container
26 Arrange container				26 Arrange container
				27 Walk 10 steps to
				storage
28 Set empty container down				28 Set empty container down
29 Release container				29 Release container
NOTE 1: Motions 1,2,3,4,5,6,24,25,26,27,28,and 29 occur every 8th cycle.				
NOTE 2: Motions 10,11, and 12 repeat 5 times per envelope or 500 times per box.				
NOTE 3: Motions 7,8,9,13, and 14 repeat 110 times per box.				
NOTE 4: Motions 15,16,17,18,19,20,21,22, and 23 occur once every box.				

Figure 7-4   (Continued).

1. Both hands start this proposed task by reaching to the container of needles. The hands are carried an estimated travel distance of 18 inches. Because the hands are empty, the symbol is the basic C-18.

2. Before the container can be picked up, it must be grabbed. The bulk storage container is, relatively speaking, a large container. Therefore, a GR-3 is appropriate.

PRE-DETERMINED TIME ANALYSIS

PROCESS: <u>Needle Packaging</u>   ,   METHOD (PRESENT / PROPOSED), STUDY № <u>1</u>
OPERATOR: <u>C. Babbage</u>   ,   ANALYST: <u>LSA</u>
DATE: <u>6</u> / <u>14</u> / <u>91</u>   SHEET № <u>1</u> of <u>2</u>

LEFT-HAND ACTIVITY	SYMBOL	ALLOW. TIME	SYMBOL	RIGHT-HAND ACTIVITY
1 Reach to bulk needles		4.2	C-18	1 Reach to bulk needles
2 Grab container		9.7	GR-3	2 Grab container
3 Pick up container		7.8	C-14-20	3 Pick up container
		17.0	10 BM-2	4 Walk 10 steps to workbench
5 Place container on bench		6.4	C-11-20	5 Place container on bench
6 Release container		1.0	LG-1	6 Release container
7 Reach for empty envelope	C-14	3.4		
8 Grasp envelope	GR-2	8.6		
9 Place envelope in work				
area	C-14	3.4		
		3.6	C-15	10 Reach for needle
		11.1	GR-1	11 Pick up needle
		3.8	C-16	12 Place needle in envelope
		1.0	LG-1	12A Release needle
13 Place envelope in box	C-26	22.8		
14 Release envelope	LG-1	1.0		
15 Reach for cover	C-8	2.8		
16 Grab cover	GR-2	8.6		
17 Carry cover to box	C-8	2.2		
18 Place cover on box	PT-A	14.0		
		29.1	C-32	19 Reach to box
		8.6	GR-2	20 Grab box

Figure 7-5 AFTWAYS Analysis of Needle Packaging Operation.

3. Once the 40-pound container has been grabbed, the packing operator must pick it up before it is moved to the work area. It is estimated that the operator will lift the container 14 inches to have it in a position that is comfortable for walking. The correct

PRE-DETERMINED TIME ANALYSIS CONTINUED

STUDY № 1   SHEET № 2 of 2

LEFT—HAND ACTIVITY	SYMBOL	Time	SYMBOL	RIGHT—HAND ACTIVITY
21 Carry box to pallet		42.6	C-30-8	21 Carry box to pallet
22 Release box		1.0	LG-1	22 Release box
23 Return to work area		44.4	C-44	23 Return to work area
24 Reach to bulk container		49.8	C-48	24 Reach to bulk container
25 Grab empty container		9.7	GR-3	25 Grab empty container
26 Arrange container		5.1	C-10-13	26 Arrange container
		170.0	10 BM-2	27 Walk 10 steps to storage
28 Set empty container down		6.6	C-14-13	28 Set empty container down
29 Release container		1.0	LG-1	29 Release container
NOTE 1: Motions 1,2,3,4,5,6,24,25,26,27,28,29 occur every 8th cycle.				
NOTE 2: Motions 10,11, and 12 repeat 5 times per envelope or 500 times per box.				
NOTE 3: Motions 7,8,9,13,14 repeat 100 times per box.				
NOTE 4: Motions 15,16,17,18,19,20,21,22, and 23 occur once every box.				

Figure 7-5   (Continued).

AFTWAYS motion is a C-14-20. The final 20 represents the 20 pounds of container weight carried by each hand. Because both hands are doing the same thing at the same time, the motion is shown on the same line of the left-hand/right-hand chart.

4. The packer next walks, with the container, 10 paces or steps to the workbench. Because 10 steps are required, the BM-2 motion is indicated and used 10 times, suggesting the AFTWAYS notation 10BM-2.

5. After the operator reaches the work area, the container is placed on the bench. The anticipated travel distance is 11 inches and the motion is C-11-20. Again, because both hands are carrying the container, the motion is shown on both the left-hand and right-hand sides of the analysis sheet.

6. Both hands simultaneously release the container. The standard Letgo is used, LG-1.

7. Beginning with this motion, the hands start to perform the independent motions required to pack the needles in the envelopes and then pack the envelopes in the boxes. The left hand reaches or is carried to the stack of empty envelopes. These envelopes are estimated, according to the workplace layout, to be about 14 inches away. The motion, naturally, is a C-14.

8. Comparatively speaking, the envelope is a medium-sized object. (Comparing the envelope to either the bulk storage container or an individual needle should make it obvious that the envelope is medium-sized.) Before the envelope can be transported, it must be grabbed using a GR-2.

9. Upon grabbing the envelope, the left hand carries it back to the work area using a C-14.

10. Meanwhile, the right hand reaches to the bulk needle storage, an average distance of 15 inches. As the supply of needles diminishes, the distance will change, but AFTWAYS allows for the use of the average distance when a highly repetitive series of motions is performed. In this case, the motions average out to a C-15.

11. The right hand next picks up, or grabs, the relatively small needle. The GR-1 is appropriate. Had there been a danger of injury to the operator, then the AFTWAYS rules would have dictated that two GR-1s be used.

12. The needle is carried to the envelope and placed inside it. This Carry averages, to the nearest inch, 16 inches. Needles are relatively light, weighing one-half an ounce each.

12a. The right hand lets go of the needle, using the standard LG-1. Motions 10, 11, 12, and 12a are performed five times per envelope and the allowed time must reflect this activity.

13. After the fifth needle is placed in the envelope, the left hand carries the envelope to the shipping box. The average distance the left hand moves is 26 inches. Each envelope weighs only ⁵⁄₃₂ of a pound, so no weight factor will be required.

14. After placing the envelope in the box, the left hand lets go. The left hand then reaches for the next empty envelope, an average distance of 14 inches. This motion is, in reality, a repeat of motion 7. As should be obvious, motions 7 through 14 are repeated 100 times per box and the allowed time must take this repetition into account.

15. Upon filling the box with 100 envelopes, the left hand, instead of reaching for another envelope, carries itself a distance of eight inches to the box cover, a C-8 motion.

16. The next logical motion is for the packer to pick up or grab the cover. The cover, again comparatively speaking, is an average-sized item requiring a GR-2.

17. The lightweight cover is then carried the eight inches back to the full box, another C-8.

18. The left hand, by placing the top on the box, is in effect putting the box together. This task is viewed, because it is performed by one hand, as being of average difficulty. The motion is a PT-A.

19. At this juncture, the right hand reaches to the covered box, a distance of 32 inches. A C-32 AFTWAYS motion is used.

20. Upon reaching the box, the right hand grabs it (classified as average-sized) with a GR-2.

21. Both hands transport the filled box an average of 30 inches to the pallet. The box, containing 100 envelopes each weighing 5/32 of a pound, has a weight of 500/32 pounds, or just over 15.6 pounds. Using the next higher weight classification of 16 pounds, a C-30-16 is specified. The time allowed, however, reflects a C-30-8 because the weight is carried by both hands.

22. The hands let go of the box, using the standard AFTWAYS Letgo.

23. Both hands return to the work area, a distance of approximately 44 inches. The entire motion pattern specified by motions 7 through 23 reflects one cycle for loading a box. Still to be considered fully is replenishing the raw material. This activity was partially done in motions 1 through 6 and will be completed in the remaining motions.

24. Once the container of needles is empty, the packer must dispose of the empty container and get a fresh one. This motion, reaching to the bulk needle container after letting go of the eighth box filled from the bulk container of needles, involves carrying the empty hands 48 inches to the container, a C-48 motion.

25. Naturally, the hands must grab the empty container before doing anything with it.

26. Having grabbed the container, the packer lifts it to a comfortable position for carrying. Because the weight is evenly distributed, the empty container is subject to a C-10-13.

27. Walking the 10 steps back to the bulk storage area is a body motion, specifically a BM-2. Because 10 steps are required, the AFTWAYS motion is a 10BM-2.

28. After arriving in the work area, the packer sets the empty container down. The anticipated distance is 14 inches and the distributed weight of the empty container requires a C-14-13 motion.

29. To complete the motion pattern, before reaching to the next full container of needles, the operator must let go of the empty container. As is always the case, an

LG-1 is the release used. Motions 1, 2, 3, 4, 5, 6, 24, 25, 26, 27, 28, and 29 occur once every eighth box.

Now that all of the motions have been described and, hopefully, fully explained, the time standard per box of 100 envelopes can be set. Motions 1, 2, 3, 4, 5, 6, 24, 25, 26, 27, 28, and 29 occur every eighth cycle, so the time allowed for these motions must, of course, be divided by eight to determine their relative contribution to the time standard. The total for these motions is 441.3 ATOMS. One eighth of this time is 55.2 ATOMS. Motions 10, 11, 12, and 12a occur five times per envelope and this time occurs 100 times per box. The total of these four elements is 41.3 ATOMS. Multiplied by 5 to determine the time per envelope yields 206.5 ATOMS. Because 100 envelopes are in a box, this time is now multiplied by 100 to determine the time per box of 100 envelopes, 20,650 ATOMS.

Elements 7, 8, 9, 13, and 14 occur 100 times per box. Their allowed time is 39.2 ATOMS. For 100 envelopes, the time is 3,920 ATOMS. Finally, motions 15, 16, 17, 18, 19, 20, 21, 22, and 23 occur once per box. The time required is 153.3 ATOMS.

The entire time per box is the sum of these times:

$$55.2 + 20,650 + 3.920 + 153.3 = 24,778.5 \text{ ATOMS}$$

Applying the PFD allowance of 15 percent, the standard becomes:

$$(24,778.5)(1) + (24,778.5)(.15) = 24,495.3 \text{ ATOMS}$$

This analysis means, of course, that 24,495.3 ATOMS will be required for each box of 100 needles. By converting the ATOMS to hours and then converting the hours to dollars, the management of the Winslow Needle Company can determine the labor cost of marketing their own needles. They will have been able to do this analysis without ever having had to produce the first package of needles. That is one of the most important points to be made about predetermined time systems.

Predetermined time systems can be used very effectively to synthesize time standards for nonexistent jobs and even jobs within an organization that have nothing in common with jobs that have rates set on them by traditional or direct measurement methods. Predetermined time systems have sets of motions, and corresponding times for these motions, that can be used to set standards for any job that can be described in terms of these basic motions.

ADVANTAGES OF PREDETERMINED TIME SYSTEMS

There are four major advantages involved in using a predetermined time system. First, predetermined time systems require a complete analysis of all motions performed in a given job. Any hidden problems and inefficiencies can be identified and dealt with—that is, improved upon or eliminated—to improve the productivity of the job. With this methods improvement comes the added feature of job standardization, which leads to uniform job procedures, a valuable training tool.

Standards set with predetermined time systems are generally considered to be more factual and less subjective than those set by time study. This "engineering" of time standards, based upon the microscopic examination of the work methods used, is much more readily accepted by organized labor because subjective performance rating is not performed.

Third, predetermined time standards can be used to synthesize time standards, that is, to develop time standards prior to beginning production on an actual job. Once work methods have been defined, the workplace layout and even the plant layout can be designed to fit the operations and methods. Tool and equipment selection and design can be directed toward the areas that are compatible with the work methods established.

Finally, predetermined time standards are extremely useful in providing information about learning time and performance rating. The development of learning curves and their subsequent application is an essential part of determining the cost of a new product or service. It is a very important facet of production planning.

SELECTING A PREDETERMINED TIME SYSTEM

As was stated earlier, there are literally thousands of predetermined time systems available for use. The major considerations in selecting the system for your use should include the following:

- How is the system to be used?
- Under what conditions is the system to be used?

Certain predetermined time systems are best suited for long-cycle, low-repetition jobs. Another consideration is the amount of methods description required. Some systems require very detailed methods analysis and are, as a result, extremely useful for training applications. Well-written motion descriptions provide a base for consistent applications.

Consideration must also be given to the availability of instructional materials. Supervisors and production workers should be introduced to the system in a manner that will give them an overall understanding of the procedures involved in applying the system. The amount of training required of an analyst is also an important consideration in selecting a system. Some of the predetermined time systems, such as MTM, require extensive training for analysts before they are certified as being proficient in the use of that system.

While there are many advantages to using a predetermined time system, some limitations are involved as well.

LIMITATIONS OF PREDETERMINED TIME SYSTEMS

The major disadvantage is the difficulty encountered with machine-paced operations. The best way to determine machine time is with a stopwatch because most predetermined time systems were not designed to replace direct measurement but rather to add another tool for work measurement. Not every system will give good standards for every type of work. Some systems, because of operations used when the original database was developed, just don't seem to be as effective when used with different types of operations. Whether or not it is a disadvantage is debatable, but a significant amount of training is required with most predetermined time systems before analysts can achieve consistent results. Krick (1962) reported that some of the systems "differ with respect to performance level embodied in the time data, by as much as an estimated 35 percent" (p. 352).

EXAMPLES OF EXISTING PREDETERMINED TIME SYSTEMS

The following section is not an endorsement of any predetermined time system. It is, rather, a brief summary of some of the better known and more frequently used systems. It is not meant to be a comprehensive summary but rather, it is a very small sampling of some of the existing (legitimate) systems.

MODAPTS

MODAPTS is a relatively easy-to-use predetermined time system. MODAPTS stands for Modular Arrangement of Predetermined Time Standards. Because it describes work in human rather than mechanical terms, it has many more potential applications than earlier work analysis systems. The application is integrated with desktop computer processing capabilities, which simplifies its use.

MODAPTS is a recognized IE technique, meeting all criteria of the U.S. Defense Department and Department of Labor to develop industrial standards. Performance times are based on the premise that motions will be carried out at the most energy-efficient speed.

MODAPTS is used to analyze all types of industrial, office, and materials handling tasks. Data from MODAPTS studies are used for planning and scheduling, cost estimating and analysis, ergonomic evaluation of manual tasks, and the development of labor standards.*

The Work-Factor System

The first predetermined time system was developed by A. B. Segur around 1925. He was one of the first to recognize the association between motion and time. He stated a principle, that, within allowances for normal variation, the time required by experts to perform a fundamental motion is consistent. Segur developed Methods Time Analysis which could be used to analyze manual and manual/machine operations. Segur emphasized that the time required for work depended on how the work was done. He stressed that a complete description of the work performed was necessary.

*Additional information about MODAPTS is available from the International MODAPTS Association, Inc. (2022 Kohrman Hall, Western Michigan University, Kalamazoo, MI 49008). The above material is reproduced with the permission of the International MODAPTS Association, Inc.

In the early 1930s, union workers in Philadelphia were dissatisfied with the quality of the stopwatch time standards set for their highly controlled incentive jobs. The results of this protest were one of the first published predetermined time systems, called Work-Factor (Karger and Bayha, 1977). Work-Factor tables are shown in Table 7-11. The Work-Factor system makes it possible to determine the normal time for manual tasks by the use of motion time data. The basic motion is defined as that motion which involves the least amount of difficulty or precision for any given distance and body member combination. Work-Factor is used as the index of additional time required over and above the basic times when motions are performed involving manual control and weight or resistance.

There are four variables that affect the time of manual motions in the Work-Factor system:

1. Body member used
2. Distance moved-measured on a straight-line basis
3. Degree of manual control required
4. Weight or resistance varies with body member used and sex of the operator.

The eight standard elements of work factor are transport, grasp, preposition, assemble, use, disassemble, mental process, and release.

Harold Engstrom, working for the General Electric Company, developed a system based primarily on "place and get." Some of Engstrom's motions required several basic motions. Critics have indicated that his system combined predetermined microscopic times with a standard data system. The time systems currently used at GE reflect Engstrom's influence (Duncan, Quick, and Malcolm, 1962).

Another system was developed by Captain John Olsen at the Springfield, Massachusetts armory. Olsen's special purpose data tried to make military production more competitive with civilian production. His system served as the basis for a form of incentive pay.

Methods Time Measurement (MTM)

The most widely publicized system of performance rating ever developed was presented in *Time and Motion Study* by Lowry, Maynard, and Stegemerten (1940). The basis of the rating was the use of four factors: skill, effort, consistency, and performance. Maynard and Stegemerten teamed with John Schwab to expand this idea into Methods Time Measurement or MTM (Maynard, Schwab, and Stegemerten, 1948). According to Robert Rice (1977) this method is the most widely used system of predetermined times. Maynard and associates performed many micromotion studies to come up with their standard elements and times. Because MTM was and is readily available, it is not surprising that it is the most frequently used of all the systems and it is also the most frequently imitated. Standard MTM data is shown in Figure 7-6. MTM is defined as a procedure that analyzes any manual operation or method into the basic operations required to perform it and assigns to each motion a predetermined time standard which is determined by nature and the

TABLE 7-11 WORK-FACTOR MOTION TIME TABLE FOR DETAILED ANALYSIS* (TIME IN WORK-FACTOR* UNITS)

(A) Arm—measured at knuckles

Distance moved, in.	Basic	Work-factors 1	2	3	4
1	18	26	34	40	46
2	20	29	37	44	50
3	22	32	41	50	57
4	26	38	48	58	66
5	29	43	55	65	75
6	32	47	60	72	83
7	35	51	65	78	90
8	38	54	70	84	96
9	40	58	74	89	102
10	42	61	78	93	107
11	44	63	81	98	112
12	46	65	85	102	117
13	47	67	88	105	121
14	49	69	90	109	125
15	51	71	92	113	129
16	52	73	94	115	133
17	54	75	96	118	137
18	55	76	98	120	140
19	56	78	100	122	142
20	58	80	102	124	144
22	61	83	106	128	148
24	63	86	109	131	152
26	66	90	113	135	156
28	68	93	116	139	159
30	70	96	119	142	163
35	76	103	128	151	171
40	81	109	135	159	179
Weight, lb: Male	2	7	13	20	Up
Female	1	3½	6½	10	Up

(L) Leg—measured at ankle

Distance moved, in.	Basic	Work-factors 1	2	3	4
1	21	30	39	46	53
2	23	33	42	51	58
3	26	37	48	57	65
4	30	43	55	66	76
5	34	49	63	75	86
6	37	54	69	83	95
7	40	59	75	90	103
8	43	63	80	96	110
9	46	66	85	102	117
10	48	70	89	107	123
11	50	72	94	112	129
12	52	75	97	117	134
13	54	77	101	121	139
14	56	80	103	125	144
15	58	82	106	130	149
16	60	84	108	133	153
17	62	86	111	135	158
18	63	88	113	137	161
19	65	90	115	140	164
20	67	92	117	142	166
22	70	96	121	147	171
24	73	99	126	151	175
26	75	103	130	155	179
28	78	107	134	159	183
30	81	110	137	163	187
35	87	118	147	173	197
40	93	126	155	182	206
Weight, lb: Male	8	42	Up	—	—
Female	4	21	Up	—	—

Note: From *Work-Factor Time Standards* by J. H. Duncan. J. H. Quick, and J. A. Malcolm. New York: McGraw-Hill, 1962, pp. 436–437. Reproduced with permission of the Science Management Corporation.

TABLE 7-11 (Continued)

(T) Trunc—measured at shoulder

1	26	38	49	58	67
2	29	42	53	64	73
3	32	47	60	72	82
4	38	55	70	84	96
5	43	62	79	95	109
6	47	68	87	105	120
7	51	74	95	114	130
8	54	79	101	121	139
9	58	84	107	128	147
10	61	88	113	135	155
11	63	91	118	141	162
12	66	94	123	147	169
13	68	97	127	153	175
14	71	100	130	158	182
15	73	103	133	163	188
16	75	105	136	167	193
17	78	108	139	170	199
18	80	111	142	173	203
19	82	113	145	176	206
20	84	116	148	179	209

Weight, lb:
Male	11	58	Up	Up	—
Female	5½	29	Up	Up	—

(F, H) Finger-hand—measured at finger tip

1	16	23	29	35	40
2	17	25	32	38	44
3	19	28	36	43	49
4	23	33	42	50	58

Weight, lb:
Male	⅔	2½	4	Up	—
Female	⅓	1¼	2	Up	—

(Ft) Foot—measured at toe

1	20	29	37	44	51
2	22	32	40	48	55
3	24	35	45	55	63
4	29	41	53	64	73

Weight, lb:
Male	5	22	Up	Up	—
Female	2½	11	Up	Up	—

(FS) Farm swivel—measured at knuckles

45°	17	22	28	32	37
90°	23	30	37	43	49
135°	28	36	44	52	58
180°	31	40	49	57	65

Torque, lb. in.:
Male	3	13	Up	Up	—
Female	1½	6½	Up	Up	—

Walk Time

Type	30-inch Paces		
	1	2	Over 2
General	Analyse from table	260	120 + 80 per Pace
Restricted		300	120 + 100 per Pace

Add 100 for 120–180° Turn at Start or Finish

Up Steps (8-inch rise—10-inch flat) 126
Down Steps 100

Head Turn:
45° 40
90° 60

1 Time Unit = 0.000 Second
= 0.0001 Minute
= 0.0000167 Hour

Work-Factor* Symbols
W—Weight or Resistance
S—Directional Control (Steer)
P—Care (Precaution)
U—Change Direction
D—Definite Stop

*Registered trademark.

TABLE 7-11 (Continued).

Work-factor grasp table				

Complex grasp from random files

Size (major dimension or length), in.		Solids and brackets [thickness over 0.0469 inch (3/64)]				Thickness of thin flat objects, in.								Add for entangled, nested or slipperty objects†	
						0.00156 (Less than 1/64)				0.0156–0.0469 (1/64–3/64)					
		Blind		Visual		Blind		Visual		Blind		Visual			
		n‡	s	n	s	n	s	n	s	n	s	n	s	n	s
0.0000–0.0625	1/16 and less	120	172	B§	B	—	—	—	—	131	189	B	B	17	20
0.0626–0.1250	Over 1/16–1/8	79	111	B	B	108	154	B	B	85	120	B	B	12	18
0.1251–0.1875	Over 1/8–3/16	64	88	B	B	102	145	B	B	74	103	B	B	12	18
0.1876–0.2500	Over 3/16–1/4	48	64	B	B	72	100	B	B	56	76	B	B	8	12
0.2501–0.5000	Over 1/4–1/2	40	52	B	B	64	88	B	B	48	64	B	B	8	12
0.5001–1.0000	Over 1/2–1	40	52	32	40	64	88	60	82	48	64	44	58	8	12
1.0001–4.0000	Over 1–4	37	48	20	22	53	72	36	46	45	60	28	34	8	12
4.0001 and up	Over 4	46	61	20	22	70	97	44	58	62	85	36	46	9	14

Size (major dimension or length), in.		Diameter of cylinders and regular cross-sectioned solids, in.													
		0–0.0625 (1/16)		0.0626–0.125 (1/8)		0.1251–0.1875 (3/16)		0.1876–0.5000 (1/2)				0.5001 and up (Over 1/2)			
		Blind		Blind		Blind		Blind		Visual		Blind		Visual	
		n	s	n	s	n	s	n	s	n	s	n	s	n	s
0.0000–0.0625	1/16 and less	S	S	S	S	S	S	S	S	S	S	S	S	S	S
0.0626–0.1250	Over 1/16–1/8	85	120	S	S	S	S	S	S	S	S	S	S	S	S
0.1251–0.1875	Over 1/8–3/16	79	111	74	103	S	S	S	S	S	S	S	S	S	S
0.1876–0.2500	Over 3/16–1/4	79	111	68	91	61	88	S	S	S	S	S	S	S	S
0.2501–0.5000	Over 1/4 1/2	62	85	56	76	56	76	44	58	B	B	S	S	S	S
0.5001–1.0000	Over 1/2 1	62	85	56	76	48	64	48	64	44	58	40	52	32	40
1.0001–4.0000	Over 1–4	56	76	48	64	40	52	40	52	36	46	37	48	20	22
4.0001 and up	Over 4	56	76	48	64	40	52	40	52	36	46	37	48	20	22

*Special Grasp conditions should be analyzed in detail.

†Add the indicated allowances when objects: (1) are entangled (not requiring two hands to separate); (2) are nested together because of shape or film; (3) are slippery (as from oil or polished surface). When objects both entangle and are slippery, or both nest and are slippery, use double the value in the table.

‡n = Non-simo; s = Simo.

§B = Use Blind column, since Visual Grasp offers no advantage; S = Use Solid Table.

condition under which it is made (Karger and Bayha, 1977). Reach is the most common or basic MTM motion. The other motions involved include:

MOVE—In which the predominant purpose is to transport an object to a destination.

TURN—In which the hand is turned or rotated about the long axis of the forearm.

TABLE 7-11 (Continued).

Work-factor assemble tables

Average number of aligns (AIS motions)

Target diameter, in.	Ratio of plug diameter to target diameter										
	To 0.224		0.225–0.289		0.290–0.414		0.415–0.899		0.900–0.931		0.935–1.000
	Closed targets										
0.875 and up	(D*)	18	(D*)	18	(D*)	18	(¼)	25	(¼†)	51	(¼‡)	59
0.625 –0.874	(D*)	18	(D*)	18	(SD*)	18	(¼)	25	(¼†)	51	(¼‡)	59
0.375–0.624	(SD*)	18	(SD*)	18	(¼))	25	(½)	31	(½†)	57	(½‡	65
0.225–0.374	(½)	31	(1)	44	(1)	44	(1½)	57	(1½†)	83	(1½‡	91
0.175–0.224	(1)	44	(1)	44	(1)	44	(1½)	57	(1½†)	83	(1½‡	91
0.125–0.174	(1)	44	(1¼)	51	(1½)	57	(1½)	57	(1½†)	83	(1½‡)	91
0.075–0.124	(2½)	83	(2½))	83	(2½)	83	(2½)	83	(2½†)	109	(2½‡)	117
0.025–0.074	(3)	96	(3)	96	(3)	96	(3)	96	(3†)	122	(3‡)	130
	Open targets										
0.875 and up	(D*)	18	(D*)	18	(D*)	18	(D*)	18	(¼†)	51	(¼‡)	59
0.625–0.874	(D*)	18	(D*)	18	(D*)	18	(SD*)	18	(¼†)	51	(¼‡)	59
0.375–0.624	(SD*)	18	(SD*)	18	(SD*)	18	(½)	31	(½†)	57	(½†)	65
0.225–0.374	(¼)	25	(½)	31	(½)	31	(¾)	38	(¾†)	64	(¾‡)	72
0.175–0.224	(½)	31	(½)	31	(½)	31	(¾)	38	(¾†)	64	(¾‡)	72
0.125–0.174	(¾)	38	(1)	44	(1)	44	(1)	44	(1†)	70	(1‡)	78
0.075–0.124	(1¼)	51	(1¼)	51	(1¼)	51	(1¼	51	(1¼†)	77	(1¼‡)	85
0.025–0.074	(1½)	57	(1½)	57	(1½)	57	(1½)	57	(1½†)	83	(1½‡)	91

*Letters indicate Work-Factors in Move preceding Assemble.

†Requires A(X)S Upright for all Ratios of 0.900 and greater.(Table value includes AIS Upright.)

‡Requires A(Y)S Upright and A(Z)P Insert for all Ratios of 0.935 and greater. (Table value includes AIS Upright and AIP Insert.)

POSITION—The motion employed to align, orient, and engage one object with another.

GRASP—Where the main purpose is to secure sufficient control of one or more objects with the fingers or the hand.

RELEASE—Motions that identify the operator relinquishing control of an object.

DISENGAGE—The motion used to identify when contact between two objects is broken.

EYE TIMES—Used when the eyes direct hand or body motions.

BODY MOTIONS—Motions made by the entire body, not just the hands, fingers, or arms.

Computerized versions of this information can convert the symbolic MTM coding to actual MTM times.

TABLE 7-11 (Continued)

Work-factor assemble tables (Continued)

Distance between targets			Gripping distance		
Distance between targets, in.	Addition to aligns, percent	Method of align	Distance from gripping point to align point, in.	Addition to aligns, percent	Length of upright motion, in.
0– 0.99	Neg.	Simo	0– 1.99	Neg.	1
1– 1.99	10	Simo	2– 2.99	10	1
2– 2.99	30	Simo	3– 4.99	20	2
3– 4.99	50	Simo	5– 6.99	30	2
5– 6.99	70	Simo	7– 9.99	40	3
7–14.99	Align and Insert first end, then Align* and Insert second end.		10–14.99	60	5
			15–19.99	80	6
15 and up	Align and Insert first end, Head Turn and Inspect second end†, then Align* and Insert second end.		20 and up	100	7 and up

*If connected, treat Second Assemble as Open Target with no Upright. †HT45° + IB$_9$S$_1$ + 2

General rules for assemble

1. When required add W and P Work-Factors to all Assemble Motions according to rules for Tansports.
2. Reduce number of Aligns by 50 percent when hand is rigidly supported.
3. Where Gripping Distance, two Targets, and Blind Targets are involved, add each percentage to original Align. Do not pyramid percentages.
4. Aligns for Surface Assemble are taken from 0.224 column and are AISD Motions.
5. Index is FIS, AIS, or FS45°S.

Blind targets

Distance from target to visible area, in.	Addition to Aligns, percent	
	Permanent (blind at all times)	Temporary (blind during assemble)
0.0– 0.49	20	0
0.5– 0.99	30	10
1.0– 1.99	40	20
2.0– 2.99	70	30
3.0– 4.99	130	50
5.0– 6.99	250	70
7.0–10.00	380	120

POST-DISENGAGE TRAVEL TABLE

Resistance to Disengage, lb.	Post-disengage Travel, lb.
0.0– 2.0	Neg
2.1– 7.0	3
7.1–13.0	6
13.1–20.0	10

Basic Motion Timestudy (BMT)

In 1951, the Canadian firm of Woods and Gordon made the first significant contribution to predetermined time system literature by a foreign source. The Canadians developed Basic Motion Timestudy which was derived from systems already available. The major advantage of BMT is its brevity. In BMT, a basic motion is defined as a single complete movement of a body member. A basic motion occurs every time a body member, being at

TABLE I – REACH – R

Distance Moved Inches	Time TMU A	B	C or D	E	Hand In Motion A	B	CASE AND DESCRIPTION
3/4 or less	2.0	2.0	2.0	2.0	1.6	1.6	A Reach to object in fixed location, or to object in other hand or on which other hand rests.
1	2.5	2.5	3.6	2.4	2.3	2.3	
2	4.0	4.0	5.9	3.8	3.5	2.7	
3	5.3	5.3	7.3	5.3	4.5	3.6	B Reach to single object in location which may vary slightly from cycle to cycle.
4	6.1	6.4	8.4	6.8	4.9	4.3	
5	6.5	7.8	9.4	7.4	5.3	5.0	
6	7.0	8.6	10.1	8.0	5.7	5.7	
7	7.4	9.3	10.8	8.7	6.1	6.5	C Reach to object jumbled with other objects in a group so that search and select occur.
8	7.9	10.1	11.5	9.3	6.5	7.2	
9	8.3	10.8	12.2	9.9	6.9	7.9	
10	8.7	11.5	12.9	10.5	7.3	8.6	
12	9.6	12.9	14.2	11.8	8.1	10.1	
14	10.5	14.4	15.6	13.0	8.9	11.5	D Reach to a very small object or where accurate grasp is required.
16	11.4	15.8	17.0	14.2	9.7	12.9	
18	12.3	17.2	18.4	15.5	10.5	14.4	
20	13.1	18.6	19.8	16.7	11.3	15.8	
22	14.0	20.1	21.2	18.0	12.1	17.3	E Reach to indefinite location to get hand in position for body balance or next motion or out of way.
24	14.9	21.5	22.5	19.2	12.9	18.8	
26	15.8	22.9	23.9	20.4	13.7	20.2	
28	16.7	24.4	25.3	21.7	14.5	21.7	
30	17.5	25.8	26.7	22.9	15.3	23.2	
Additional	0.4	0.7	0.7	0.6			TMU per inch over 30 inches

TABLE II – MOVE – M

Distance Moved Inches	Time TMU A	B	C	Hand In Motion B	Wt. (lb.) Up to	Dynamic Factor	Static Constant TMU	CASE AND DESCRIPTION
3/4 or less	2.0	2.0	2.0	1.7				
1	2.5	2.9	3.4	2.3	2.5	1.00	0	
2	3.6	4.6	5.2	2.9				A Move object to other hand or against stop.
3	4.9	5.7	6.7	3.6	7.5	1.06	2.2	
4	6.1	6.9	8.0	4.3				
5	7.3	8.0	9.2	5.0	12.5	1.11	3.9	
6	8.1	8.9	10.3	5.7				
7	8.9	9.7	11.1	6.5	17.5	1.17	5.6	
8	9.7	10.6	11.8	7.2				B Move object to approximate or indefinite location.
9	10.5	11.5	12.7	7.9	22.5	1.22	7.4	
10	11.3	12.2	13.5	8.6				
12	12.9	13.4	15.2	10.0	27.5	1.28	9.1	
14	14.4	14.6	16.9	11.4				
16	16.0	15.8	18.7	12.8	32.5	1.33	10.8	
18	17.6	17.0	20.4	14.2				
20	19.2	18.2	22.1	15.6	37.5	1.39	12.5	
22	20.8	19.4	23.8	17.0				
24	22.4	20.6	25.5	18.4	42.5	1.44	14.3	C Move object to exact location.
26	24.0	21.8	27.3	19.8				
28	25.5	23.1	29.0	21.2	47.5	1.50	16.0	
30	27.1	24.3	30.7	22.7				
Additional	0.8	0.6	0.85					TMU per inch over 30 inches

Figure 7-6 MTM Times. (Copyright 1973 by the MTM Association for Standards and Research. Permission to reprint must be obtained from the MTM Association, 16–01 Broadway, Fairlawn, NJ 07410.)

TABLE III A — TURN — T

Weight	Time TMU for Degrees Turned										
	30°	45°	60°	75°	90°	105°	120°	135°	150°	165°	180°
Small — 0 to 2 Pounds	2.8	3.5	4.1	4.8	5.4	6.1	6.8	7.4	8.1	8.7	9.4
Medium — 2.1 to 10 Pounds	4.4	5.5	6.5	7.5	8.5	9.6	10.6	11.6	12.7	13.7	14.8
Large — 10.1 to 35 Pounds	8.4	10.5	12.3	14.4	16.2	18.3	20.4	22.2	24.3	26.1	28.2

TABLE III B — APPLY PRESSURE — AP

FULL CYCLE			COMPONENTS		
SYMBOL	TMU	DESCRIPTION	SYMBOL	TMU	DESCRIPTION
APA	10.6	AF + DM + RLF	AF	3.4	Apply Force
			DM	4.2	Dwell, Minimum
APB	16.2	APA + G2	RLF	3.0	Release Force

TABLE IV — GRASP — G

TYPE OF GRASP	Case	Time TMU	DESCRIPTION	
PICK-UP	1A	2.0	Any size object by itself, easily grasped	
	1B	3.5	Object very small or lying close against a flat surface	
	1C1	7.3	Diameter larger than 1/2″	Interference with Grasp
	1C2	8.7	Diameter 1/4″ to 1/2″	on bottom and one side of
	1C3	10.8	Diameter less than 1/4″	nearly cylindrical object.
REGRASP	2	5.6	Change grasp without relinquishing control	
TRANSFER	3	5.6	Control transferred from one hand to the other.	
SELECT	4A	7.3	Larger than 1″ x 1″ x 1″	Object jumbled with other
	4B	9.1	1/4″ x 1/4″ x 1/8″ to 1″ x 1″ x 1″	objects so that search
	4C	12.9	Smaller than 1/4″ x 1/4″ x 1/8″	and select occur.
CONTACT	5	0	Contact, Sliding, or Hook Grasp.	

EFFECTIVE NET WEIGHT			
Effective Net Weight (ENW)	No. of Hands	Spatial	Sliding
	1	W	W x F_c
	2	W/2	W/2 x F_c

W = Weight in pounds
F_c = Coefficient of Friction

Figure 7-6   (Continued).

250

TABLE V – POSITION* – P

CLASS OF FIT		Symmetry	Easy To Handle	Difficult To Handle
1—Loose	No pressure required	S	5.6	11.2
		SS	9.1	14.7
		NS	10.4	16.0
2—Close	Light pressure required	S	16.2	21.8
		SS	19.7	25.3
		NS	21.0	26.6
3—Exact	Heavy pressure required.	S	43.0	48.6
		SS	46.5	52.1
		NS	47.8	53.4
SUPPLEMENTARY RULE FOR SURFACE ALIGNMENT				
P1SE per alignment: >1/16 ≤1/4''		P2SE per alignment: ≤1/16''		

*Distance moved to engage—1'' or less.

TABLE VI – RELEASE – RL

Case	Time TMU	DESCRIPTION
1	2.0	Normal release performed by opening fingers as independent motion.
2	0	Contact Release

TABLE VII – DISENGAGE – D

CLASS OF FIT	HEIGHT OF RECOIL	EASY TO HANDLE	DIFFICULT TO HANDLE
1—LOOSE—Very slight effort, blends with subsequent move.	Up to 1''	4.0	5.7
2—CLOSE—Normal effort, slight recoil.	Over 1'' to 5''	7.5	11.8
3— TIGHT—Considerable effort, hand recoils markedly.	Over 5'' to 12''	22.9	34.7
SUPPLEMENTARY			

CLASS OF FIT	CARE IN HANDLING	BINDING
1— LOOSE	Allow Class 2	————
2— CLOSE	Allow Class 3	One G2 per Bind
3— TIGHT	Change Method	One APB per Bind

TABLE VIII – EYE TRAVEL AND EYE FOCUS – ET AND EF

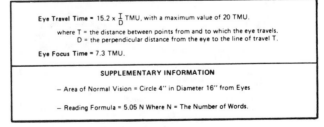

Eye Travel Time = $15.2 \times \frac{T}{D}$ TMU, with a maximum value of 20 TMU.

    where T = the distance between points from and to which the eye travels.
           D = the perpendicular distance from the eye to the line of travel T.

Eye Focus Time = 7.3 TMU.

SUPPLEMENTARY INFORMATION

— Area of Normal Vision = Circle 4'' in Diameter 16'' from Eyes

— Reading Formula = 5.05 N Where N = The Number of Words.

Figure 7-6 (Continued).

TABLE IX – BODY, LEG, AND FOOT MOTIONS

TYPE		SYMBOL	TMU	DISTANCE	DESCRIPTION
LEG–FOOT MOTION		FM	8.5	To 4"	Hinged at ankle.
		FMP	19.1	To 4"	With heavy pressure.
		LM__	7.1	To 6"	Hinged at knee or hip in any direction.
			1.2	Ea. add'l inch	
HORIZONTAL MOTION	SIDE STEP	SS__C1	*	<12"	Use Reach or Move time when less than 12". Complete when leading leg contacts floor.
			17.0	12"	
			0.6	Ea. add'l inch	
		SS__C2	34.1	12"	Lagging leg must contact floor before next motion can be made.
			1.1	Ea. add'l inch	
	TURN BODY	TBC1	18.6	——	Complete when leading leg contacts floor.
		TBC2	37.2	——	Lagging leg must contact floor before next motion can be made
	WALK	W__FT	5.3	Per Foot	Unobstructed.
		W__P	15.0	Per Pace	Unobstructed.
		W__PO	17.0	Per Pace	When obstructed or with weight.
VERTICAL MOTION		SIT	34.7	——	From standing position.
		STD	43.4	——	From sitting position.
		B,S,KOK	29.0	——	Bend, Stoop, Kneel on One Knee.
		AB,AS,AKOK	31.9	——	Arise from Bend, Stoop, Kneel on One Knee
		KBK	69.4	——	Kneel on Both Knees.
		AKBK	76.7	——	Arise from Kneel on Both Knees.

TABLE X – SIMULTANEOUS MOTIONS

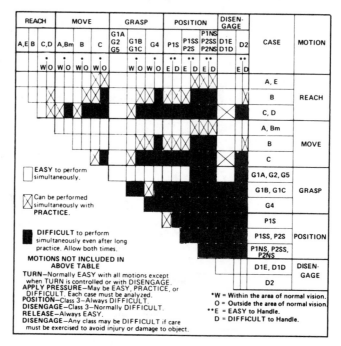

Figure 7-6 (Continued).

SUPPLEMENTARY MTM DATA

TABLE 1 — POSITION — P

Class of Fit and Clearance	Case of † Symmetry	Align Only	Depth of Insertion (per ¼")			
			0 >0≤1/8"	2 >1/8≤¾	4 >¾≤1¼	6 >1¼≤1¾
21 .150" — .350"	S	3.0	3.4	6.6	7.7	8.8
	SS	3.0	10.3	13.5	14.6	15.7
	NS	4.8	15.5	18.7	19.8	20.9
22 .025" — .149"	S	7.2	7.2	11.9	13.0	14.2
	SS	8.0	14.9	19.6	20.7	21.9
	NS	9.5	20.2	24.9	26.0	27.2
23* .005" — .024"	S	9.5	9.5	16.3	18.7	21.0
	SS	10.4	17.3	24.1	26.5	28.8
	NS	12.2	22.9	29.7	32.1	34.4

*BINDING—Add observed number of Apply Pressures.
DIFFICULT HANDLING—Add observed number of G2's.

†Determine symmetry by geometric properties, except use S case when object is oriented prior to preceding Move.

TABLE 1A — SECONDARY ENGAGE — E2

CLASS OF FIT	DEPTH OF INSERTION (PER 1/4")		
	2	4	6
21	3.2	4.3	5.4
22	4.7	5.8	7.0
23	6.8	9.2	11.5

TABLE 2 — CRANK (LIGHT RESISTANCE) — C

DIAMETER OF CRANKING (INCHES)	TMU (T) PER REVOLUTION	DIAMETER OF CRANKING (INCHES)	TMU (T) PER REVOLUTION
1	8.5	9	14.0
2	9.7	10	14.4
3	10.6	11	14.7
4	11.4	12	15.0
5	12.1	14	15.5
6	12.7	16	16.0
7	13.2	18	16.4
8	13.6	20	16.7

FORMULAS:
A. CONTINUOUS CRANKING (Start at beginning and stop at end of cycle only)

$$TMU = [(N \times T) + 5.2] \cdot F + C$$

B. INTERMITTENT CRANKING (Start at beginning and stop at end of each revolution

$$TMU = [(T + 5.2) F + C] \cdot N$$

C	=	Static component TMU weight allowance constant from move table
F	=	Dynamic component weight allowance factor from move table
N	=	Number of revolutions
T	=	TMU per revolution (Type III Motion)
5.2	=	TMU for start and stop

Figure 7-6   (Continued).

rest, moves and comes to rest again. Basic Motion Timestudy takes the following five factors into consideration in determining times:

1. Distance moved
2. Visual attention needed to complete motion
3. Degree of precision required in grasping or positioning
4. Amount of force needed in handling weight
5. Simultaneous performance of two motions.

The motions of BMT fall into one of three classifications:

Class A: Stopped without muscular control by impact with a solid object.
Class B: Stopped entirely by use of muscular control.
Class C: Stopped by use of muscular control both to control the slow-down and to end it in a grasping or placing action.

Recognition of a force factor is made. Whenever a heavy object must be handled or when friction must be overcome, added muscular effort is required.

MTM Variations

Universal Standard Data was born in Sweden in the early 1950s. An offshoot of MTM, Universal Standard Data is best suited for long-cycle operations.

Another variation of MTM is Master Standard Data. This system was developed in the 1950s by engineers working for Serge A. Birn Associates. As was the case with Universal Standard Data, Master Standard Data was concerned with speed of application without loss of accuracy.

The large number of systems developed during the 1950s and 1960s prompted the MTM Association to develop several derivatives of the standard Methods Time Measurement System. The basic system became known, officially, as MTM-1. MTM-2, synthesized from MTM-1, uses a number of motion combinations. With MTM-2, speed of application is increased but accuracy and method description are sacrificed.

Another synthesized system is MTM-3, in which there are only four basic motions and only 10 possible time values. Application of MTM-3 is gauged to be seven times as fast as MTM-1 and three times as fast as MTM-2 (Maynard, 1971). Again, as was the case with MTM-2, accuracy is given up for speed of application.

SUMMARY

"One problem with predetermined time systems is that they are not automatic; that is, analyst judgment is required. Different analysts get different times for the same job because of interpretations of the various rules" (Konz, 1979, p. 140). Predetermined time systems are not intended to replace time standards set by time study. They are but another of the tools used to measure and hopefully, increase productivity. They can accomplish this goal however, only when each system is applied by an analyst trained in the rules, methods, and procedures of that system. Significant harm in terms of invalid standards, inconsistent standards, and disbelief of the entire analysis procedure may result if these tools are misused. Their proper use, however, can be a tremendous aid to management.

REVIEW QUESTIONS

1. What is a *predetermined time system?*
2. What are the major uses of predetermined time systems?
3. What are the four major advantages to using predetermined time systems?
4. What are the limitations of using predetermined time systems?
5. Why are methods important when using predetermined time systems?
6. Describe the procedure for setting standards with a predetermined time system.
7. What are some major systems commonly used?
8. Predetermined time systems are believed to be useful in settling grievances. Explain.
9. Predetermined time systems are believed to be useful in methods analysis and improvement. Explain.

PRACTICE EXERCISE

Proper application of any predetermined time system requires considerable training with the system. It is not the intent of this book to create expert applicators of any predetermined time system. Therefore, no practice problems are provided.

CHAPTER 8

Work Sampling

OBJECTIVES

After completing this chapter, the reader should be able to

- Define *work sampling*.
- Describe the standards set with work sampling.
- Describe the types of work that work sampling is most effective in setting standards for.
- Describe the procedure involved in conducting a work sampling study.
- Establish activity standards based on work sampling studies.

INTRODUCTION

For many jobs, it is impractical to set performance standards or productivity measures by the methods described thus far in this text. There are many jobs where time study, standard data, and predetermined time standards just are not useful, productive ways to set standards. These methods are most useful in setting time standards for jobs that have relatively low- or short-cycle times and a large number of repetitions. Studies can be made relatively quickly and standards established, either based on actual work performed, as with time study, or based on elemental descriptions and synthesized as with predetermined time systems. Some jobs, however, do not have short-cycle times or high repetition rates. Jobs, such as those in maintenance, materials handling, and many office-type jobs tend to have rather long cycle times. The tasks performed are generally not repeated on a regular basis.

When these jobs are studied to determine their contribution to the firm's productivity, a different method is required. Direct observation of these jobs, with sufficient measurements to present a reliable picture of the work performed, would take a tremendous amount of time. As a result, the cost of making such studies would be tremendous. Therefore, jobs that have long cycles and low frequency of repetition require a different type of analysis.

DEFINITION OF WORK SAMPLING

Work sampling, or activity analysis, is the process of making sufficient random observations of an operator's activities to determine the relative amount of time the operator spends on the various activities associated with the job. Although it is not the express purpose of work sampling to determine how long specific tasks should take, work sampling data, coupled with historical production data, can be used to develop standards. The major goal of work sampling, however, is to determine how long, or how much of the work day, is spent on specific types of work. Work sampling may identify the fact that certain operators spend a large portion of their time waiting for work, or performing paper work tasks, or even performing activities that are not included in their job descriptions. This analysis of the productivity of various job classifications can provide tremendous insight into ways to increase an organization's overall productivity.

Although the definition of work sampling sounds straightforward enough, it needs careful examination. The first key word in our definition is *sufficient*. One of the basic foundations of statistical sampling theory is the concept that the larger the sample size, the better, or more accurate, the results will be. Obviously, if we observed a worker all the time, our summary of the way the worker utilized his or her time would be perfectly accurate. But, if we spent all of our time measuring or observing the performance of the worker, our time would not be spent very productively. In work sampling, a sufficient number of observations must be made to be confident that the results accurately summarize the work performed. There are statistical formulas to help determine how many observations should be made. This procedure will be explained later in this chapter.

Our explanation of a seemingly straightforward concept seems to contain two ambiguous terms. We use the term *accurate*. Accuracy is relative. An archer is called accurate if every arrow shot hits the bull's eye. However, this accuracy depends on the bull's eye's size. The bigger the bull's eye, the easier it is to hit. Realistically, we recognize that a bull's eye that is too large threatens the credibility of the archer to call himself accurate. Common industrial levels of accuracy that are used when estimating the percentage of time spent on particular activities are ± 10 percent, ± 5 percent, and ± 1 percent of the time spent on the activity.

Our explanation also includes the word *confident*. If our archer wanted to be 100 percent confident of hitting the bull's eye, he would shoot at targets containing nothing but bull's eyes, or better yet, he would shoot first and provide the bull's eye after the fact. If we wanted to be 100 percent confident that we were 100 percent accurate we would observe an operator 100 percent of the time. As we have already emphasized, this idea is impractical. Instead, we acknowledge ahead of time that there is a chance we will make a mistake. There is a chance we will develop an analysis that is not within the desired accuracy levels. By using statistical sampling theory, we can specify ahead of time how often we will be correct and how often we will be incorrect in our estimate. Common confidence levels are 90, 95, and 99 percent. So, based on our analysis of the worker's activities, we believe that our results will be accurate 90, 95, or 99 times out of 100. As can readily be surmised, the more confident we want to be, the more often we will have to observe the job. The more we observe the job, the higher the cost of analysis.

The number of observations that an analyst must make of a particular job also depends on how much time is devoted to a particular task. The less time an operator spends doing a particular task, the more observations that will be required to ensure that the task is measured properly relative to its contribution or use of the operator's time. If only three percent of the time is really devoted to a task, then many more observations would be required to pin down the three percent contribution than if 30 percent of the time were spent on the task. If, in conducting a work sampling study, we knew ahead of time the relative amount of time spent on each task there would be no need for the study. Instead, we would establish the time we believe will be spent on the shortest of the major tasks performed. Never would we expect to measure a task that occupied less than one percent of an operator's time. As a practical matter, the number of observations required when the proportion of time is less than about 15 percent would probably make the cost of the study prohibitive.

Again, there are two ways to look at accuracy, in terms of work sampling. One is relative accuracy, where the range is a percentage of the observed proportion of time spent on an activity. For example, if an activity required 10 percent of the day, 48 minutes would be spent performing the activity. Plus or minus 10 percent accuracy would mean plus or minus 4.8 minutes. On the other hand, if an activity required 25 percent of the day, or 120 minutes, the same plus or minus 10 percent would be plus or minus 12 minutes. Thus the same relative accuracy gives a larger acceptable range. Absolute accuracy establishes the range in minutes. To be within plus or minus 4.8 minutes on all estimates requires differing sample sizes, depending on the estimated proportion of time spent on the activity.

Statistically, sufficiency depends on the accuracy, relative or absolute, and confidence desired. Tables 8-1(a), (b), and (c) tabulate the required number of observations for relative accuracy, confidence, and time spent on a job. Table 8-2 shows the reading required for absolute accuracy.

These tables have been calculated using standard statistical sampling formulas whose use is illustrated at the end of this chapter.

Collecting sufficient observations of a worker's activities in and of itself is not enough to guarantee that an accurate representation of the kinds of work done is available. The sampling must be representative of all work performed and the samples must be taken from all the possible conditions that the operator faces on the job. If some activity occurs only at certain locations or certain times of the year, this activity must not be permitted to bias the results. Neither too much nor too little emphasis should be placed on these events. The "famous" *Literary Digest* poll, which up until 1936 predicted, fairly accurately, U.S. Presidential elections by sampling the population, predicted President Roosevelt's defeat. The usual reason given for the incorrect projection is that, in sampling the 1936 electorate, the magazine only contacted households with telephones. In 1936, the sample of homes with telephones was not representative of the population as a whole and the results did not accurately reflect the thinking of the electorate as a whole. Only part of the population was sampled because not everyone had a telephone in 1936.

When work sampling is employed, it is imperative that the operator perform all activities normally found on the job. Just as importantly, the operator must not perform tasks that are alien to the job. Many companies, in the few days of work just preceding Christmas, sanction, although unofficially, employee parties. A work sampling study conducted during the week just prior to Christmas might incorrectly conclude that the operator spends much of the working time munching Christmas goodies.

Thus, it is most important that the work sampling study be conducted when working conditions are truly *representative* of the normal working conditions found in a given job.

TABLE 8-1(a)   SAMPLE SIZES FOR 90% CONFIDENCE

P	1%	5%	10%
1	2,678,960	107,159	26,790
2	1,325,950	53,038	13,259
3	874,948	34,998	8,749
4	649,446	25,978	6,494
5	514,145	20,565	5,141
6	423,944	16,957	4,239
7	359,515	14,381	3,595
8	311,193	12,448	3,112
9	273,609	10,944	2,736
10	243,542	9,742	2,435
15	153,341	6,133	1,533
20	108,241	4,330	1,082
25	81,181	3,427	812
30	63,141	2,526	631
35	50,255	2,010	503
40	40,590	1,624	406
45	33,074	1,323	331
50	27,060	1,082	271
55	22,140	886	221
60	18,040	722	180
65	14,571	583	146
70	11,597	464	116
75	9,020	361	90
80	6,765	271	68
85	4,775	191	48
90	3,007	120	30
95	1,424	69	14
99	273	11	3

In addition to making sure the sample is representative, it is just as important that the sample be taken at random times throughout the work day. If the analyst makes all of the observations at the same time every day, the subjects will come to expect the observation at that time and adjust their behavior accordingly. When this adjustment is made, the sample is no longer representative and the results obtained are not valid. A random sample can be generated or used to ensure that the time observations that are made are truly random.*

CONDUCTING A STUDY

It is recommended that a uniform procedure be followed each time a work sampling study is performed. Although there is no fixed procedure, the following guidelines may be helpful (Barnes, 1957).

---

*The April 1981 issue of *Industrial Engineering* magazine includes a minicomputer program that aids in scheduling work sampling observations.

TABLE 8-1(b) SAMPLE SIZES FOR 95% CONFIDENCE

P	1%	5%	10%
1	3,803,180	152,127	38,032
2	1,882,380	75,295	18,824
3	1,242,120	49,684	12,421
4	921,984	36,879	9,220
5	729,904	29,196	7,299
6	601,851	24,074	6,019
7	510,384	20,415	5,104
8	441,784	17,671	4,418
9	388,428	15,537	3,884
10	345,744	13,830	3,457
15	217,691	8,708	2,177
20	153,664	6,147	1,537
25	115,248	4,610	1,152
30	89,637	3,585	896
35	71,344	2,854	713
40	57,624	2,305	576
45	46,953	1,878	470
50	38,416	1,537	384
55	31,431	1,257	314
60	25,611	1,024	256
65	20,685	827	207
70	16,464	659	165
75	12,805	512	128
80	9,604	384	96
85	6,779	271	68
90	4,268	171	43
95	2,022	81	20
99	388	16	4

Establish the Purpose

First, the objective of the study should be established. Work sampling can be used to determine an overall perspective on the work done. It can be used to determine a more precise analysis of the time spent on various work elements or it can be used in conjunction with production records to set performance standards. The analyst must establish the use of the results before making the study.

Identify the Subjects

Second, the people performing the task under consideration must be identified. If general office work is being studied with the objective of determining overall productivity, the appropriate employees should be specified; in larger companies, specific job classifications should be identified. Likewise, if a study of machine tools utilization is to be performed, the specific tools that will be studied should be specified. The workers and supervisors involved must naturally be informed of the nature of the study.

TABLE 8-1(c)   SAMPLE SIZES FOR 99% CONFIDENCE

P	1%	5%	10%
1	5,356,170	214,247	53,562
2	2,651,040	106,041	26,510
3	1,749,320	69,973	17,493
4	1,298,470	51,939	12,985
5	1,027,950	41,118	10,280
6	847,610	33,904	8,476
7	718,794	28,752	7,188
8	622,182	24,887	6,222
9	547,039	21,882	5,470
10	486,925	19,477	4,869
15	306,582	12,263	3,066
20	216,411	8,656	2,164
25	162,308	6,492	1,623
30	126,240	5,050	1,262
35	100,477	4,019	1,005
40	81,154	3,246	812
45	66,126	2,645	661
50	54,103	2,164	541
55	44,266	1,771	443
60	36,069	1,443	361
65	29,132	1,165	291
70	23,187	927	232
75	18,034	721	180
80	13,526	541	135
85	9,548	382	95
90	6,011	240	60
95	2,848	114	28
99	546	22	5

Identify the Measure of Output

The third step in making the study is the identification of the measure of the output produced or the types of activities performed on the jobs being studied. This step is especially important if the objective of the study is to measure productivity with the intent of setting a standard. For example, in an insurance office, working time might be compared with the number of claims processed per day.

Establish a Time Period

Fourth, the time period during which the study will be conducted must be established. Starting and stopping points for the study must be defined as well. The longer the period the better, but this constraint will be counterbalanced by the cost of making the study. Whatever period is specified, the time allowed should be sufficient to be representative of

TABLE 8-2(a) SAMPLE SIZE FOR 90% CONFIDENCE WITH ABSOLUTE ACCURACY SHOWN

P	±.01	P	±.05	P	±.10	P	±.25
1	268	1	11	1	3	1	0
2	530	2	21	2	5	2	1
3	787	3	31	3	8	3	1
4	1039	4	42	4	10	4	2
5	1285	5	51	5	13	5	2
6	1526	6	61	6	15	6	2
7	1762	7	70	7	18	7	3
8	1992	8	80	8	20	8	3
9	2216	9	89	9	22	9	4
10	2435	10	97	10	24	10	4
11	2649	11	106	11	26	11	4
12	2858	12	114	12	29	12	5
13	3061	13	122	13	31	13	5
14	3258	14	130	14	33	14	5
15	3450	15	138	15	35	15	6
16	3637	16	145	16	36	16	6
17	3818	17	153	17	38	17	6
18	3994	18	160	18	40	18	6
19	4165	19	167	19	42	19	7
20	4330	20	173	20	43	20	7
21	4489	21	180	21	45	21	7
22	4644	22	186	22	46	22	7
23	4792	23	192	23	48	23	8
24	4936	24	197	24	49	24	8
25	5074	25	203	25	51	25	8
26	5206	26	208	26	52	26	8
27	5334	27	213	27	53	27	9
28	5455	28	218	28	55	28	9
29	5572	29	223	29	56	29	9
30	5683	30	227	30	57	30	9
31	5788	31	232	31	58	31	9
32	5888	32	236	32	59	32	9
33	5983	33	239	33	60	33	10
34	6072	34	243	34	61	34	10
35	6156	35	246	35	62	35	10
36	6235	36	249	36	62	36	10
37	6308	37	252	37	63	37	10
38	6375	38	255	38	64	38	10
39	6438	39	258	39	64	39	10
40	6494	40	260	40	65	40	10
41	6546	41	262	41	65	41	10
42	6592	42	264	42	66	42	11
43	6632	43	265	43	66	43	11
44	6668	44	267	44	67	44	11
45	6697	45	268	45	67	45	11
46	6722	46	269	46	67	46	11
47	6741	47	270	47	67	47	11
48	6754	48	270	48	68	48	11
49	6762	49	270	49	68	49	11
50	6765	50	271	50	68	50	11

TABLE 8-2(b) SAMPLE SIZE FOR 95% CONFIDENCE WITH ABSOLUTE ACCURACY SHOWN

P	±.01	P	±.05	P	±.10	P	±.25
1	380	1	15	1	4	1	1
2	753	2	30	2	8	2	1
3	1118	3	45	3	11	3	2
4	1475	4	59	4	15	4	2
5	1825	5	73	5	18	5	3
6	2167	6	87	6	22	6	3
7	2501	7	100	7	25	7	4
8	2827	8	113	8	28	8	5
9	3146	9	126	9	31	9	5
10	3457	10	138	10	35	10	6
11	3761	11	150	11	38	11	6
12	4057	12	162	12	41	12	6
13	4345	13	174	13	43	13	7
14	4625	14	185	14	46	14	7
15	4898	15	196	15	49	15	8
16	5163	16	207	16	52	16	8
17	5420	17	217	17	54	17	9
18	5670	18	227	18	57	18	9
19	5912	19	236	19	59	19	9
20	6147	20	246	20	61	20	10
21	6373	21	255	21	64	21	10
22	6592	22	264	22	66	22	11
23	6803	23	272	23	68	23	11
24	7007	24	280	24	70	24	11
25	7203	25	288	25	72	25	12
26	7391	26	296	26	74	26	12
27	7572	27	303	27	76	27	12
28	7745	28	310	28	77	28	12
29	7910	29	316	29	79	29	13
30	8067	30	323	30	81	30	13
31	8217	31	329	31	82	31	13
32	8359	32	334	32	84	32	13
33	8494	33	340	33	85	33	14
34	8621	34	345	34	86	34	14
35	8740	35	350	35	87	35	14
36	8851	36	354	36	89	36	14
37	8955	37	358	37	90	37	14
38	9051	38	362	38	91	38	14
39	9139	39	366	39	91	39	15
40	9220	40	369	40	92	40	15
41	9293	41	372	41	93	41	15
42	9358	42	374	42	94	42	15
43	9416	43	377	43	94	43	15
44	9466	44	379	44	95	44	15
45	9508	45	380	45	95	45	15
46	9543	46	382	46	95	46	15
47	9569	47	383	47	96	47	15
48	9589	48	384	48	96	48	15
49	9600	49	384	49	96	49	15
50	9604	50	384	50	96	50	15

TABLE 8-3 RANDOM NUMBERS BETWEEN 0700 AND 2000

1395	1038	1807	0948	1235	1477	0929	1928
1294	1638	1444	1190	1447	1009	1657	1240
1545	1891	0730	0729	1777	1317	1598	1149
1205	1025	0859	0833	1689	1556	0812	1254
0704	1563	1549	0721	1299	1829	1955	1338
1621	0727	1565	1137	1186	0766	1458	0718
1581	1795	1108	0839	1106	1645	1746	0851
1310	1778	1023	1152	1413	0812	1344	1954
1456	0954	1452	1523	1585	1735	1686	1057
0837	1306	1779	1447	1051	1538	1617	1414
0735	1876	1088	0732	1552	0724	1401	0727
1630	1036	1703	0804	1360	1372	1218	1940
1721	1334	0849	1765	1427	0734	1413	1723
1859	1607	1012	1266	0922	1602	1038	1103
0955	0970	0831	0917	1877	1649	0842	1624
0916	0941	1338	0802	1750	1536	1116	1259
1498	1529	1992	1725	0817	1143	1663	1547
1318	0834	1812	1593	0819	1494	1673	1658
1133	1655	1502	1261	1877	1806	1649	1887
1801	1232	1124	1373	1928	1546	0740	1985
1934	1102	1982	1918	1953	0885	1018	1787
1324	1395	1623	1627	1387	1017	1893	1481
1833	1735	1097	1979	1358	1156	1092	1150
0870	1475	0751	1383	0759	1869	1689	1199
1503	1650	1720	1341	1404	1563	1857	0816
1326	1422	1638	1627	1878	1417	0804	1691
1532	0707	0728	1763	1513	1001	0721	1881
0938	1237	1214	0702	1403	1784	1487	1021
1394	1133	1632	1784	0936	1812	1691	1656
0731	1119	1459	0752	1845	1723	1849	0810
1957	0906	1500	0768	1144	1712	1994	1363
0822	1840	1748	1628	0734	1755	1399	1824
1470	1777	1015	1833	0742	1429	1462	1531
0957	0752	1335	1551	0855	1977	1581	1555
1994	1700	1296	0859	1351	1295	0930	1427
0751	1549	1111	1219	1282	1859	1350	1716
0713	0747	1163	1189	1557	0758	0952	2000
1178	0830	1350	0914	1332	1065	1935	0917
1280	1404	1038	1028	1354	0916	0989	1591
1648	0801	1319	1833	0726	1932	1989	1717
0874	0835	1538	1641	0713	1119	0963	1367
0810	1222	1944	1060	1554	1462	0941	0839
1507	0748	1985	0997	0817	1638	1291	1448
1992	1888	1156	1448	1112	0929	1672	0711
1612	0853	1710	1916	1646	1138	1349	1259
1556	0902	1689	0737	1717	1251	1245	1002
1523	1940	1574	1550	0922	1642	0754	0862
1872	1227	1833	1345	0972	0944	0948	0918
1510	1034	1352	1718	1268	1818	1605	1581

the work normally performed on the type of job that is being studied. For example, if a payroll clerking operation is being studied, more activity might be required, or at least a different type of activity be performed, when it is close to payday.

Define the Activities

This step involves defining the activities that are performed by the people under study. This specification may be a very broad definition, such as the definition used in a machine utilization study, including only the categories of working, idle—not working, and idle—mechanical breakdown. Or, it might include a listing of 10 or more specific work activities such as repairing drill presses in the production area or checking out tools at the tool crib.

Determine the Number of Observations Needed

After the work elements are defined, the number of observations for the desired accuracy at the desired confidence level must be determined. The sample size, remember, is dependent on the percentage of time believed to be spent on the major work element requiring the smallest portion of the operator's time. If a reasonable guess cannot be made, then a trial study of perhaps 20 to 40 observations should be made to get an estimate of this portion. These initial observations should be included with the rest of the observations taken during the remainder of the work study.

Schedule the Observations

Once the number of observations required has been determined, either from appropriate statistical calculations or from tables such as those provided in Tables 8-1 through 8-3, the actual observations must be scheduled. Typically, the analyst will assign an equal number of observations each day during the course of the study. For example, if 800 observations are required and 20 work days are established as an appropriate observation time, 40 observations should be recorded each day. A random number table can be used to establish the random times for each observation.

Inform the Personnel Involved

Before the study is actually performed, the personnel involved should be informed about the objective of the study and the methodology that will be employed. As in any productivity measurement study, this part of the procedure is very important. Workers and their supervisors might think that they personally are being measured rather than the work they are doing. As we shall see, there are some special problems that may arise during a work sampling study and potential problems should not be compounded by ignoring this part of the process.

Record the Raw Data

The next and perhaps the easiest part of any work sampling study is the actual recording of the raw data. Although this recording can be performed by anyone, it is desirable that a trained analyst be employed. It is also imperative that the observations be made at exactly the same location every time. Failure to be consistent in this manner may bias the results.

Summarize the Data

After the data have been collected, they must be summarized. This process simply involves totaling the observations made for each work element and calculating the percentage of time actually spent on that particular task. (This step may be easily adapted to computer analysis.) If a standard is to be set, this percentage of time is compared with the output for the time of the study and the time per unit of output is calculated. For example the summarization of work sampling data may result in the statement that a machine under study is utilized or productive 87 percent of the time. The formal statement of the result might be as follows: "We are 95 percent confident that, within ±5 percent, the machine is productive 87 percent of the time." Or it may yield a statement that is much more specific, such as, "We are 90 percent confident that the typist spends 18 percent of his time, within one percent, typing thank you letters in response to orders placed," or it may be expressed as a standard, "We are 99 percent confident that the time required to type a thank you note is .072 hours." Regardless of how the results of the work sampling study are expressed, the relative number of observations made of the particular activity divided by the total number of observations is the basic measure of the work performed during the work sampling study.

The preceding paragraphs represent the general procedure for performing a work sampling study. Before examining the advantages and limitations of this style of work measurement, we shall look at some examples. The first is a work sampling study that was used to set very broad or general productivity standards; the second is a study used to set much more definitive standards on a precisely defined set of work elements; and the third is the setting of standards based on work sampling and historical production records.

Table 8-3 shows random numbers between 0700 and 2000. It contains four-digit numbers representing times in a 24-hour clock that occur in completely random order. The use of this random number table will be illustrated in the following examples.

Example: Cash Register Operation

The cash register operator at the ABC Store was always complaining that he was overworked. He felt that another clerk had to be employed by ABC to ease the workload. Before consenting to his demands, the IE for the ABC Store decided it would be appropriate to measure the utilization of the cash register operator. The objective of the study would be to determine the utilization or the percentage of time the cash register operator spent on productive activities as opposed to the time spent not doing productive, i.e., "checking-out" work.

The lone operator was to be observed. The only activities that seemed to be pertinent, or productive, were if the operator was working, that is, ringing up prices, bagging purchases, or collecting money, or if the operator was idle, waiting for the next customer to visit the check-out station.

As a practical consideration, the analyst selected a one-month analysis period for conducting the study. This time period, she believed, would be representative of the conditions normally found in the business. The peaks of activity around paydays would be balanced by the valleys of activity during the days just before paydays.

The analyst wanted to be 99 percent confident that her time estimate was within five percent of the true time spent on each activity, and, more specifically, she wanted her estimate to be within five percent of the actual time the cash register operator was working. Using Table 8-2 for these conditions, coupled with an estimate that the check-out counter was really working 75 percent of the time (based on an initial trial set of observations), a sample size of 721 was specified. Based on the 20 working days typically found in a four-week month, and subtracting the 21 observations already made in determining the initial time estimate, 35 observations would have to be made to have a meaningful set of results. A random number table, such as Table 8-4, can, when used properly, be used to identify the 35 observation times made during each working day.

At the ABC Store, the normal work hours are from 10:00 a.m. until 6:00 p.m. In 24-hour time, these times would be read as 1000 and 1800. Reading random numbers from Table 8-3, beginning at the randomly selected first row and third column, the first 34 observation times are shown in Table 8-4. The 34 times scheduled for this day are similar to, but are not exactly the same as the observations scheduled for each of the 20 days of the study.

Once the analyst develops the schedule and identifies exactly when each of the 700 observations is to be made, she must explain the procedure to the operator and the supervisor. In this case, the explanation would probably follow the approach of assuring the operator that sufficient data were needed to document (or refute) the operator's claim that

TABLE 8-4   25 SAMPLE TIME OBSERVATIONS
(From Table 8-3, First Row/Third Column)

In order, from Table 8-3			
1444	1124	1335	1152
1549	1623	1111	1523
1108	1720	1350	1447
1023	1638	1038	1725
1452	1214	1319	1627
1703	1632	1538	
1012	1459	1156	
1338	1500	1710	
1502	1748	1352	
	1015	1137	

TABLE 8-5 SAMPLE DAY'S SCHEDULE
AND OBSERVATION RECORDING

Time	Activity	Time	Activity
1012	Working	1500	Working
1015	Idle	1502	Idle
1023	Working	1523	Working
1038	Working	1538	Working
1108	Working	1549	Working
1111	Idle	1623	Working
1124	Working	1627	Idle
1137	Working	1632	Idle
1152	Working	1638	Working
1156	Idle	1703	Working
1214	Working	1710	Working
1319	Working	1720	Working
1335	Working	1725	Working
1338	Working	1748	Idle
1350	Working		
1352	Idle		
1444	Working		
1447	Working		
1452	Working		
1459	Working		

extra help was necessary. Following the explanation, the next 20 days would be spent making observations.

Each observation made was recorded when the analyst was at the same location in the store. Table 8-5 shows the tabulated data for the first day. Table 8-6 shows the summary of the observations for all 721 viewings of the check-out operator's performance during the study. Also shown in Table 8-6 is the percentage of time spent on each activity. As can be seen, the cash register operator was busy 79 percent of the time and idle 21 percent of the time. Whether these data provide sufficient justification to hire another worker and purchase the required equipment is a decision that must be made. But at least the decision can be made based on reliable data, not merely the supposition of a very biased operator!

A common area where work sampling is used to set standards is in maintenance (Candy, 1982).

TABLE 8-6 SUMMARY OF WORK
SAMPLING DATA

Activity	Number of observations	Percent of time
Working	570	79
Idle	151	21
TOTAL	721	100

Blivet Manufacturing wants to measure the productivity of its maintenance workers. (These are part of Blivet's indirect labor costs.) A common belief among Blivet employees is that the maintenance worker spends most of his or her time going back to the tool room to obtain the correct tool. The Industrial Engineering Department wants to document just how the maintenance workers really spend their time.

Blivet's IE has easily identified the objective of the proposed study and the people doing this type of work. After consultation with the appropriate supervisory personnel, it is decided that a quantifiable measure of output for a maintenance worker is number of jobs completed. The type of jobs the Blivet maintenance employees perform are defined as follows:

- *Minor Repair*—Performing minor repair and/or replacement tasks such as replacing burned-out light bulbs, changing air conditioning filters, or replacing various restroom supplies.

- *Preventive Maintenance*—Performing routine maintenance tasks on various production equipment prior to (or without) the equipment breaking down. Much of this work is performed at specified intervals. Examples include changing oil and lubricating fork lift trucks after each 2000 hours of operation or rebuilding the control system of a machine after 10,000 hours of operation.

- *Emergency Repair*—Performing nonroutine repair tasks that occur when a piece of production equipment breaks down. Breakdowns are not predictable, nor are the causes of breakdowns foreseeable. Production breakdowns have a high priority for use of available maintenance employees.

- *Housekeeping*—Performing routine tasks associated with the general upkeep of the facility. Examples of this activity include floor sweeping, aisle painting, and trash removal.

- *Record Keeping*—Maintenance employees must process the appropriate paperwork when preventive maintenance and emergency repair tasks are completed. This paperwork includes material requisitions as well as records of work performed.

- *Tool Keeping*—Each maintenance worker is responsible for maintaining his or her own supply of tools. These tools must be in working order and available if needed for an emergency repair.

- *Personal*—Each maintenance worker at Blivet spends some time each day taking care of personal needs.

- *Idle*—Not all of a maintenance worker's time is spent performing maintenance activities. Not all time is or necessarily should be spent on routine tasks. There are times when the maintenance workers are waiting for job instructions, and other times when they are literally sitting around and "waiting for something to break."

Having defined the eight maintenance activities, the Blivet analyst must determine how many observations to make and when to schedule them. The analyst wants to complete the study within 30 working days and to be 90 percent confident that the results are accurate to within 10 percent of the true time spent on the activities. Table 8-1 shows the appropriate number of observations to make, when the estimated proportion of time spent on the shortest of the major tasks is eight percent, to be 3112. To complete this many readings during the required time span means that 104 observations must be made for each of the 30 days. Therefore, in an eight-hour work day, 12 or 13 observations must be made every hour. When trying to pin down activities of such relatively short duration with work sampling, a high number of observations must always be made. In this case, by the time the analyst makes the required visits to the work area, virtually the entire day will be scheduled. Unless a number of workers can be observed simultaneously, the economy of making such a study may be difficult to justify. Presuming that this study is justifiable, the random numbers for the first hour's observation are

```
0849    0804    0819
0831    0802    0855
0833    0859    0812
0839    0817    0842
```

These numbers were taken from Table 8-3, beginning in the fifth row and third column. Similar schedules must be developed for each hour for each of the 30 days of the study.

Once all of this preliminary work has been completed, the IE for Blivet must explain, to the maintenance employees that will be affected, just exactly what will be happening during the study. After the study has been explained and the natural "spying" fears have been allayed, the actual study is performed. Table 8-7 shows the number of observations made for each. Because output units of jobs completed was selected, it is noted that work activities that led to completed jobs were tallied and that each separate job was tallied only one time. From the results shown in Table 8-7, it can be seen that the Blivet Company's maintenance employees spent more of their time in preventive maintenance than in any other activity. As a result, it should be expected that the bulk of the job tickets turned in as complete should be for jobs of this type, thus squelching the rumor that all of the maintenance workers' time is spent getting the right tool.

TABLE 8-7   MAINTENANCE WORK SAMPLING RESULTS

Activity	Frequency	Percentage
Minor Repair	812	26
Preventive Maintenance	1107	36
Emergency Repair	214	07
Housekeeping	444	14
Record Keeping	129	04
Tool Keeping	44	02
Personal	162	05
Idle	200	06

Example: insurance claims

Earlier, we mentioned that a particular insurance office measured its output in the number of claims filed and processed. During the past year, which had 250 working days, this particular insurance office processed 30,000 claims. The office employed four clerks whose primary responsibility was to process claims. A preliminary analysis of the data might indicate that each clerk was supposed to process 30,000/250/4 or 480 claims per day. However, to assume this number might be unreasonable because it would assume that those clerks did nothing but process claims. It might be much more realistic to assume that a clerk could process more than 480 claims in a day if all the clerk really did was process claims. To determine a better estimate of the standard time required to process a claim, a work sampling study was performed.

The following work activities were identified for the clerks under study:

Work on Claims
Answer Telephone
Converse with Customers
Wait on Researchers
Idle

A work sampling study over a period of 50 working days was performed. The percentage of time spent on each activity was determined to be:

Work on Claims	60%
Answer Telephone	14
Converse with Customer	10
Wait on Researchers	10
Idle	6

As the study indicates, the clerk spends only 60 percent of the time working on claims. Despite this relatively low amount of time, the typical clerk still processes an average of 480 claims per eight-hour work day. Because only 60 percent of the clerk's time, or 4.8 hours each day, are spent processing claims, the average time required for each claim is only

$$4.8 \text{ hours}/480 \text{ claims per 8 hours} = .01 \text{ hours per claim}$$

meaning that, in an eight-hour day, if a clerk did nothing else, 800 claims would be processed. This information might help the office manager to better, more effectively, or more productively, utilize his clerical staff.

ADVANTAGES OF WORK SAMPLING

Work sampling has many advantages over some of the more traditional direct measurement techniques used for setting time standards. Some of the claims presented in favor of work sampling include (Heiland and Richardson, 1957):

- Work sampling provides a procedure that can be used to measure the productivity contribution of a number of tasks that might not be measurable by other means. The high-cycle time and low-repetition-rate jobs are very suitable for this type of analysis.

- Work sampling studies can be performed on a number of different operators simultaneously. Proper planning of the path followed to make observations can reduce the number of analysts required for a particular study. In direct observation methods, one analyst is required for each job studied.

- When determining time utilization, as work sampling often does, it is more economical to randomly sample the work performed than it is to continuously observe the work done. Theoretically, the analyst can perform other work between observations.

- A work sampling study usually is conducted over a longer period of time than a direct observation study. A longer period of time spent making observations helps to ensure that there is no "faking it" for the sake of the study. The results are likely to be more representative of the work actually performed.

- Work sampling studies, because they are based on random observations, can be completed at the discretion of the analyst. There is no need to finish the study while a particular job is being run. The studies can be interrupted with no loss of validity if some other more pressing need for the analyst comes along.

- Establishing sampling risks ahead of time, such as 95-percent confidence and accuracy within five percent, give work sampling results a degree of reliability that is quantified much easier than a direct observation procedure.

- Work sampling avoids the tediousness of time study and is, therefore, much easier on the observer.

- Many people feel very uncomfortable when they are watched for a continuous period of time. These self-conscious feelings are not only uncomfortable, but sometimes even distract from the work being performed. Because work sampling requires brief observations, it is often preferred by analysts making the study.

LIMITATIONS OF WORK SAMPLING

Although work sampling has many purported advantages, there are also a number of drawbacks. Some of the drawbacks often suggested include the following (Davidson, 1960):

- The results of a work sampling study are not quantifiable in the same sense that direct measurement results are. A work sampling observation, while it can be used in conjunction with historical production figures, is generally acknowledged to not give as good a standard set by a direct observation method such as time study. A work sampling study usually can only describe the general characteristics of operator performance, such as working or not working, or the general type work being performed. There is no way, in this method of study, to determine whether or not an operator is doing the proper work, working in the appropriate way, or using correct procedures.

- The economics of maintaining a study are questionable. Theoretically, anyone can make sample observations once the observation schedule has been established. On a practical basis, the observations are usually made by the engineer or the technologist who designed the study. The assertion that the analyst will use the in-between observation times productively has to be questioned. The analyst has to be thinking about when to make the next observation and, once the thought process is centered on another task, it is almost impossible to believe that the analyst will use the in-between time productively.

- Theoretically, a large number of operators performing different operations can be studied concurrently. However, as a practical matter, it is not cost effective for one analyst to try to observe operators who are located all over the plant. The analyst can spend the entire working day journeying from one observation site to another.

- Work sampling cannot provide the detailed analysis of work performed that the elemental analysis prepared in time study can, nor can it compete in detail with the descriptions prepared when a predetermined time system is used to set the standard. Although work sampling may tell us that a person is working 86 percent of the time, it does not precisely define what the work is nor does it serve as much of a description as to the general type of task completed. Even if a standard, such as .01 hours per form, is set, the work sampling standard does not contain much descriptive information.

- The statistics used in work sampling, while clear to someone with a basic knowledge of probability and statistics, are more than a little confusing to the layman. Convincing most people that random observations of a particular set of activities can result in a reliable prediction of how time is spent can be difficult in itself. Tie in a possible personal evaluation of performance and there will be such an outcry of disbelief that it may be most difficult to sell work sampling results. If the people being directly affected, the operator and supervisor, don't believe the results, the value of the study is minimal.

- Work sampling identifies the large components of specific jobs. No record is kept, however, of how the job is done or how the job should be done. When time study is used to set rates, each time some small part of the method changes, the standard must be re-evaluated. When productivity is analyzed using work sampling, the standard methods are not well-defined. Any change in method can have an unknown effect on the time required to complete a job.

- Sometimes, the analyst gets sloppy or lazy. Although this limitation is possible in any work measurement system, it is far more likely to occur when work sampling is used. For work sampling to be effective and reliable, the proper sample size must be observed. As a quick examination of Tables 8-1, 8-2, and 8-3 shows, when the conditions are right, a sample size in the hundreds of thousands, or even in the millions, might be specified. This requirement may not be economically practical. Although the time study analyst may know when enough observations have been made, work sampling does not offer that same feeling of confidence based on experience that other methods do. When economics are permitted to override the study itself, the results become suspect.

SUMMARY

In some instances, work sampling can be used to set standards at a lower cost and with more reliability than some other methods. Work sampling can be used to set standards and it can be economical.

Work sampling is effective, however, in determining the relative utilization of employees' time and in developing performance measures or standards for low-cycle and long-duration jobs. Work sampling can be used very successfully to set standards for jobs such as those found in maintenance.

As is the case with all methods of work measurement, the proper use can make each an effective tool. When used improperly, they can be barriers to competent productivity analysis.

SAMPLE SIZE DETERMINATION

The reliability of work sampling data is dependent on the sample observations made. In turn, the reliability of these observations depends on the number of observations made. The number of observations made depends on the following:

1. *Confidence*—The confidence associated with any sampling procedure depends on how often we desire the results to be truly indicative of the time spent on an activity. Generally, the greater the accuracy desired, the larger the sample size will be.

2. *Accuracy*—The accuracy associated with work sampling depends on how close we desire our observed estimates to be to the relative time actually spent performing the activity being observed. The closer we desire to be, the more observations we will be required to make.

3. *Percentage of Time Allocated*—The number of observations required for a particular task also depends on how much of the day is normally spent performing the task. The more time spent on an activity, the fewer the observations that should be required to identify that activity as a significant part of the job. A preliminary estimate of the activity that requires the least amount of time, or the smallest percentage of time, is required before the sample size can be determined. A preliminary study or sampling often is required to estimate this sample size.

After the confidence, accuracy, and percent utilization have been estimated, it is necessary to determine the precise sample size necessary to give the sampling results statistical significance. Although tables, such as Tables 8-1 and 8-2 are available, they were constructed only by careful use of the appropriate statistical formulas.

If we call Z the representative of the confidence level, A the accuracy desired, and P the preliminary estimate of the percentage of time spent performing a particular activity, then we can calculate n, the sample size, from the following formula:

$$n = Z^2(1 - P)/(P)(A^2)$$

Typical values for Z are

CONFIDENCE	Z
99.9%	3.250
99.0	2.575
95.0	1.960
90.0	1.645
80.0	1.245
75.0	1.151
50.0	.675

For 95-percent confidence, Z is equal to 1.96. If 10-percent accuracy or .10 accuracy is desired for a job believed to take 15 percent of the time as a minimum, the sample size is calculated as

$$n = (1.96)^2(.85)/(.15)(.10)^2$$
$$n = 2,177$$

For 95-percent confidence and 10-percent accuracy, we should take a random sample of 2,177 observations.

REVIEW QUESTIONS

1. Define *working sampling*.
2. What is meant by *random?*
3. What is meant by *confidence?*
4. What is meant by *accuracy?*
5. Describe the limitations of work sampling.
6. Describe the advantages of work sampling.
7. What types of activities should work sampling be used with to set standards?
8. What types of activities would work sampling never be used with to set standards?

PRACTICE EXERCISES

1. A work sampling study requires 95-percent confidence with five-percent accuracy. How many observations must be made?
2. Using the first row and first column in Table 8-3, determine the observation schedule for the first day of a 100-day schedule as described in Exercise 1 above.

3. Determine the utilization factor for the following tabulated work sampling observations:

ACTIVITY	NO. OF OBSERVATIONS
Working	852
Absent	116
Idle	22

4. Determine the utilization factor for the following tabulated work sampling observations:

ACTIVITY	NO. OF OBSERVATIONS
Typing	247
Filing	211
Dictation	100
Telephone	156
Coffee Break	18
Idle	45
Absent	18

CHAPTER 9

Physiological Work Measurement

OBJECTIVES

Upon finishing this chapter, the reader should be able to

- Describe the general procedure used to physiologically measure the work performed by an individual.
- Describe some of the current uses of data derived from such studies.
- Describe the productivity improvement implications for which physiological work measurement studies can be responsible.

INTRODUCTION

The work measurement procedures we have examined so far—time study, standard data (based on time study), predetermined motion time systems, and work sampling—have all evaluated worker productivity based solely on the time required to complete an operation or finish a job or certain number of jobs. This basis is, after all, the definition of *time standard*. It has been assumed that work performed can be translated into a production count or at least a summary of how much time was spent productively. The traditional time standard has been, for all intents and purposes, the most easily observed measure of work performed. From the time of Taylor, productivity has been measured this way. When standards are set in this traditional manner, an allowance is made for fatigue. Usually, this allowance is an estimate made by the analyst or, in some cases, it is negotiated between a labor union and company.

There is another way to measure the work performed by an individual. "If the human body is considered as a system in which chemical energy can be converted into mechanical energy, the output is muscular work. This work,* however, is not held in high regard in Western countries, and it is a general philosophy that it should be eliminated" (Singleton et al., 1971, p. 93). This alternate method of measuring work involves measuring the chemical energy consumed by the worker in performing the work. This measurement is done through a variety of physiological measures. The chemical energy used by a worker can be used to predict the amount of mechanical work that a worker is capable of doing when performing a specific task. Technology has advanced so that oxygen consumption,

---

*The *work* referred to here is physical work.

heart rate, body temperature, and other similar human physiological characteristics can be measured. Furthermore, analysis of the results of these types of measurements have shown that there is a strong relationship between some of these measurements and work performed. Additionally, Brouha (1960) determined that, "when mechanical work stops, physiological work continues above the resting rate until recovery is complete" (p. 3), meaning that the body keeps working even though actual physical work has ceased.

Although considerable research has been performed in this area, it is not the intent of this chapter to present a complete treatment of this topic. Rather, a brief overview of some of the methods used to physiologically measure work, a summary of some potential uses of data collected in this manner, and a discussion of some productivity implications will be presented. The reader who is truly interested in delving further into physiological work measurement is urged to use the bibliography at the end of the book as a starting point for additional research.

MEASURING PHYSIOLOGICAL WORK

Whatever measurement of physiological work is used, the objective of the measurement is to determine when a person is working and how hard the person is working. When people are resting, they have steady-state physiological characteristics. When physical work is performed, the observed characteristic, whether oxygen consumption, heart rate, or body temperature, increases. When work is completed, these characteristics take some time to return to normal or steady state. This period is commonly called *recovery*. Figure 9-1 shows a typical recovery curve.

For workers to produce more effectively, it is necessary for them to either work at such a pace that the physiological measures stay close to normal or steady-state rates or that they be given sufficient time to recover, physiologically speaking, once a task has been completed. Physiological studies can be performed to determine the standards required to meet these conditions. "There are a number of useful indices of physical effort: ventilation volume, energy consumption, heart rate, electromyography, and output of a force platform" (Konz and Cahill, 1971, p. 1).

The three most frequently used measures are heart rate, oxygen consumption, and metabolic measurement systems (Barnes, 1980). Figure 9-2 shows a piece of equipment typically used to perform such measurements. Whichever of these measurement devices is used, the general energy expenditure pattern resembles the one shown in Figure 9-1.

Heart Rate Measurement

Heart rate, or pulse measurements, are recorded electrically via electrodes affixed to the human body. The rate is recorded at regular intervals and a picture of heart rate as a function of time is obtained. Beats per minute is the common measure of heart rate. According to studies performed and documented by Konz and others, "For most tasks now used in industry, heart rate can be used to evaluate the physical cost associated with the task. If a person is 'calibrated' it is possible to estimate energy consumption from heart rate"

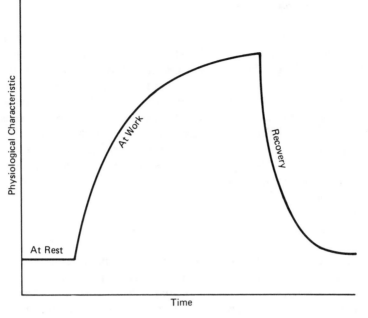

Figure 9-1 Recovery Curve.

(Konz and Cahill, pp. 2–10). In other words, pulse can be converted to an indication or measurement of work performed. Naturally, the actual process is more complicated but this is the goal of the study.

Figure 9-2 Example of a Physiological Measuring Device—The Metabolic Measurement Cart.

Oxygen Consumption

Oxygen consumption is also a common way to physiologically measure the energy expended by a person at work. The amount of oxygen used indicates how hard the individual is working. This amount is determined by measuring the oxygen content of the air the worker is breathing and then measuring the oxygen content of the air that the worker expels. The difference is used to determine the oxygen consumption rate, which is generally expressed in terms of number of liters of oxygen consumed per minute. Samples of expired air can be taken in specially designed air bags connected to face masks and the relative amount of oxygen present can be determined. One problem of gathering data in this way is in the actual gathering of the sample data. The masks are somewhat burdensome to wear and may interfere with the worker's normal performance of the task.

Metabolic Measurement

Metabolic measurement, performed similarly to the measurement of oxygen consumption, checks the expired air for other indicators of energy used such as carbon dioxide content. Again, as with the other measures, the purpose of this type of measurement is to determine the energy expended by a worker while performing a task. Ideally, the worker can be trained to perform the task at near to normal or resting levels. If this level cannot be accomplished, tasks that require a large consumption of energy can be identified and appropriate recovery periods can be specified to permit the worker to recover, physiologically, before proceeding with additional work.

Not surprisingly, some people are much better at muscular or physical work than others. According to Barnes (1980) studies were made of 2000 healthy college students. The ability of these students to withstand physiological stress varied by a factor of 10. A number of reasons, including physical fitness, age, and sex, were identified as potential sources for the variations.

When the muscles of the body have been overworked, they respond by becoming fatigued. This sensation is not the feeling of fatigue or of being tired that many people have at the end of a day; rather, it is characterized by aching and even pain in the muscles concerned (Edholm, 1967). Static work is much more fatiguing than rhythmic work. When muscles contract and expand, they relax and recover somewhat. When static work is performed there is little, if any, relaxation. Hence, little recovery and actual physical fatigue results (Edholm, 1967).

USES OF PHYSIOLOGICAL WORK MEASUREMENTS

Davis, Faulkner, and Miller (1971), in work performed at Eastman Kodak, identified several potential uses of work physiology in the industrial setting. Some of the applications included:

1. *Determining whether a particular job is within the physical capabilities of particular people.* This application might be considered the extreme case of fitting the individual to the job. In this case, the physiological requirements of a job must be established, hopefully in terms of one of the common measures. It also requires that all potential workers have their physiological abilities and capabilities measured. Although a significant amount of measurement must be performed, once the data are collected and used in the manner described, the required work can be performed more productively.

2. *Identifying the best method to perform a job.* Other things being equal, it is better to select a method for performing a job that uses less energy than some comparable method that requires the expenditure of additional physiological energy. Of course, the traditional concepts of work design must be followed in designing alternative ways to perform the job under consideration.

3. *Evaluating the work requirements of new or proposed jobs.* Simulation of new jobs within a laboratory, while the jobs are being performed under a variety of controlled test conditions, can be helpful in determining the physical demands that may be placed on the worker in an actual production situation. These data can then be used in fitting the job to the best-suited person.

4. *Ranking the jobs in terms of actual physical difficulty.* This application can provide the basis for evaluation when wage and salary criteria are evaluated. Most job evaluation systems include a factor that pays at least lip service to the concept of evaluating jobs based on difficulty. Physiological measures can provide quantitative support to various assertions about jobs and their relative physiological ranking. They are based on the assumption that performing more difficult jobs requires more physiological energy. This assumption may be especially true of jobs that are categorized as straight physical work.

Davis, Faulkner, and Miller (1971) went on to say, "The importance of obtaining quantitative data in the physiological responses of operators as they are actually performing their job is sometimes underestimated. It is practically impossible to estimate accurately the difficulty of a job merely by observing it" (p. 1173). Salvendy and Stewart (1975) substantiated this statement with some research results that showed that, ". . . biological measures . . . predicted significantly more effectively the production performance . . ." (p. 385).

Example: Carton Stapling

Davis et al. (1971) reported on an application at Eastman Kodak. There is a job that involves stapling corrugated shipping cartons. The packed cartons arrive at the staple operation at an uneven rate; sometimes several arrive almost simultaneously, sometimes none arrive for several minutes. When a carton arrives, the operator usually uses four staples, applied with a power stapler, to secure the top. The cartons are then placed on a pallet for later disposal. As needed, the operator removes loaded pallets with a hand truck and secures empty pallets for future use. Additional staples are procured as needed.

Based upon heart rate data collected, it was discovered that large energy expenditures were required by the operator. The heart rate also tended to increase during the day without sufficient recovery. One operator's heart rate was even measured as high as 140 beats per minute late in the work day. Evidently, a large amount of work was being performed although it was not obvious to the casual, or even to the trained, observer.

Two recommendations came out of the study. They fit the general types of information use suggestions presented earlier. First, only workers who were physically fit were assigned to this type of job. (Thus fitting the job to the worker and the worker to the job. Specific suggestions were made as to the conditions of physical fitness.) Second, because of the build-up of fatigue during the work day workers were rotated to this job throughout the day, providing ample opportunity for each individual worker to recover, physiologically speaking, at least (Davis et al., 1971).

Example: Pre-employment Screening

Ayoub (1982) has suggested, and some companies such as Reynolds Metals Company have adopted, a pre-employment physiological screening program to match job demands with worker abilities. Within a one-hour pre-employment examination, potential employees can be classified as physiologically acceptable, marginal, or unacceptable.

Studies of many different jobs by physiologists over the years have led to a body of knowledge that makes assumptions like the following about job demands and physiological responses reasonable:

> (1) At submaximal work, heart rate, oxygen consumption, and ventilation rates are linearly related to each other regardless of the type of work done. (2) The presence or absence of thermal stress has no effect on the oxygen consumption associated with job performance. (3) Heart rate is linearly related to the increase in thermal stress . . . (4) Maximum heart rate that can be reached by an individual is independent of the state of health or the degree of fitness and depends only on age . . . (5) Physiological responses of unfit or unhealthy individuals are markedly higher than those of normal individuals performing comparable jobs (Ayoub, 1982, p. 42).

These and other similar assumptions, along with the data accumulated from physiological studies, led to the development of certain "standards" for use in determining a potential employee's suitability. For example,

- An employee is judged fit for employment if energy expenditure, as measured by oxygen consumption while performing a task, is 40 to 50 percent of that individual's predicted capacity.
- An employee is judged fit for employment if heart rate is no more than 75 percent of that individual's predicted maximum.

To be effective, this type of a screening program requires that, "a given organization studies and catalogs the demands of the major jobs ahead of time . . ." (Ayoub, 1982, p. 46). When in place, these physiological screening programs are effective in fitting the person to the job. They have the distinct advantage of basing the placement of the individual on the individual's own characteristics, not generalized height-weight-age statistics (Ayoub, 1982).

IMPLICATIONS FOR PRODUCTIVITY

"Our general problem can be stated thus. Any task makes demands on the man required to carry it out. What . . . we find out about these demands will lead to better selection and training, better machine design, better environment designing" (Singleton et al., 1971, p. 15). Physiological measures can help.

However, the techniques of work physiology do not provide a satisfactory analysis of all industrial jobs. For certain types of jobs, measures of this sort, such as heart rate and energy expenditure, fail to reflect all of the demands placed on the operator. While these measures are effective for jobs that are primarily physical, they are not nearly as effective when other aspects of the job, such as mental demands, become important. There is always the danger of, as is the case with any methodology for doing anything, forgetting what your objective is and forcing every problem to fit your solution (Davis et al., 1971). It is very easy to get wrapped up in the technology of physiological work measurement and forget that it is a measure of performance that is desired and not necessarily a *physiological* measure of performance.

Even with this limitation, "the use by industry of these techniques to analyze and improve production is now an established trend in Europe. There can be little doubt that this approach will gain increasing importance in the United States" (Davis et al., 1971, p. 1178). "The work designer must know about the laws that govern the capabilities and limitations of the human body as an energy converter, and that the designer must have some understanding of the biological organism" (Singleton et al., 1971, p. 94). Some elablorate studies (Singleton et al., 1971) have indicated the following generalizations:

1. The longer the body has to work without rest, the lower is its ability to convert chemical energy into the mechanical energy necessary for work.
2. The older the body, the less mechanical energy it can convert from chemical energy.

SUMMARY

It must be remembered that, although physical activity is normal activity, the human body is an "unsatisfactory and expensive source for mechanical energy" (Singleton et al., 1971 p. 91). Although we can measure the physical work that people perform, there are better ways to use people than in uncomplicated and repetitive work.

REVIEW QUESTIONS

1. Define and describe *physiological work measurement*.
2. What are the most common methods of physiologically measuring work?
3. What are the three most frequently used measures of physiological work performed?
4. What is the basic measurement unit of heart rate?
5. What is the basic measurement unit of oxygen consumption?
6. What are some uses of physiological work measurement?
7. What are the limitations of physiological work measurement?
8. What are the productivity implications of measuring work with physiological measures rather than with more traditional measures?

PART IV

Improving Productivity

Many methods have been suggested for improving worker productivity. These suggestions have ranged, over the years, from cracking the whip to coddling the worker. Thus far, no magical solution to the productivity problem has been developed. However, a number of methods have been developed and implemented with some degree of success.

Among these methods are the concepts of effective job design, financial incentives, and participatory management. More specifically, job design is based on the principle of designing the job to fit the person. Ergonomics, or human factors, based on the idea of designing work consistent with human capabilities and limitations, is the subject of Chapter 10. Ergonomics, using an interdisciplinary approach, emphasizes physical work design concepts that should lead to increased productivity.

Money seems to be at the root of almost every human motivation. Incentive systems, based on sound time standards, reward productive performance with monetary prizes. Although it has limitations, money seems to work as a motivator. Chapter 11 examines traditional incentive pay systems as well as profit sharing and productivity sharing plans. Some results have been significant, as in the case of Lincoln Electric.

Although money is a known motivator, there are other aspects of work that motivate workers. High on the list of these other factors is what is known as *participative management*. This concept involves letting the worker take part in the decision-making process. Known by a variety of names, such as "quality of work life," these nontraditional motivational procedures have brought about significant productivity improvements when they have been properly administered. Chapter 12 describes the most common of these methods.

CHAPTER 10

Introduction to Ergonomics

OBJECTIVES

Upon completing this chapter, the reader should have

- An understanding that work can be viewed as a man-machine system and can be studied like other systems.
- An understanding of the scope of the formal study and application of human factors engineering, or ergonomics.
- Knowledge of the major accomplishments possible with a human factors program.
- Knowledge of the major components of an ergonomics program.
- An understanding that ergonomics is an interdisciplinary study combining physiology, anthropometry, and classical industrial engineering in work design and improvement.
- An awareness of the availability of ergonomic work design guidelines and checklists and their applicability to productivity improvement.

INTRODUCTION

Since its inception, classical industrial engineering has passed through many different phases. Initially, what we might call "quick and dirty" methods were used to try to maximize the use of production factors. Unfortunately, this emphasis often led to maximizing machine contributions while minimizing the human effort. Many of the contributors minimized or literally denied the need for the worker to be considered in the work improvement process.

From its beginning however, there has been a need to recognize that people are a very important part of the work system. Management is aware that productivity tends to improve as work is made safer, healthier, less repellent, and less arduous. Human factors, or ergonomics, "is making it possible, in good conditions of safety and health and comfort, to combine and pursue successfully, the twin objectives of higher productivity and greater well-being for men and women involved in industrial work" (*International Labour Review*, January 1961, p. 5).

Human factors engineering attempts to enhance and facilitate the use of physical objects such as tools and equipment and the actual workplace provided to people while they are performing specified tasks.* Human factors design is concerned with what we might call man-machine systems. There are arrangements of people and equipment that can achieve a high level of productivity when it is recognized that humans do certain

---

*This discussion should sound very familiar to work methods improvement. There is a strong relationship or overlap between the two.

things well. By complementing these abilities with machines, overall effectiveness or productivity can be optimized (Van Wely, 1969).

APPROACH TO HUMAN FACTORS DESIGN

Human factors engineering in work design is an impressive title. It sounds so broad and encompassing that it could cover almost anything relating to people. Human factors, or ergonomics, involves a study of entire work systems and strives, through the application of industrial engineering, psychological, and physiological principles to design jobs that are the most productive possible. But, in the context of our discussion, the term *ergonomics* is concerned with fitting physical and mental capabilities and limitations of people into the work environment.

Gustafson and Rockway, as cited in Hutchingson (1981), offered several questions that are important regarding design and improvement of the workplace environment.*

1. Do we need the human at all? Can the human, as a component of a productive system, contribute something to the functioning of the system that is to be designed and built? Would it not be better to design the human right out of the system from the start, if that is possible?

2. If the human is needed in the system, what functions should the human perform? What functions should be automated? Which functions should the human monitor? Which functions should the human serve as the backup to the machine for and be able to over-ride the machine if necessary?

3. How should the human function be organized into tasks and jobs? What parts of the job should be done at which time? What sequence should the human and the machine parts of the job be done in? Who should do this organizing?

4. What human-machine equipment is required? How should it be designed and arranged to provide for maximum system productivity? How should it be designed and arranged for maximum human productivity? How can it be designed and arranged for maximum machine productivity?

5. How can the operator best be protected from the working environment? What is required to sustain the operator's performance in terms of basic human physiological needs? Are there any special needs that just cannot be designed around?

6. What kind of a person is needed for the various jobs? The task or job defined above, in number 3, must be divided into the skills and the knowledges required. Then the individual who best fits these requirements should be placed in the job.

7. How can we best teach the user of the system to use the system to its greatest advantage? Where, and how, and with what equipment?

8. How valid are our decisions we made regarding the above? Human factors programs ideally include a test program to evaluate the system as well as an ongoing feedback system to continually monitor performance (pp. 6–7).

*Reproduced by permission of McGraw-Hill.

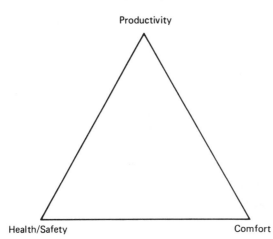

Productivity

Health/Safety   Comfort

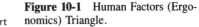

Figure 10-1   Human Factors (Ergonomics) Triangle.

As viewed by the ergonomics specialist, there seem to be three major components to most jobs. These components considered may be as shown in Figure 10-1, as the three points of a triangle.

There are two kinds of productivity. The first is actual work efficiency which is output divided by input. The second is physiological efficiency, which is a measure of the physical work required to complete a task. Both parts of efficiency or productivity have been discussed at length earlier in this book. The objective in both parts is for the worker to be able to work "smarter, not harder." It means more work and more production with less human effort.

The second point on the triangle is employee health and safety. The human factors specialist tries to determine the relationship between worker health (and safety) and productivity. Theoretically, high productivity should go hand-in-hand with improved health and safety. The health of a worker is dependent on the presence or absence of certain factors that might cause some physiological problems, such as breathing poisonous fumes. Often, there is a temptation to short-change the safety and health area because doing so will improve current productivity figures. It is imperative that the human factors specialist always be on guard for such short-sighted savings.

Health and safety are also concerned with areas of physiological measurement of the work performed. Work physiology should tell us what the human body is best suited for in the man-machine system. Standard measures, made under close supervision, can be most useful in ensuring that a safe job is designed.

The third area, or third point on the triangle, is worker comfort. Workers, quite simply, tend to be able to produce more when they are comfortable. Human factors studies have helped to determine what the various comfort zones are, both real and imagined.

Within the broad framework shown as the three points on the triangle—productivity, health, and comfort—the following have been identified as major components of the human factors system (*International Labour Review,* January 1961, pp. 9–10).

TABLE 10-1 STATISTICAL VALUES FOR U.S. ARMY WOMEN, 1946 AND 1977

Measurement	N	Mean	SD	Min.	Max.	Total	Stature ratio
Weight (kg.)							
1946	8107	59.63	9.00	39.0	111.6	72.6	
1977	1331	59.97	8.69	39.9	125.1	85.2	
Stature							
1946	8121	162.14	6.00	141.0	184.0	43.0	1.000
1977	1331	162.96	6.52	142.6	183.8	41.2	1.000
Sitting Height							
1946	8119	83.66	3.19	71.0	97.0	26.0	.516
1977	1331	85.08	3.59	73.1	96.2	23.1	.522
Bust Circum.							
1946	8115	88.91	7.68	68.0	128.0	60.0	.548
1977	1331	88.21	6.43	68.9	128.4	59.5	.541
Waist Circum.							
1946	8115	67.00	6.24	52.0	110.0	58.0	.413
1977	1331	71.01	6.90	56.5	117.5	61.0	.436
Hip Circum.							
1946	8113	95.09	6.70	74.0	126.0	52.0	.586
1977	1331	95.52	6.39	77.4	134.6	57.2	.586
Age (years)							
1946	8118	27.30	5.57	16.0	52.0	36.0	
1977	1331	23.10	5.40	17.0	60.0	43.0	

Note: From *Proceedings of the Human Factors Society, 22nd Annual Meeting, 1978.* Copyright © 1978 by the Human Factors Society. Reprinted with permission.

1. *Physiological Measurements.* The human factors or ergonomics charge in this area is to determine the physiological limits of human muscular action. Basic measures such as energy expenditure per unit time, effect of climatic conditions such as humidity and air flow on performance and effort of rest pauses on human performance during all kinds of work have to be determined.

2. *Anthropometry.* The correct design of the workplace, including the design of workbenches and seats and the proper positioning and arrangement of tools and equipment in light of studies of human body measurements (anthropometry) constitute this phase of ergonomic emphasis. Tables 10-1 and 10-2 show some typical anthropometric data. There are large amounts of this type of information available in reference books.* It is constantly updated as the physical characteristics of our population change. Note that there are separate data for male and female workers.

3. *Engineering Psychology.* Layout and presentation of all types of indicating and control devices and instrument display panels to take advantage of human capabilities

*See the Bibliography for a listing of some of these guidelines.

TABLE 10-2 STATISTICAL VALUES FOR U.S. ARMY MEN, 1966, AND WOMEN, 1977

Measurement	N	Mean	SD	Range Min.	Range Max.	Range Total	Stature ratio
Weight (kg.)							
Men	6677	72.23	10.60	45.2	128.7	83.5	
Women	1331	59.97	8.69	39.9	125.1	85.2	
Stature							
Men	6682	174.52	6.61	151.8	199.7	47.9	1.000
Women	1331	162.96	6.52	142.6	183.8	41.2	1.000
Sitting Heights							
Men	6682	90.69	3.66	77.2	103.2	26.0	.520
Women	1331	85.08	3.59	73.1	96.2	23.1	.522
Chest/Bust							
Men	6682	93.77	6.69	71.8	124.2	52.4	.537
Women	1331	88.21	6.43	68.9	128.4	59.5	.541
Waist Circum.							
Men	6682	80.29	8.18	58.8	127.7	68.9	.460
Women	1331	71.01	6.90	56.5	117.5	61.0	.436
Hip Circum.							
Men	6682	94.21	6.25	77.2	134.2	57.0	.540
Women	1331	95.52	6.39	77.4	134.6	57.2	.586
Age (years)							
Men	6682	22.17	4.64	17.0	55.0	38.0	
Women	1331	23.10	5.40	17.0	60.0	43.0	

Note: From *Proceedings of the Human Factors Society, 22nd Annual Meeting, 1978.* Copyright © 1978 by the Human Factors Society. Reprinted with permission.

and limitations regarding assimilation of information and reaction to this information is another of the tasks facing the human factors specialist.

4. *Workplace Design.* The design, positioning, and direction of operation of measurement and control devices so that the action taken by the operator is in accordance with normal human expectations is an important part of the job. For example, simply being aware of, and including in the workplace layout the fact that, when a worker turns a wheel-type device clockwise, he or she expects whatever is being controlled or operated to increase, is an important part of designing jobs that will be productive. It is possible to have a clockwise motion decrease a controlled operation but, because this activity is not consistent with human expectations, it will require that the operator do some thinking. This activity is best likened to some of the nonproductive Therbligs discussed earlier in this book. Also, in a critical situation, the operator might not pause to think before operating and cause a serious problem.

5. *Working Environment.* Three major components comprise this part of human factors consideration in job design. The first is providing adequate illumination. This factor ensures that the worker will be able to complete the task under conditions of comfort, efficiency, and safety.

The second is control of noise levels. This element has two subparts. The first of these parts is controlling noise so that no damage to hearing occurs. The second is the psychological effect of noise and the interference it has with effective performance and communication on the job. Human factors considerations in the work design process seek to eliminate these noise considerations or, if it is impossible to eliminate them, at least to protect the operators from them. Closely related to noise are problems caused by vibration. The problems and solutions often are quite similar to noise problems.

The third environmental component which is of extreme importance in an ergonomic sense deals with the effects of extremes in temperature and humidity on the work capacity of the individual. Human factors try not only to establish the optimums but also identify what happens to the individual who must be subjected to extremes in all of these areas. The need for rest to permit the body to recover and the methods of controlling either the level or amount of exposure are important aspects of this part of ergonomics.

6. *Biological Factors.* Biological factors in the arrangement of working periods is also part of the ergonomic design of work. This section should have or could have been called ''Physiological Factors'' in light of what we know of the physiological factors that influence worker fatigue. Factors such as duration and intensity of work performed, the frequency and duration of rest periods, and the effect of shift work are all part of the process.

As should be noted and re-emphasized, ergonomics is a cross-disciplinary endeavor. The main scientific disciplines concerned with fitting the job to the worker and the worker to the job are industrial engineering, anthropometry, physiology, and psychology. The contributions of these areas to ergonomics will now be discussed in more depth.

Increasing productivity is the main focus of this text. The problem we have set out to solve is the productivity problem. The reader should, at this point, be quite aware of how industrial engineering relates to productivity improvement and the work design process.

Anthropometry is the study of basic features of the human which are most important in laying out or designing the work area or equipment used. It is concerned with measurements of the parts of the body such as height, weight, arm reach, and so on, taking into account the range of variation of individuals around the average. This information can be used in the design of equipment and the layout and location of controls and assembly parts based on the consideration of the relationship between the operator and his or her physical capabilities (McCormick and Sanders, 1982).

We have discussed the physiological aspects of work in the chapter on physiological work measurement. Factors such as strength and the human body's adaptability to environmental conditions such as heat stress are important aspects of this part of ergonomics. As was shown in Chapter 9, the main industrial application of this type of information is placing the individual in the job that he or she is capable of performing.

The final major area of human factors is engineering psychology. This part of ergonomics is concerned primarily with the design and location of controls based on information gained from experimental psychology. Proper applications in this area can often increase the amount produced, decrease worker effort, increase job safety, and increase product quality.

WORK DESIGN CONSIDERATIONS

Most of the ergonomics studies that have been completed are concerned with jobs that have considerable psychological physical demands. Improvement in performance concerns itself with trying to find the optimum compatibility between sensory tasks, mental tasks, and motor tasks (Yoder, 1972). Perhaps as a result of the industrial studies, or perhaps as a prelude to the studies, certain generalizations to work design have been suggested. Van Wely (1969) stated these best as the "Eleven Commandments."*

I. *Aim at movement.* Muscles should be used, but not for holding or fixation. Movement reduces monotony.

II. *Use optimum movement speeds.* Movements that are too quick or too slow are fatiguing and inefficient. Try to find the specific optimum speed of each movement and have the operator perform the motion at that speed.

III. *Use Movements Around the Middle Positions of the Joints.* Long duration or frequent use of extreme positions of a joint, especially under load, are harmful and have poor mechanical advantage; yet occasional extreme positions, while not loaded, are desirable.

IV. *Avoid Overloading of Muscles.* Keep the dynamic forces on the muscles to less than 30 percent of the maximum force that the muscle can exert. If this must be exceeded, and it sometimes must be in order for a job to be done, the analyst must recognize that the human can exert a dynamic force up to 50 percent of the muscle capacity, but that this can be done for no more than 5 minutes. Static muscular load should be kept at less than 15 percent of the maximum force that the muscle can exert.

V. *Avoid Twisted or Contorted Postures.* Pedals should not be operated by standing operators. Arm supports should be used only when the upper arms cannot relax and they should be perpendicular to the floor. Bending the head backwards causes glare in the field of vision as well as sore necks.

VI. *Vary the Posture.* Any fixation causes problems in the muscles, joints, skin, and blood circulation. Do not use pedals in microscopic work. (Ayoub [1973] offers specific suggestions for this consideration.)

VII. *Alternate Sitting, Standing, and Walking.* Sitting for more than one hour and continuous standing for more than one half hour are, in the long run, too fatiguing. Standing for more than one hour a day is fatiguing and causes physical abnormalities.

*Reproduced with permission of the Kansas Engineering Experiment Station at Kansas State University.

Concrete floors are very fatiguing. Elastic supports, such as wood, rubber, or carpet are better.

VIII. *Use Adjustable Chairs.* When the operator must sit for more than one half hour continuously or longer than three hours a day, use adjustable chairs and footrests. When adjusting these, remember,

    a. The height of the seat should be such that the elbows are about the height of the working plane.

    b. The footrest should be adjusted so that no pressure occurs in the back of the knees.

    c. The backrest should be such that the lower back is supported. The bottom of the backrest should be 4 to 8 inches above the seat.

IX. *Make the Large Man Fit and Give Him Enough Space; Let the Small Woman Easily Reach.* For standing it is essential that the working place be adjustable. A platform for the short worker is a simple solution. The working plane should be within two inches of elbow height.

X. *Train Workers in Correct Use of Equipment.* People must be instructed and trained in good working postures. Sitting, standing, and especially lifting are often done the wrong way.

XI. *Load People Optimally.* Neither maximum nor minimum are optimum. Optimum physical and mental loads give better performance, more comfort, less absenteeism, and do less harm to the worker.

Applying ergonomic principles is as much a matter of following the above commandments as it is breaking new ground. The successful analyst or work designer should be cognizant of human capabilities and limitations. Murrell (1969) suggested an "ergonomic checklist" not all that far removed from the principles of motion economy. It might provide the groundwork for designing work or improving work using the principles of human factors engineering. It is not the only checklist, but is one of the more reasonable lists suggested. The following questions, based on Murrell's checklist, should be asked about any job that is being studied.

    1. *What is the operator expected to do?* The specific nature of the task, whether it is mental or physical, and the end result must be defined. Looking at the work objective is the first step in any job analysis task.

    2. *Will the best use be made of human capabilities?* There are certain talents people have that machines, as of this writing, just don't have, such as adapting to unexpected changes. Jobs should be designed to make optimum use of human capabilities.

    3. *Will the worker be required to perform functions that are not well-suited to human capabilities?* Just as humans have capabilities that make them more valuable in certain situations or doing certain tasks, machines have certain activities, like performing repetitive tasks reliably over extended periods of time, that people are not as good at.*

---

*Chapter 4 gives a more complete listing of the capabilities and limitations of both people and machines.

4. *Can these functions be transferred to machines?* Tasks, like the repetitive one mentioned, might be better performed by machines. The ergonomics specialist is charged with designing work that will have people do what they do best and have machines do the rest.

5. *Will the equipment anthropometrically fit the operator?* Common characteristics in which sizes have been compiled for significant portions of the population include height, sitting height, knee height, buttock-knee length, and so on. Studies have also shown that these characteristics tend to change with age.

6. *Has the existing equipment been designed for this particular application or is it being used because it has always been used?* Sometimes, economic considerations may answer this question. Where possible however, the long-term economic gains should be compared with short-term costs before changes are ruled out.

7. *Does the operator sit or stand?* As the "Commandments" showed, there are fatigue considerations for standing and sitting. The work designer should be aware of these considerations.

8. *Is the operator's posture satisfactory?* Poor posture can increase fatigue and lead to occupational illness. Human factors can suggest the proper posture and design of work stations to try to encourage workers to use correct posture.

9. *Is the equipment likely to be operated by women?* There are measured differences between men and women, as Tables 10-1 and 10-2 showed. For example, women are, on the average, shorter than men. If it is likely that women will operate the equipment, then it should be designed to accommodate them.

10. *Has equipment that is likely to be operated by women been anthropometrically designed for women?* Again, this question refers to the observed and measured physical differences between the sexes.

11. *What information does the operator need to do the task?* The human has a limited capacity to discriminate between different types of information. Most people, studies have shown, can absolutely tell the difference between from 4 to 9 or 10 individual inputs (*International Labour Review,* January 1961). It is important not to overload operators with excessive information.

12. *How should the operator receive information?* Whenever possible, it is desirable to have people process information so that they have to make relative judgments about it. It is far easier to differentiate between two successive sounds, for example, if the operator just has to determine whether one sound was louder or higher pitched than the other.

13. *What type of display should be used to transmit information to the operator?* Many types of displays are available to transmit information. Humans have five senses: vision, hearing, touch, smell, and taste. They can all be used to transmit information to the operator. Certain types of displays transmit certain types of information better than others. For example, flashing lights are appropriate for danger.

14. *Where are the instruments located relative to the importance of the information they transmit to the operator?* The most important information has to be most accessible.

Information that is important to a pilot, such as altitude, must be readily available at all times. A pilot should not have to search a display panel to determine such a critical piece of information. Likewise, a machine operator should not have to search for some similarly critical information.

15. *What controls and what type of controls will the operator need to perform the job?* The type of control, whether it is a lever, joystick, or steering wheel may be dictated by what is being controlled. Direction, for example, might best be controlled by a wheel-type of control, similar to a car's steering wheel. Making the controls operate consistently with human expectations can be a very important contribution of the ergonomics specialist to the design of a work station.

16. *If the operator is standing, have foot motions been eliminated?* People cannot comfortably operate foot controls while standing. It is very difficult to do physically, especially if it is a regular part of the job cycle.

17. *If force is required of the operator, is the force mechanism enhanced?* People are not very strong, especially when compared with machines. People tend to get tired very easily and take time to recover their strength. Machines, however, can be built to apply considerable force, apply the force consistently, and literally never get tired.

18. *Will operators have to communicate with other operators or with supervisors on a regular basis?* Communication may be required as part of a job. If it is, the work environment must be designed to permit the necessary communication. It is obviously ridiculous to require a jack hammer operator to speak while operating the tool.

19. *If communication between operators is required, what format must it take?* Communication is not necessarily verbal. Just as people can receive input through several senses, they can also transmit information through more than just verbal communication. Arm motions are often just as effective, if not more so, than verbal communication in transmitting certain kinds of information.

20. *How much physical work is expected of the operator?* Physical work is fatiguing. If significant physical work is required, then adequate steps for recovery must be provided.

21. *Can the operator perform the physical work expected?* More specifically, can the operator continue, on a sustained basis, to perform the work required? Most people can occasionally make extraordinary effort but they cannot continue this effort on a regular basis. It is important that these "super" efforts not normally be required as part of a job.

22. *What type of work environment is anticipated?* Special conditions may arise in working environments. Factors such as heat may fatigue a worker more quickly or noise may make communication difficult as well as increase fatigue. If special environmental conditions are present, a job should be designed with them in mind.

23. *Is excessive noise expected?* Although Federal regulations, such as those of the Occupational Safety and Health Act, set maximums for noise exposure, workers are not as productive when they work in very high noise environments. Certain frequencies, even at lower decibel levels, are more disturbing than others. Inter-

mittent noise is much more disturbing than continuous noise. The length of time a worker is exposed to higher levels affects the fatigue the noise causes (*International Labour Review*, January 1961). Because in most industrial settings it is impossible to eliminate noise completely, steps should be taken to control it. These steps can include mufflers, sound absorbers or even personal protective devices. Physical problems such as hearing loss are often the result of long-term exposure to noise. It is very important that noise exposure be kept to a minimum when a job is designed.

24. *Is excessive heat expected?* The effects of heat are much more quickly recognized than those of noise and can be much more physiologically damaging. People get tired faster in the heat. Jobs should be designed to minimize heat exposure or maximize operator protection from heat. If this protection is impossible, productivity requirements should be adjusted. It is important to provide adequate recovery facilities for workers who are exposed to extreme heat.

25. *How much illumination is necessary for the type of work to be performed?* Inadequate lighting can contribute to fatigue. This factor is relatively easy to remedy so there should be no excuse for providing too little illumination. Also important to the illumination issue is the distribution of lighting in the work area. The Illuminating Engineering Society has prepared guidelines for required illumination for different types of work.

26. *Will the mental or physical demands of the job overload the operator?* Although humans are good at making decisions, there are definite limits to the number of decisions that can be made. Machines can and should make routine decisions, permitting the individual to make critical decisions.

27. *If the operator is overloaded, what steps can be implemented to reduce the load?* There have been considerable advances in technology that allow workers' abilities to be enhanced by machines. Computers can fly jet airplanes without pilots. Computers also aid the sometimes difficult and stressful task of air traffic control.

28. *Are there any maintenance requirements?* Although maintenance may not be performed frequently, all the human factors considerations should be applied to maintenance tasks as well as to normal working procedures, both in work design and work station layout. Can maintenance be performed easily? Maintenance time is usually nonproductive time. If it can be easily performed, then it can be kept to a minimum and productivity can be increased.

29. *Has the equipment been designed to facilitate repairs?* Everything breaks down once in a while. Even preventive maintenance won't prevent occasional failures. Ease of repair or replacement can keep productivity losses to a minimum.

While no means complete, the above listing and discussion of questions about work design address the relationship of man (or woman) and the work that must be accomplished. Note that the questions refer to all three points of the ergonomics triangle. The more care given to the relationship, the more productive the system will be. The following section briefly describes some industrial applications of the above principles and then describes a more complete example of the application of some of the ergonomics principles.

INDUSTRIAL APPLICATIONS OF ERGONOMICS

Although it is a relatively new discipline, there have been many ergonomics applications made over the years. Some companies have invested heavily in ergonomics. Eastman Kodak, Eli Lilly and Company, Lockheed, and Liberty Mutual have all contributed significantly to the area. Projects that have been documented include applications in many different areas. Some of these industrial applications include:

- Studies of maximum acceptable weights and work loads for manual handling tasks
- Studies of the effects of heat stress on maximum acceptable work loads
- Studies of the physical work capacity of females
- Studies of lower-back injuries
- Studies of automobile driver fatigue and the development of effective corrective measures
- Studies of the performance and acceptance of respirator facial seals
- Design of aircraft escape hatches
- Environmental design: lighting, temperature, and atmosphere; noise, vibrations, and acceleration
- Design for speech communication
- Data entry systems design
- Designing for the handicapped: visually, physically, and hearing
- Reduction of safety hazards found at work stations in a textile plant (Kahlil and Ayoub, 1976)
- Selection of workers to determine physiological capacities and to attempt to calibrate their heart rates and oxygen consumption responses to physical work.
- A visual sorting task involving the scanning, detection, and removal of defective empty gelatin capsules was made at one company with the objective of optimizing the man-machine relationship at the interface of the inspector and the inspection point. Types of lighting, single versus multiple work stations, and job rotation were some of the factors analyzed.
- Proofreading of label and literature documents for errors at various stages of the process of creating large quantities of these products. The objective was to optimize man-machine interfaces to minimize error possibilities.

Example: Industrial Application of Ergonomics

The following case study, showing an in-depth industrial application of ergonomics, was described by T. M. Kahlil.*

*Reprinted with permission from the *1976 Annual AIIE Conference Proceedings*, Copyright © American Institute of Industrial Engineers, Inc., 25 Technology Park/Atlanta, Norcross GA 30092.

In the electronics industry workers used to assemble transistors, capacitors, amplifiers, and other small components in an integrated circuit board according to a specific design. The task involves selection of the proper electronic component from a bin, searching for the proper location in the board, assembling the component according to a sequence provided by the circuit design. This task is a very complex one and requires exceptional perceptual and psychomotor capabilities. It overburdens the visual senses and requires a special mentality to keep track of the proper sequence for assembling items without any errors. Productivity of the workers was measured based on the number of assemblies produced per day per worker.

A careful review of the task indicates that the task is overburdening the operator with functions which can be performed more effectively and at a faster rate by a combined man-machine system. Delegating the tedious and repetitive task of searching and arranging according to a predesigned sequence to a machine, and concentrating human effort on manual manipulation needing high level of manual dexterity, would create a more effective man-machine system.

Since the technology is available to do this, the solution has been implemented in one factory. A programmable machine was installed which provides the operator with components according to the design sequence of assembly. It also sheds light on the proper location where the component should be inserted.

With the new man-machine system in operation, production output per worker increased by more than one third. Errors made resulting in board malfunctioning were reduced from 15 percent to less than one percent and inspection plans were reduced from 100 percent to only spot inspections. A cost analysis of the old and new man-machine systems well justified the expenditures resulting from the change and dramatically increased the plant's productivity (Kahlil, 1976, p. 58).

SUMMARY

Much of the human factors or ergonomics work in this country is directed at the design of equipment, layout, and procedures used to maintain relatively complex work systems. One of the major goals of the human factors specialist is to provide information to the work designer to try to assure that the development of systems, equipment, and procedures of work are compatible with the *capabilities and limitations* of the people who will use them. The emphasis has been on, and continues to be needed on, "designing equipment and procedures that a man (or a woman) can use reliably" (Swain, 1973, p. 129). "The human factors field received its initial impetus in the military services, and it is probably that the primary systematic application of human factors is still within the military services. Over the years, however, considerations of human factors has been extended, in varying degrees, to other areas of application such as the design of transportation equipment, production machinery, and processes, communications equipment, . . . , and consumer products and services. However, it should be emphasized that the application in most of these areas has so far been sporadic and limited . . . (but) as one looks ahead, the potential range of applications of human factors principles and data is tremendous . . ." (McCormick and Sanders, 1982, p. 9).

PARTIAL LISTING OF DESIGN GUIDE BOOKS

CHAPANIS, A. *Ethnic Variables in Human Factors Engineering*. Baltimore, MD: Johns Hopkins Press, 1975.

DAMON, A., STOUDT, H., & McFARLAND, R. *The Human Body in Equipment Design*. Cambridge, MA: Harvard University Press, 1966.

DREYFUS, H. *The Measure of Man*. New York: Whitney Library of Design, 1960.

VAN COTT, H. P., & KINKADE, R. G. (Eds.). *Human Engineering Guide to Equipment Design*. Superintendent of Documents, U.S. Government Printing Office, Washington, D.C., 1972.

WOODSON, W. E. *Human Engineering Guide for Equipment Designers*. Berkeley: University of California Press, 1964.

REVIEW QUESTIONS

1. What is *human factors engineering* (or ergonomics)?
2. What is the main objective of human factors in work design?
3. What are the three major components of the human factors program?
4. What are the human factors questions that should be raised concerning the design and improvement of the working environment?
5. What are the three major disciplines that combine to make up human factors?
6. What is *anthropometry*?
7. What is *engineering psychology*? How does it differ from "regular" psychology?
8. What types of tasks are most amenable to improvement with human factors or ergonomics studies?
9. In general terms, what are the "Eleven Commandments"?
10. What is an *ergonomic checklist*?
11. How can a checklist be used in productivity improvement?

CHAPTER 11

Incentives to Increase Productivity

OBJECTIVES

After completing this chapter, the reader should

- Understand what a wage incentive plan is and how it relates to productivity improvement.
- Understand the relationship between incentive systems and time standards.
- Be able to develop incentive pay plans and calculate some relatively straightforward situations.
- Have an appreciation for the success of the Scanlon plan and the Lincoln Electric incentive or productivity-sharing plans.

INTRODUCTION

Time standards are defined as the time the typical operator, working at a normal pace under normal working conditions, requires to complete a specified task, with sufficient time allowed for personal needs, fatigue, and delay. Standards reflect expected productivity. If a worker can complete the specified task in less time than expected, it is logical to presume that the worker is more productive than the typical worker who would perform that task.

It is also logical to assume that, if the standard is accurate, the most productive worker should be able to produce the product in less time than the standard allows. When this situation occurs, it is also logical to assume that management should be willing to reward the more productive worker with more than typical wages for the job in question. It should also be logical to pay a bonus to motivate the typical worker to produce above-standard, or in less time than required by the standard.

Much has been written and said about wage incentive plans. At one time, they were unquestionably used as a means of regulating production and payment. They were a device management used to make more money for themselves and, incidentally, their employees. Barnes (1980) explained how they work. "That means that more work can be put through the machine or process in a given time, and the overhead cost of operating this machine will be pro-rated over a greater number of pieces and so the unit cost will be lower" (p. 483). This relationship has been substantiated by research. Productivity was 42.9 percent higher in plants with incentives than in plants without incentives and furthermore, it was 63.8 percent higher than in plants with no work measurement system at all (Fein, 1973).

Although substantiated by research, using incentive wages to motivate workers has some built-in problems. Watmough (1965), in a critical look at incentive systems, stated, "Most reasons stem, not from physical disagreement with the basic incentive idea, but from endlessly recurring mistakes—very, very serious mistakes—in the installation, administration, and maintenance of the many incentive systems presently in operation" (p. 356).

Workers are worried that if they show they can produce at a rate that is consistently above the standard, management will either raise the standard or, perhaps worse, reduce the work force to compensate for the higher productivity of some workers. This problem is very real. During the installation of an incentive plan, great care must be taken to ensure that a fair standard is established. A single time study will not do. Instead, a system like Cox (1959) described, using better data, is necessary. "Rates were not set based on a single time study of a particular job, but rather time values were established for various functions after accumulating perhaps hundreds of studies on similar equipment" (p. 356).

Traditionally, incentive plans have been accepted only with some reluctance by organized labor. According to a position paper issued by the AFL-CIO in 1969, the general feeling among labor leaders was that incentive plans offered many problems, especially regarding the consistency of the standards used as the basis for the incentive system. Therefore, organized labor has been, for the most part, opposed to traditional wage incentive plans.

Although there have been some abuses in the past (Sloane and Whitney, 1977), some unions, such as the United Auto Workers (UAW) and International Ladies Garment Worker's Union (ILGWU), have learned to live with the systems. As a matter of fact, both the UAW and the ILGWU have their own industrial engineering staffs to monitor the standards used as the base for the incentive pay systems.

Among the union objections to wage incentives are the legitimate charges of abuse that occurred during the past, a belief that speed-ups follow the installation of incentive plans, and the belief that incentives cause worker disharmony through competition and disrupt the harmony of a work situation. Until very recently, unions have fought to keep incentive systems out of their organized plants. As recently as 1976, their efforts were fairly successful. During that year, "a Bureau of Labor Statistics study of 1711 major collective bargaining agreements, covering more than 7.6 million workers (but not railroad, airline, or government employees), 1504 contracts covering 5.5 million workers had one or more jobs for which workers were paid on an hourly basis. Only 467 contracts allowed some form of incentive pay" (Scheuch, 1981, p. 485). Recent economic conditions have encouraged many unions, such as the United Auto Workers in its Landmark 1982 contract with Ford Motor Company, to accept a form of incentive pay—profit-sharing. Economic conditions forced the unions to accept this form of incentive pay in return for job security.

Rather than being problems of incentive systems, many of the unions' objections to incentive systems are problems of labor and management not trusting each other. Not only must an incentive plan have a strong database behind it, but it must also be accepted and understood by the employees who are working under it. The operators need to have faith that the incentive will not suddenly be changed because of one outstanding performance.

One of the best sources of instruction about the plan is from the worker's immediate supervisor. The line supervisor can be the bridge between operators and management. First however, the supervisors must have a thorough understanding of all aspects of the rate setting process.

INCENTIVE SYSTEM DESIGN

The following are necessary conditions for a successful incentive system:

1. The program must be technically sound. That is, the program must be based on accurate standards.
2. The program must guarantee some reasonable incentives for all workers. No upper limits should be placed on earnings.
3. The program must be simple enough so each employee can determine his or her own wage payment.
4. The program must indicate procedures to be followed when the work method and hence, the standard, changes.

Incentive plans are based on the following principle: "Employees are compensated for increased productivity above an acceptable productivity level, at a predetermined participation ratio, in proportion to the increase in productivity or in accordance with an established plan" (Fein, 1973, p. 18). To be technically sound, it is of utmost importance that valid time standards exist. These standards may be set by time study, predetermined time systems, standard data applications, work sampling, or historical records. Management often forgets that even the most routine job is constantly changing. "It cannot be over-emphasized that incentive systems are imperfect tools that require constant attention and improvement" (Lokiec, 1966, p. 313). As methods change and new tools are introduced, either officially or unofficially, any job changes from when it was originally studied. Although it is a big job to install a system initially, it is only half of the task. "The other half is supervising the application to insure maintenance of the prescribed level of productivity and earnings of the operator" (Lokiec, 1966, p. 313).

It must be emphasized that to be technically sound it is important that valid time standards exist. As long as the parties involved—workers and management—agree that the standards accurately and fairly reflect the current expected productivity of the typical worker on the job as it presently exists, then the incentive plan will have a good chance of succeeding.

When standards are valid and accurate, they will guarantee that each employee participating in the incentive plan, regardless of the job performed, has the same opportunity to earn an incentive wage. Extra effort on Job A should produce the same increase in productivity as the same extra effort on Job B. These equal increases in productivity must be rewarded with equal increases in compensation.

Once the technical aspects of a plan are addressed, the focus of attention shifts to administration of the plan. Workers will generally support an incentive plan when the following conditions occur.

1. After putting in a fair day's work, employees must discover that they can earn incentive wages. Although it is not the result of any formal policy, current surveys indicate that American workers average an incentive earning of 30 percent in excess of their base earnings (Krick, 1962).

2. A second condition that will cause the workers to support incentive wage plans is guaranteed minimum daily earnings. Although incentives are based on the belief or desire to reward performance above standard, thereby reducing costs and increasing productivity, standard performance, or even occasional substandard performance, cannot be penalized monetarily. Remember, standard performance is the performance of an average operator and some workers who, by the very nature of that definition, must perform below standard. Continued substandard performance shows a need for corrective action.

3. There should be no need for an upper limit or incentive earnings if the standards are technically sound and reflect the current work method. Although it is unlikely, if a worker can product at 200 percent of standard and the rates are sound, then the worker should be paid a corresponding incentive wage. Occasionally, there is a worker whose individual abilities are such that they can legitimately double expected production. These "superstars" deserve their reward because they have indeed increased productivity. Although once in a while we may see a superstar, we should not normally expect to see an entire production department of these outstanding operators.

4. Each worker must be able to determine, based on the daily or weekly production count, the amount of money earned for the period. (This element is important for the success of an incentive program.) It is most important for the worker to know what he or she has earned. Although people "trust" their employers, it is human nature to double-check wage payments. Also, when workers are able to determine and verify that extra effort indeed leads to extra income, the incentive plan's motivational value is reinforced.

5. A well-administered wage incentive plan will be able to accommodate any changes that are made in the production method. A change in method will usually result in changing the standard. When a standard changes, the potential earnings of an employee are affected. If the method reduces the time required to complete the job and a new standard is not set to reflect this change, the rate becomes *loose,* meaning that the worker can earn a larger incentive on that job than on a comparable job with a valid time standard.

   Similarly, if more time is required after a methods change than the standard allows, then the rate is *tight.* When a rate becomes tight, it is most difficult for employees to earn incentive pay. Care must be taken to ensure that the same direct relationship between productivity and the reward for increasing productivity remains. The pay rate may be negotiated, but the standard itself must remain as the analyst set it. Horror stories abound about companies that negotiated time standards to adjust the earnings of the employees on an incentive plan. The correct procedure is to negotiate the pay rate.

6. For an incentive plan to be accepted, the job should be at least in part operator-paced rather than machine-paced. Machine-paced portions of a job cannot be placed on an incentive pay plan.

Incentive systems, when properly developed, can result in significant increases in productivity. Historically, individual incentives have been successful (Maynard, 1970). In the past, people have not liked to depend on others for their financial rewards; however, we are currently seeing substantial success with group incentives, which are more commonly known as *productivity-sharing plans.* Traditionally, incentive systems have been used to pay support personnel, such as materials handlers, based on the production of the people they support. Whether individuals or groups, the incentive system must be structured so that product quality is not sacrificed for production quantity. Paying only for acceptable products and penalizing defective production is a way to handle this potential difficulty. In summary, for an incentive plan to be effective, it must be *fair* to all concerned.

TYPES OF INCENTIVE PLANS

Although all incentive plans tend to pay workers more when they produce more, there are a variety of plans in use. Some are more sophisticated than others. Some are based on factors other than hourly production. This section will examine a sampling of individual incentive plans. Two relatively famous group plans, the Scanlon plan and the Lincoln Electric plan, will be examined in more depth in a later section of this chapter. The first plan we will look at is not really an incentive plan; it is a plan that provides a convenient baseline, or point of reference.

Daywork Rate Plan

This reference plan is called *daywork.* In daywork, the base rate of pay is always paid to the worker regardless of productivity. the worker is paid for time spent on the job. In daywork, it is assumed that this time will be spent working at 100 percent of standard. Daywork is included in the discussion of standards because there are often situations when a worker's productivity is controlled by events that the worker cannot influence. For example, if an operator's task is to operate a machine, there will be certain parts of the job that the worker cannot speed up, such as the time that the machine is operating. If the worker is required to tend to the machine while it is operating, then there is no way the worker could be expected to produce at more than 100 percent during the machine-controlled portions of the job. An equitable pay plan for this portion of the job is to pay daywork.

Example: Machine-Controlled Job

A worker must tend a machine 100 percent of the time. The negotiated hourly payrate for this job is $6.50. If the worker is on the job for 40 hours in a given week, the gross pay would be

$$\text{Pay} = (\text{Hourly Rate})(\text{Number of Hours})$$

And, in this example

$$\text{Pay} = (6.50)(40) = \$260.00$$

Straight Piece-rate Plan

The straight piece-rate plan is the most basic of all incentive plans. In this plan, wages are based strictly on the production of the worker. Each unit of output completed earns the worker additional compensation. The higher the productivity, the higher the wages. The lower the productivity, the lower the wages. In the straight piece-rate plan, there is no minimum pay guaranteed. Because this idea is not consistent with the conditions established earlier for an effective incentive plan, we will modify the plan and describe a similar plan, but one that guaranteed a minimum or "fair day's pay."

When the straight piece-rate plan is modified to the *straight piece-rate with 100-percent guarantee* incentive plan it guarantees a normal day's pay to every worker regardless of productivity. If the worker does not produce at 100 percent of the standard, the pay is calculated as if the worker were being paid daywork. Although theoretically there should be no reason for an employee not to make the standard production, that is, there is no theoretical reason why a worker should not produce to standard, there may be circumstances beyond the worker's control, such as a machine breakdown, or there may be extenuating conditions that prohibit the worker from producing as expected. If a worker consistently does not make production at standard, then retraining or other appropriate action should be taken.

Example: Straight Piece-rate Pay

The hourly pay rate for a particular job classification is $5.00 per hour. The standard for each completed unit on this job is .0083 hours. The worker is guaranteed 100-percent pay regardless of productivity, although production in excess of standard is paid an incentive based on a straight piece-rate plan. In other words, based on .0083 hours per piece, the worker would be expected to produce

$$(1 \text{ hour})/(.0083 \text{ hours per piece}) = 120.5 \text{ pieces per piece}$$

The value of each piece, based on the hourly rate of $5.00 per hour is

$$(\$5 \text{ per hour})/(120.5 \text{ pieces per hour}) = \$.041 \text{ per piece}$$

1. If the worker produced 1200 pieces in a day, determine the pay. If the worker received daywork pay, the gross pay would be

$$(8 \text{ hours})(\$5 \text{ per hour}) = \$40$$

Incentive pay for this production job would be

$$(1200 \text{ pieces})(\$.041 \text{ per piece}) = \$49.20$$

The worker would have earned an extra \$9.20 this day for producing above the expected production, this represents a bonus of (9.20)/(40) or 23 percent of the day's work. This example is certainly reasonable motivation and reward for producing over standard.*

2. If the worker on this job produced 800 pieces in a given day the incentive pay would be only

$$(\$.041 \text{ per piece})(800 \text{ pieces}) = \$32.80$$

Because this pay is less than the daywork rate, the worker would, because of the 100-percent guarantee, receive the daywork rate of \$40.00 for this day's production.

Example: Process-Controlled Job

The hourly pay rate for a particular job classification, with a piece-rate incentive and 100-percent guarantee, is \$7.50 per hour. This job is 40 percent machine-process-controlled. During an eight-hour day, determine the worker's pay for producing (a) 350 units and (b) 500 units. Before doing any specific calculations, we must perform two more generalized ones. First, we must determine the daily guarantee. If 40 percent of the time is process-controlled, the operator, regardless of production, is guaranteed 40 percent of the daywork rate. The eight-hour straight daywork pay would be

$$(8 \text{ hours per day})(\$7.50 \text{ per hour}) = \$60.00$$

Forty percent of this amount is

$$(.4)(60) = \$24.00$$

Second, we must determine the piece rate, or the actual value of each piece, based on the 14 hours per thousand standard.

$$14 \text{ hours/thousand} = .014 \text{ hours/piece}$$
$$.014 \text{ hours/piece} = 71.4 \text{ or } 71 \text{ pieces/hour}$$

Using the pay rate of \$7.50 per hour, the value of each piece becomes

$$\$7.50 \text{ per hour}/71 \text{ pieces per hour} = \$.105 \text{ per piece}$$

*This motivation may be blunted by the income tax bite. If incentive earnings push the taxable income up into another tax bracket, the worker may be less inclined to produce at higher levels. However, the lure of extra dollars now often offsets worry about paying taxes later.

a. If the daily production is 350 units, the piece-rate pay will be (350)(.105)=$36.75. The total pay for the day will be this amount plus the amount guaranteed by the process-controlled contribution of $24.00, or $60.75. Because this total exceeds the dayrate pay for eight hours, the pay will be $60.75.

b. If production is 500 units, then a larger incentive can be earned. The piece-rate earned is ($.105)(500) or $52.50. Coupled with the process-controlled time paid at the straight $24.00, the earnings for this day are $76.50.

Standard Hours Plan

Not all incentive plans use the piece-rate for determining the wage payment. A *standard hours plan* converts the production rate to the hours allowed for producing a certain quantity. This plan then pays the hourly rate for "earned" hours. A typical standard hours plan will guarantee pay for eight hours each work day.

Example: Standard Hours Rates

A worker is paid incentive wages, using a standard hours plan with an eight-hour guarantee. The hourly rate is $6.42. The standard is 3.4 hours per thousand. How much would a worker earn if production were 2,700? The production count must be converted to earned hours. The standard of 3.4 hours per thousand is used, combined with the production of 2,700. Based on the standard, production of 2,700 units should take

$$(2700 \text{ units})(3.4 \text{ hours per } 1000 \text{ units}) = 9.18 \text{ hours}$$

Paid at the hourly rate of $6.42, earnings for this worker are

$$(9.18 \text{ hours})(\$6.42 \text{ per hour}) = \$58.94$$

Compared with straight daywork pay,

$$(8 \text{ hours})(\$6.42 \text{ per hour}) = \$51.36$$

The worker earns more than standard, but this pay is fair because the worker produced 9.18 hours worth of production in eight hours. By using the same resources as the typical worker, more product was produced. This amount represents a productivity improvement and management should be happy to reward this level of performance.

Example: Standard Hours Rates
with Process Control

A worker is paid incentive wages, using a standard hours plan with an eight-hour guarantee. The job is 25 percent process-controlled. The hourly rate is $5.66. The standard is 11.6 hours per thousand. How much would a worker earn if she produced 623 units in a day? She would automatically be paid for 25 percent of the work day, or two hours of process-controlled time at the daywork rate.

$$(2 \text{ hours})(\$5.66 \text{ per hour}) = \$11.32$$

The allowed hours for producing 623 units are

$$[(623 \text{ units})(11.6 \text{ hours}/1000)] = 7.23 \text{ hours}$$

The pay for the earned 7.23 hours is

$$(7.23 \text{ hours})(\$5.66 \text{ per hour}) = \$40.92$$

The total pay for the day is $40.92 plus the guaranteed $11.32 for the process-controlled time, or $52.24. This amount compares favorably with the guaranteed minimum of (8 hours) ($5.66/hour) or $45.28.

All of the incentive plans we have looked at thus far have had two characteristics in common. First, the bonus paid for production over standard was the same as that paid for production up to standard. Second, when a guarantee was specified, it was a 100 percent of standard guarantee, or a guarantee of paying daywork regardless of production.

The Taylor Plan

The Taylor plan, while similar in design to a straight piece-work incentive system with a 100-percent guarantee, is typical of a variation of these plans. The Taylor plan pays a different, higher rate for every unit produced above the standard. This plan provides a greater incentive for producing over standard. Workers who increase productivity can receive substantial bonus payments, depending on the rate differential.

Example: Taylor Plan Incentive

A worker is paid incentive wages using a Taylor plan incentive with a 100-percent guarantee. Normal pay is $4.73 an hour and incentive pay is $6.15 an hour. The standard for the job is .28 hours. Calculated earnings for this job are based on the fact that 175 acceptable parts are produced in a 40-hour work week.

For production up to 100 percent of the standard, the piece rate is determined. The expected production per hour is

$$1/.28 \text{ hours per piece} = 3.57 \text{ pieces per hour}$$

Each piece is worth

$$\$4.73 \text{ per hour}/3.57 \text{ hours per piece} = \$1.33 \text{ per piece}$$

In an eight-hour day, we would expect (8 hours)(3.57 pieces per hour) or 28.56 pieces per day or (28.56 pieces per day)(5 days/week) = 142.8 pieces per week. This amount would, to simplify the wage payment procedure, be rounded down to 142 units per week. Payment for production over 142 units per week is paid at the higher rate of $6.15 an hour.

$$\$6.15 \text{ per hour}/3.57 \text{ pieces per hour} = \$1.72 \text{ per piece}$$

In this example, the recorded production of 175 units exceeds standard by

$$175 - 142 = 33 \text{ units}$$

The week's pay is

$$(142 \text{ pieces per week})(\$1.33 \text{ per piece}) +$$
$$(33 \text{ pieces over standard})(\$1.72 \text{ bonus per piece}) =$$
$$\$188.86 + \$56.76 = \$245.62$$

Example: Taylor Plan Incentives
with Process Control

A worker is paid incentive wages using a Taylor plan incentive with 100-percent guarantee. The job has 20-percent process-control time. Regular pay is \$3.85 per hour and incentive pay is \$4.40 an hour. The standard is 3.6 hours per thousand. An operator's weekly production was 11,600. How much would the earnings be? Generally, in plans like this one, the machine or process-controlled time is paid at the standard rate. In a 40-hour week this amount is

$$(.20)(40 \text{ hours per week}) = 8 \text{ hours}$$

Eight hours will automatically be paid as daywork. The standard hourly production rate is

$$(3.6 \text{ hours per thousand}) = .0036 \text{ hours per unit}$$

Each hour would have an expected production of

$$1/.0036 \text{ hours per unit} = 277.8 \text{ units per hour}$$

In a 40-hour week, this amount would be (277.8 units/hour) (40 hours/week) or 11,112 units per week. The piece-rate at standard rate is

$$\$3.85 \text{ per hour}/277.8 \text{ pieces per hour} = \$.014 \text{ per piece}$$

A weekly production at standard would earn

$$(\$.014 \text{ per piece})(11,112 \text{ pieces per week}) = \$155.57 \text{ per week}$$

Added to this amount would be the process-controlled time paid as daywork, or

$$(\$3.85 \text{ per hour})(8 \text{ hours}) = \$30.80$$

In this example, actual production exceeded standard by

$$11600 - 11112 = 488 \text{ units}$$

These units of extra production are paid at the bonus rate of

$$\$4.40 \text{ per hour}/277.8 \text{ units per hour} = \$0.16 \text{ per unit}$$

Bonus earnings are

$$(488 \text{ units})(\$.016 \text{ per unit}) = \$7.81$$

Total earnings for the week are

$$\$7.81 + \$30.80 + \$155.57 = \$194.18$$

The Gantt Plan

The second area of similarity that we mentioned before was guaranteeing a 100-percent base pay. The Gantt plan only promises to pay the worker 80 percent of standard for production under that rate. Production above 80 percent of standard is paid an incentive directly relating to the quantity of acceptable product produced.

Example: Gantt Plan Incentives

A worker is paid using a Gantt plan incentive system with an 80-percent guarantee and a straight piece-rate incentive for production above this amount. The quantity produced in a day is 817 units. The standard is 10 hours per 1000. The hourly pay rate is $5.75 per hour. Earnings for this day are determined by the now familiar process of first calculating the expected eight-hour production. With a standard of 10 hours per thousand, each unit requires

$$10 \text{ hours}/1000 \text{ units} = .01 \text{ hours/unit}$$

The expected hourly production is

$$1 \text{ hour}/.01 \text{ hours per unit} = 100 \text{ hours per unit}$$

The expected standard production for an eight-hour day would be 800 units; 80 percent of this standard is

$$(.8)(800) = 640$$

Any production in excess of 640 will be paid at the straight piece-rate incentive. This rate is

$$\$5.75 \text{ per hour}/100 \text{ units per hour} = \$.0575 \text{ per unit}$$

The guarantee of 80 percent is

$$(640 \text{ units})(\$.0575) = \$36.80$$

The incentive pay for production over this 80-percent rate is the production in excess of the 80-percent level

$$817 - 640 = 177$$

These "extras" are paid at the straight piece-rate of $.0575 per piece, or

$$(177)(\$.0575) = \$10.18$$

The total pay for this day is

$$\$36.80 + \$10.18 = \$46.98$$

Because production exceeded the standard, payment by the Gantt plan was the same as if a straight piece-rate plan was used. Pay would be affected only if production had been between the 80-percent guarantee and the expected eight-hour production at standard; for example, if the operator in this example had produced 740 units, which is less than the 800 units specified by the standard for a typical eight-hour work day. Again the 80-percent guarantee is $36.80. Production over the 80-percent level is

$$740 - 640 = 100$$

These 100 units are paid at the straight piece-rate of

$$(100)(\$.0575) = \$5.75$$

The daily earning then become

$$\$36.80 + \$5.75 = \$42.55$$

This rate compares less than favorably with the amount that would have been paid under a 100-percent guarantee plan, or

$$(800)(\$.0575) = \$46.00$$

Sometimes a higher bonus rate is established, as in the Taylor plan. It is also possible to have a different minimum guaranteed. These parts of incentive plans are often negotiated between labor unions and management when incentives are part of the contract. When the conditions of the incentive payment are negotiated, remember that it is of utmost concern that the time standard *never* be subject to negotiation. If the rates are disputed by the union, then a mechanism can and should be defined whereby the standards are audited by an individual whose interest lies with making sure that the rates are fair and consistent, not on one side or the other of the dispute. It is recommended that an impartial third party handle this problem should it arise. Sometimes, however, it is a difficult clause to negotiate.

Example: Gantt Plan for Process-Controlled Job

A Gantt plan with an 80-percent guarantee is used on a job that is 30 percent process-controlled. The plan uses a standard hourly rate of $4.80 and a bonus rate of $6.00 per hour. The standard for the job is .05 hours per piece. Dayrate is paid at the standard pay rate. Determine the earnings for a worker if 187 pieces are produced on a given day.

Generally, in a plan like this one, the dayrate is paid for the entire expected output, although it may be negotiated otherwise. The dayrate for this job, which is 30 percent machine- or process-controlled, is

$$(.30)(8 \text{ hours})(\$4.80 \text{ per hour}) = \$11.52$$

The standard rate is paid for the guaranteed 80 percent of the expected daily output. This output is

$$(1)/(.05 \text{ hours per piece}) = 20 \text{ pieces per hour}$$
$$(20 \text{ hours/hour})(8 \text{ hours})(.8) = 128 \text{ units}$$

The rate for each of these pieces is

$$\$4.80 \text{ per hour}/20 \text{ units per hour} = \$.24 \text{ per piece}$$

Thus, the guaranteed pay, including the pay for the process-controlled part of the job, is

$$(128 \text{ units})(\$.24 \text{ per piece}) + \$11.52 = \$42.24$$

The bonus rate is paid for production in excess of 128 units. This bonus rate is

$$\$6.00 \text{ per hour}/20 \text{ pieces per hour} = \$.30 \text{ per piece}$$

In this example, the production to be paid at the bonus rate is

$$187 - 128 = 59 \text{ units}$$

The incentive pay is

$$(\$.30)(59) = \$17.70$$

And the total earnings for the day would be

$$\$42.24 + \$17.70 = \$59.94$$

Had the employee been paid by the straight daywork rate of $4.80 per hour, the daily pay would only have been

$$(\$4.80)(8) = \$38.40$$

By paying the worker an incentive, the company was able to secure higher production from the same resource. It cost the company more, $21.54 to be exact, but the company got more for their money than they would have had they been required to invest in a second work station and hire a second operator. This example shows productivity improvement—getting more for less.

All of the incentive systems we have looked at so far, actually prototypes of the many incentive plans, have been concerned with providing more money to the worker in return for the worker producing more units of product. They have assumed that the sale of these extra units is possible and that the sales of these units will automatically bring a beneficial return to the company. Although these assumptions are reasonable, they don't always hold true. Sometimes, there is no benefit to the company and sometimes there is a large benefit to the company. Sometimes, productivity can be improved by workers through methods other than simply producing more units in the same amount of time. When this situation happens, workers might think that it is fair to share in the gains of the increased productivity.

There are a number of productivity-sharing plans that have been around for a long time.* These plans, the earlier mentioned Scanlon plan and the Lincoln Electric plan, don't immediately pay the worker for increasing productivity as the individual incentive or the basic group incentive plans do. Rather, all of the workers share, with the company,

---

*Productivity-sharing plans differ from profit-sharing plans in that productivity-sharing plans show a very rapid sharing of the gains, whereas profit-sharing plans usually are long-term.

the gains from increased productivity. This sharing of the fruits of increased productivity occurs on a regular basis. The workers generally receive a share of the productivity gain commensurate with their individual contribution to that gain.

THE SCANLON PLAN

The Scanlon plan is a company-wide incentive plan rather than an individual incentive plan. There are three major components usually found in this type of plan:

1. A high degree of labor-management cooperation
2. Bonus pay for increases in productivity
3. A suggestion system designed to increase efficiency and reduce costs.

Labor-Management Cooperation

The philosophy behind labor management cooperation is basic to the Scanlon plan. It involves a sharing between labor and management of problems, goals, and ideas. In effect, Scanlon plan users say, "We're all in this together." People work more productively so that the company will benefit and so, in turn, everyone associated with the company will benefit. In individual incentive plans, all the employees are looking out for themselves. In these plans, the philosophy is that if each individual benefits, then the company will as well. There is an important distinction between traditional incentive plans and productivity-sharing plans.

It must be noted that the Scanlon plan is not a substitute for collective bargaining or a tool to drive a potential union away. According to the National Commission on Productivity and Work Quality Report of May 1975, the installation of a Scanlon plan was becoming a relatively common and appropriate topic for collective bargaining agreements. There are many successful applications of Scanlon plans in organizations where the employees are represented by a labor union.*

The key element to the success of the Scanlon plan is a basic confidence and trust that the sharing of information and knowledge will benefit all concerned. Exchange of information forms a foundation for mutual efforts to improve productivity. This collaboration, the "we're all in it together" feeling gives the worker an increased feeling of value, both to himself and the company. It also provides the worker with the chance to see his company and himself earn more money. Management doesn't abdicate its responsibility, but the worker is given an active role in the operation of the company.

Bonus Pay for Increased Productivity

The second component of a Scanlon plan is bonus pay for increased productivity. Basic labor costs, as represented by time standards or any other measures, must be determined

---

*The recent UAW agreements with GM, Ford, and Chrysler include profit-sharing plans, as have many other recently negotiated contracts.

before the Scanlon plan is installed. When the base ratio of labor costs compared with sales decreases, it means that productivity has increased. If labor costs remain relatively constant and sales increase, there should be an increased income, and profit, for the company. This increased profit is shared between the workers and company. Although the split is usually negotiated, one of the more common distributions of the extra profit generated by increased productivity is 50-50. Thereby, the workers and company share equally in the gains resulting from increased productivity. This bonus is paid on a regular basis (for example, quarterly) so that workers can see some tangible and fairly rapid returns for their increases in productivity. The 50-50 ratio has evolved because it is all most managements will give to their employees. Although, according to the base ratio of the Scanlon plan, the savings are directly attributable to the labor factor, management has traditionally shown a reluctance to give more than 50 percent. Unions have recognized this threshold and have been willing to accept an even split with management.

The Suggestion System

The final component of a Scanlon plan is the suggestion system. In most cases, companies implement the philosophy of participation with a committee system made up of departmental committees and an overall steering committee. The committees, made up of "production" workers, supervisors, and technical support personnel, meet to discuss suggestions for improvements. These committees meet regularly, usually once a month, although there are some that meet as often as once a week. The committees, regardless of the frequency of their meeting, make sure that all suggestions and resultant actions have been recorded. The committee processes all suggestions and attends to previous suggestions on which action has not been completed. Although the majority of these suggestions are usually approved by the committee, the departmental committee does not have the right to accept or reject the ideas presented.

The minutes of the production committee are forwarded to the screening committee, which meets regularly, usually just after accounting makes figures available reporting the results of the previous period. This committee, chaired by a top company official but also including union officers, employees, and other company representatives, analyzes the previous period's performance. A second function is to take up any company problems or matters of interest that management wants to communicate to all employees. This committee, because it is representative of all employees, provides a useful vehicle for communications. Third, the screening committee discusses suggestions that have not been resolved at the first committee level. The screening committee makes the final decision on the implementation of all suggestions (*Harvard Business Review*, September 1969).

The Scanlon plan has been used successfully to increase productivity. In one successful application of the plan, a company broadened the duties of the employee members of the committees to make them responsible for chasing down the results of the pending suggestions. Making this responsibility the employee's, who has a vested interest in seeing about the suggestion, greatly facilitated getting rapid action on employee suggestions.

In addition to greater productivity, the plan, due to its worker participation emphasis, tends to give the workers a greater sense of satisfaction. This byproduct is very significant and cannot be overlooked. As a matter of fact, this sense of satisfaction is an important element in the Quality of Work Life programs that will be discussed in the next chapter.

THE LINCOLN ELECTRIC PLAN

One of the most cited of all incentive plans is the Lincoln Electric plan. One of the best descriptions of this plan is found in Henderson's *Compensation Management.**

> Upon being made general manager of the Lincoln Electric Company in 1914, James Lincoln established an incentive compensation plan. After 20 years of effort, the incentive plan evolved substantially into the plan that exists today. Its principal features are as follows:
>
> 1. The company guarantees 30 hours of work 50 weeks a year to each employee who has at least two years of service. (It guarantees no specific rate of pay, and the worker must be willing to transfer from one job to another and work overtime during periods of peak demand. The company reserves the right to terminate the agreement providing six months advance notice is given.)
> 2. Standard job evaluation procedures set the base wage, suing the six compensable factors of mentality, skill, responsibility, mental application, physical application, and working conditions to determine the importance of each job. The combination of job evaluation and labor market requirements then sets the actual dollar worth of the job.
> 3. The majority of employees are on a piecework incentive plan. The factory products—arc welding equipment and electric motors—lend themselves to standardized operation and the setting of rates. Both the workers and management, however, recognize labor's opportunity to improve both the quality and the quantity of output. Every possible job that can be standardized has a piece rate. Rates are set through the use of normal time-study procedures. The jobs that are not on piece rates include clerical work, tool room operations, maintenance and repairs, and experimental work. New employees and employees on new jobs receive a temporary exemption from piecework standards. Employees in a few small assembly operations work on a group piecework plan.
> 4. All employees may participate in the suggestion program with the exception of department heads and members of the engineering and time-study departments (suggestions for improvements are a fundamental part of their jobs). Any suggestions that lead to organizational progress (e.g., improved manufacturing methods, sales, procedures, waste reduction, or new or improved products) are considered during merit rating.
> 5. Twice a year, a merit-rating program appraises the actual work performance of each employee. This appraisal program uses four report cards. Each card rates the work performance according to one of the four following work variables: (1) *dependability* (the ability to supervise oneself, including one's work safety performance, orderliness, care of equipment, and effectiveness in the use of one's skills); (2) *quality* (one's success in eliminating errors and reducing scrap and waste); (3) *output* (one's willingness to be

*Reprinted with permission of Reston Publishing Company, a Prentice-Hall Company, 11480 Sunset Hills Road, Reston, Virginia 22090.

productive, not to hold back work effort or output, and recognize the importance of attendance); (4) *ideas and cooperation* (initiative and ingenuity in developing new methods to reduce costs, increase output, improve quality, and effect better customer relations). The supervisor doing the rating informs subordinates of their scores. The individual scores of each group are posted by number only. It is possible, through the process of elimination, to identify the score of a specific employee. Many employees openly state their scores. Managers at levels above that of immediate supervisor responsible for appraising performance take an active role in reviewing all merit ratings.

6. Each employee annually has the opportunity to purchase from ten to 25 shares of company stock. Upon the employee's retirement or termination of employment, the company has an option to repurchase the stock. Currently, about 25 percent of the employees own 45 percent of the stock.

7. Employees elect representatives to an "advisory board." This board has the opportunity to suggest changes in policies and operation; however, the final decision on all changes is made by management.

8. Independent work groups or "subcontractor shop" operations, in which employees have the opportunity to earn specified piece-work rates, perform their own quality control and develop their own production procedures in completing subassembly operations within given cost, quantity, and quality parameters.

9. All profits of the business are split three ways: (a) the corporation retains a certain share for capital improvement and financial security; (b) shareholders receive approximately 6 percent to 8 percent dividends based on the book value of the two types of company stock; and (c) employees receive all remaining profits.

10. The annual cash bonus earned by the employees closely approximates their annual earnings. The actual distribution an employee earns is a function of the employee's annual earnings as a percentage of the total labor cost, individual performance appraisal merit rating, and total amount of profit earned by Lincoln Electric.

The year-end bonus plan was initiated when James Lincoln turned down a request for a 10-percent increase in wages in 1934 because he felt the profit picture would not warrant such an increase. The workers then responded with the request for a year-end bonus if, through increased productivity and lowered costs, the year-end profits were larger. After some deliberation, Lincoln agreed to this efficiency-oriented proposal. To everyone's surprise (including Lincoln's), the bonus amounted to $350 instead of the $35 to $50 expected by Lincoln. In 1980, over 2,600 employees shared a bonus of $46 million. Over the past several years, the annual bonus has ranged from a low of 88 percent to a high of 115 percent of annual earnings.

By many standards, Lincoln Electric is not an easy place to work. There is no room for the "goof off" or "I don't care" worker. The success of the entire business depends on a high level of contribution by each member. There is a mutual understanding of need and a mutual respect based on democratic principles espoused and lived by James F. Lincoln.

The democratic principles go much deeper than the basic elements of the previously described incentive system. For example, there are no reserved employee parking spaces; there is one cafeteria (with excellent food) and all employees—workers and managers

321

alike—sit wherever spaces are available. In addition, there is a policy of promotion from within that requires all promotional opportunities to be posted (including many senior positions). There are no definite lines of promotion, and promotions are given by qualification only. The benefits program includes a two-week paid vacation for employees with one year of service, three weeks for 13 years of service, four weeks for 19 years of service, and five weeks for 25 years of service. There is a paid medical, surgical, and hospital plan; life insurance; and a retirement plan beginning at age 60 with pension based on years of service and total earnings excluding bonus. Other benefits include an annual picnic, company dinner, and a Quarter-Century Club.

In addition, employees may challenge any time study. If a time study results in a lowering of a rate, the involved employee may request transfer to a job that pays an equal or higher rate. Piece-rate is not a tool of speedup but rather a tool of fair or equitable distribution of rewards for the effort of a motivated employee. There is no limit to earning, and no rate can be changed unless there has been a change in method, design, or tooling. Employees challenge less than one-fifth of 1 percent of all rate changes. There is a periodic review of all rates.

The principles of incentive compensation are a fundamental part of the democratic process. Lincoln Electric has the *highest-paid* factory workers in the world and, measuring in units of work produced, the *lowest-cost* workers in any factory in the world in a similar line of work.*

SUMMARY

As noted consultant Mitchell Fein said, "To achieve high productivity, management must manage effectively and enhance the will of the employees to work. The will to work is enhanced by removing restraints to, by assuring employees' motivation, and providing motivation" (Fein, 1970, p. 35).

In this chapter, we have examined incentive plans. Chapter 12 will survey some of the better known or better publicized motivational procedures that have been suggested for improving productivity. Included among these are Quality of Work Life programs, quality control circles, and some alternative working plans such as flexitime. These motivational techniques don't have all the answers, but an awareness of them is important when trying to improve an organization's productivity.

REVIEW QUESTIONS

1. What is an *incentive wage payment plan?*
2. How do incentives relate to productivity improvement?
3. What four characteristics must be present for an incentive pay system to work?

*J. F. Lincoln, "Incentive Compensation: The Way to Industrial Democracy," *Advanced Management*, February 1950, pp. 17–18.

4. What is meant when an incentive system is *(said to be)* technically sound?
5. What is a "loose" rate?
6. What is a "tight" rate?
7. What consequences might result from a loose rate?
8. What consequences might result from a tight rate?
9. What conditions must be present for workers to really support an incentive plan?
10. What is the average (reported) of incentive earnings in this country?
11. How much of an upper limit should be placed on incentive earnings?
12. What are the advantages of individual incentive plans?
13. What are the disadvantages of individual incentive plans?
14. What are the advantages of group or productivity-sharing incentive plans?
15. What are the disadvantages of group or productivity-sharing incentive plans?
16. What is *daywork?* Why is it part of incentive plans?
17. What is meant by the *guaranteed minimum wage* in an incentive plan?
18. Describe the straight piece-rate incentive plan with 100-percent guarantee.
19. Describe the standard hour incentive plan with 100-percent guarantee.
20. Describe the Taylor-type incentive plan.
21. Describe the Gantt-type incentive plan.
22. What is process-controlled time? How does an incentive plan handle this situation when it occurs?
23. What is the recommended way to change the rate at which incentive pay is calculated and paid?
24. When and how should a time standard be changed? What effect would changing a standard have on an incentive pay system?
25. Describe the major components of the Scanlon plan.
26. Describe the major features of the Lincoln Electric plan.

PRACTICE EXERCISES

1. A worker is paid on a daywork rate of $6.13 per hour. The standard on the job is 7.5 hours per 1000 units. During a given 40-hour work week a worker produces 5541 units. Calculate the wages for the week.

2. A worker is paid on a straight piece-rate incentive plan with 100-percent guarantee. Hourly pay is $5.64. The standard on the job is 4.8 hours per 1000 units. (a) During a given week, a worker produced 9100 units. Calculate the wages for the week. (b) During a second week production was only 7200 units. Calculate wages for this week.

3. A worker is paid on a standard hour plan with 100 percent of standard guaranteed. The job is 35 percent process-controlled. Hourly wages are $7.88. The standard is .125 hours per unit. In a given day, a worker produced 75 units. Calculate wages for the day.

4. A worker is paid on a Taylor-type of incentive with a 100-percent guarantee. The process is 22 percent machine-controlled. Hourly wages for production up to guarantee are $4.75. Bonus wages are $5.50 per hour. The standard for the job is 18 hours per 1000. Weekly production was 2250. Calculate the wages for the week.

5. A worker is paid on a Gantt-type incentive plan with a 75-percent guarantee. The process is 13 percent machine-controlled. Hourly wages are $7.05. The standard is .64 hours per unit. On five consecutive days, the production was as follows: Monday—10; Tuesday—14; Wednesday—7; Thursday—11; and Friday—13. Calculate wages for the week (a) on a daily basis and (b) on an overall weekly basis.

6. A worker is paid on a standard hour plan with 100 percent of standard guaranteed. Hourly wages are $11.55. The standard for this task is 6.4 hours per 1000. If, in a given week, production was 6633, determine the wages for the week.

7. A 17-percent process-controlled job is paid using a straight piece-work incentive at an hourly wage of $4.87. Standard time per piece is .075 hours. 100 percent of standard is guaranteed. (a) If production is 1168 units on a given day, what would the wages be for that day? (b) At what production level does the incentive start paying the worker?

8. A Gantt plan has been modified to guarantee 85 percent of standard at a regular rate of $6.00 per hour. For the production over the guarantee, the rate increases to $6.50 per hour. If the standard is 2.6 hours per thousand and the daily production is 4000, what will the wages for this day be?

9. A Taylor plan has a base rate of $3.75 an hour and a bonus rate of $4.75 per hour for a job that is 100 percent of standard guaranteed with a standard of .06 hours per unit. If, in a given week, there are 7450 units produced, calculate the weekly wages. What would the pay be if these 7450 units were produced in equal amounts each day during a five-day week and the pay calculated every day instead of once a week? Would it ever make a difference if the pay were calculated in this way rather than weekly?

CHAPTER 12

Alternative Methods for Increasing Productivity

OBJECTIVES

After completing this chapter, the reader should have an under-standing of the following methods of motivation and productivity improvement:

- Quality of work life
- Quality control circles
- Flexitime
- Job enlargement/job enrichment
- Job rotation.

The reader should also have an appreciation of the problems involved when one of these, or any other alternative productivity im-provement method, is employed.

INTRODUCTION

This chapter could have been subtitled, "If Japan Can, Why Can't We?" but that title was taken by the award-winning NBC Television White Paper of 1980 that examined productivity. Traditionally, our country has led the world in productivity and productivity gains. However, as we saw earlier in this book, this tradition is rapidly changing. Although our traditional measurement and analysis tools are effective, there are, perhaps, even better or at least complementary methods to use that will permit us to work smarter, not harder.

Thus, this chapter is called Alternative Methods for Increasing Productivity. The topics of this chapter, however, are more like a theme with variations. Revolving around what is known as Quality of Work Life, topics such as formal Quality of Work Life Programs, Quality Control Circles, Flexitime, and so on address productivity improvement from a different perspective than the one that has been emphasized in this text. There are no magic formulas. If there were, they would be used universally and there would be no productivity problems. However, they do hold some promise that, if implemented and administered properly, they will lead to increases in productivity.

> The best, and the most productive employer-employee relationships are formed where there is open communication and a sense of worth and mutual respect between participants. The least productive are those based on paternalism, where the workers are treated like unruly children, or the "license plate shop" theory where the workers are treated like prisoners (Donohue, 1980, p. 13).

Each of the following sections is presented as a separate entity, although the reader will easily discover the common thread that binds them together. The sections have been written to provide the reader with an appreciation and understanding of the concept and to cite some instances where the procedure has worked and, just as importantly, to illustrate some of the method's improper uses. Hopefully, the reader will be able to identify some potential applications and will delve more deeply into appropriate methods.

QUALITY OF WORK LIFE

A relatively new concept in the productivity improvement process is the Quality of Work Life, or QWL program. Quality of Work Life programs are industry's attempt to improve productivity while allowing employees a more enjoyable work environment. General Motors and the United Auto Workers are leading this country into this style of participatory management. American Telephone and Telegraph, along with the Communications Workers of America and American Steel Firms and the United Steel Workers have joined GM and the UAW in nationwide QWL programs.

The need for the programs was caused by the changing social values in this country, by the higher education levels achieved by our citizens, and by the stagnation of productivity growth that was described earlier in this book. The majority of our work force belongs to the "baby-boom" generation. These workers have had it relatively good, never experiencing the "Depression" or having to stand in soup lines like many older workers. Howe and Mindell (1979) summarized some of the outstanding differences between traditional (older) and contemporary (younger) employees. These differences are as follows:*

- The traditional employee displays more loyalty and commitment to the company than the contemporary employee.
- Compensation has always been important to employees, but tends to be valued more as a consequence of performance by contemporary employees.
- Contemporary employees are more concerned with organizational recognition than are traditional employees.
- In contrast to traditional employees, contemporary ones have more desire to participate in decisions that affect them.
- Traditional employees are more concerned with job security.
- Contemporary employees value communication from management as to what's going on in the company.
- Contemporary employees have more short-term goal orientation, as opposed to traditional employees, whose is long-term.
- More than traditional employees, the contemporary group desires work to be challenging.

*Reprinted, by permission of the publisher, from "Motivating the Contemporary Employee," by Howe and Mindell, *Management Review*, September, 1979. Copyright © 1979 by AMACOM, a division of American Management Associations, page 52. All rights reserved.

'. . . *You've Gotta Be Kidding!* . . .'

Figure 12-1 (Reproduced with permission from the Atlanta *Constitution*, September 7, 1982, p. 4A.)

- Contemporary employees are more concerned that their work be worthwhile.
- Contemporary employees desire development opportunities.
- Contemporary employees want their work to be interesting.
- Contemporary employees want their work to be creative.
- Traditional employees usually put their work before family and leisure; contemporary employees tend to place their priorities first, with leisure, then family and work.

Two significant events of the 1960s and 1970s, namely Vietnam and Watergate, led to a strong demand by these contemporary workers for personal rights (Guest, 1980). Younger workers do not like the authoritarian (Taylor) nor the bureaucratic (Weber) approach to management. Managers brought up in the classic American traditions generally do not like younger workers. Managers at the Tarrytown, New York, plant of General Motors went on record describing their young workers as, "a bunch of hippies who only wanted to work a half day for full pay" (Guest, 1980, p. 311).

Higher educational levels are another reason for needing QWL programs. More people in general, and more women and minorities in particular, are getting college degrees. The mass communication media daily educates workers about economic and political issues. Workers believe they can more profitably contribute to their jobs than they have been allowed and these workers are not at all hesitant about speaking up.

The "productivity problem" is another factor that has shown the need for American industry to consider QWL programs. As has been noted, foreign countries such as Japan and Germany have, for the most part, been beating U.S. companies in the marketplace. According to Jerome Roscow, President of the Work in America Institute, the belief has sunk in that companies and unions, working together, can increase productivity and improve profit while achieving worker satisfaction through the implementation of QWL programs (*Business Week,* May 11, 1981).

There is a need for increased productivity and quality. Some strong claims have been made that QWL programs are the means to that end. How this program works, some success stories, and further discussion follow.

QWL Implementation

Quality of Work Life programs require a redefinition of what has been traditionally found in our employer-employee and labor-management relations. Three significant changes must be made for QWL to work (*Business Week,* May 11, 1980). It must be pointed out that it is far easier to describe what has to be done than it is to actually do it.

Nonadversarial Relationship First, a nonadversarial relationship must be developed on the production floor. Supervisors and workers have to learn to work together, to mutually solve problems. The "us against them" feeling that is prevalent in American industry must be eliminated. This item represents a major challenge and a major change in philosophy.

> Change must start with management taking the first steps, unilaterally and without *quid pro quo* . . . Management must provide the basic conditions which will motivate workers to improve productivity: job security, good working conditions, good pay, and financial incentives. There must be a diminution of the win-lose relationship and the gradual establishment of conditions in which workers know that both they and management gain and lose together (Fein, 1974, p. 87).

Collective Bargaining Second, the collective bargaining process must be reformed. The contracts negotiated must be based on the mutuality of interests developed on the shop floor. Without this change, the old adversarial relationships will not die. When no union is present, management must carefully explain the change in philosophy, although a good argument can be made that, if there is no union and no threat of a union, management and the workers are well on their way to successfully completing these first two requirements. The existence of a union doesn't necessarily mean that these two requirements cannot be met if both sides want them to be met, however.

Participatory Management Third, management must be willing to change. Management must be willing to move from the classic hierarchical form, as suggested by Taylor, where everything flows down to a broader participatory base. As in the case of the

Scanlon plan, employees must be permitted to help make decisions and management must abide by the jointly made decisions or be willing to completely explain why it can't or won't.

Adoption of this participatory management program is often delayed or even prevented by obstacles. It is one thing to say, "change your philosophy;" it is quite another to actually achieve this change. Three obstacles have typically blocked the adoption of QWL programs: union (worker) acceptance or lack thereof, first level supervision's acceptance or lack thereof, and top management's willingness or unwillingness to establish realistic long-term goals for QWL programs. Thus far, only one of these obstacles has been remedied to any great degree. Unions are accepting QWL programs but work still needs to be done on the other two obstacles.

The first documented union reaction to QWL programs was that management was trying to increase production without doing anything else. QWL will not succeed if it is only used to increase productivity (Guest, 1980). That productivity will increase when the program is implemented has been documented, but the programs have not worked when increased productivity was the primary objective.

The United Auto Workers was the first union to deal with quality of work life. Top management of Harmon International Industries, a Tennessee-based mirror manufacturing company, invited the UAW to work with company officials on a QWL program. The objectives of the initial program were twofold: to improve the job quality and increase productivity. The second part of the objective is easy enough to evaluate. The output of some assembly lines, during the QWL program, doubled (*Business Week,* May 11, 1981). However, quality of the job is harder to quantify. Starting with a traditional factory environment, the work environment seemed to have changed so that there was a spirit of cooperation between the workers and supervisors. The old, authoritarian relationship seemed to disappear. The reliance on a rigid set of rules disappeared. There was, on the surface, an openness and informality rather than a rigid and fixed system of interrelationships. And, at top management's insistence, interpersonal relationships seemed to improve. Workers were viewed as people, not just as elements of the production process.

All of these characteristics are typical of QWL programs. They are what the programs are supposed to evolve into. Unfortunately, the success at Harmon was short-lived. The owner sold the plant and the new top management team did not support the concept of QWL (Macy, 1980). However, the concept was not lost on the UAW. Although the Harmon plan did not have the support of management, it seemed to flourish while it had the complete support of top management. It failed only after the new management team withdrew that support.

General Motors began to use participatory management in its nonunion plants, perhaps as a means of circumventing union organization at these facilities. However, the UAW knew first-hand of the potential benefits of QWL programs and endorsed the concept in their collective bargaining with GM. The most famous use is at GM's Tarrytown, New York, facility. At one time, this facility was producing the worst quality of any General Motors plant. In lieu of closing the plant, GM, with the cooperation of the UAW, installed the QWL program. After this effort, Tarrytown ranked among the best of GM's manufacturing plants (CBS News, 1977; *Time,* May 5, 1980). Today, the UAW has a

National Joint Committee to Improve the Quality of Work Life, with 50 QWL programs in UAW-GM bargaining units (Bluestone, 1980).

The unions have accepted QWL. First-line supervisors remain the largest hurdle to undisputed acceptance of QWL. Many supervisors achieved their position because of long years of hard work. QWL threatens their perceived position of control. Proper training and assurance from top management is required before QWL can be implemented. The QWL programs must be permitted to evolve naturally and a true belief must be developed in their value. Remember that, at the first QWL installation, QWL failed when the owner sold the company. When the driving force left, the program left as well. Had it really been established, really been supported by all of the employees, supervisors, and managers, it might have survived the change in company ownership.

Long-term Objectives Setting long-term objectives is the third major problem facing QWL programs. Patience would, perhaps, be a better term to use. QWL marks a fundamental shift in attitudes and thinking. These changes don't come overnight. Expecting too much too soon can sabotage any QWL program. Once unions, supervisors, and top management accept this patience and realize that QWL is a reality, QWL can work as advertised and programs can start to achieve the long-term objectives of increased worker satisfaction, higher productivity, and improved product quality. This realization takes time. The programs that have failed have often made obvious mistakes in this respect.

Studies made at locations where QWL has existed long enough to be an indicator of success have revealed, "a more constructive collective bargaining relationship, a more satisfied work force, and improved product quality . . ." (Bluestone, 1980, p. 41). The jury is still out, however. The ultimate success of QWL is still questionable. Unions are endorsing the programs while companies are experimenting with this new form of participatory management. Workers want a change in their work environment and they want a say in how they work; QWL programs provide it. The byproducts of quality and productivity improvement certainly make it worthy of investigation.

QUALITY CONTROL CIRCLES

Quality control circles or quality circles is a type of QWL program. It involves the use of a company's most valuable and important resource, its workforce. According to Donald L. Dewar of the International Association of Quality Circles, the concept is a way of "capturing the creative and innovative power that lies within the workforce" (Dewar, 1979, p. 161). A *quality circle* is defined as a group of volunteer workers, "from the same group who usually meet for an hour each week to discuss problems, investigate causes, recommend solutions, and take corrective action when authority is in their purview" (Rieker, 1976, p. 90). Quality circles aim to effect improvements in product quality, productivity, and motivation of the workforce. Quality circles are effective because they give workers the opportunity to influence decisions about their work. When this opportunity is offered, the workers are then more likely to become interested and concerned with the output. Again, according to Dewar, "when fully implemented, quality circles create in the individual a sense of participation and contribution'' (Dewar, 1979, p. 161). This statement almost is similar to our description of the desired objectives of a QWL program.

Quality Circles in Japan

Quality circles began in Japan, which had to overcome a terrible worldwide reputation. At one time, the phrase ''made in Japan'' was virtually synonymous with ''junk.'' To make a drastic change in world opinion, a major change had to be made in Japan. Japanese industrial leaders recognized this need and were committed to making the required changes. The first step in this massive change was training. The Japanese turned to the country with the best expertise and knowhow, the acknowledged world leader at the time, the United States. Two of the Americans who provided this training were W. Edwards Deming, in statistical quality control, and Dr. Joseph M. Juran, in quality management. These gentlemen made their initial contributions to the Japanese in the early 1950s (Gryna, 1981).

While Americans traditionally concentrated on training technical personnel such as engineers, the Japanese were so committed to implementing productivity and quality improvements that they trained not only engineers and managers, but also workers. The desired results, when the workers were able to apply these procedures, were fewer communications problems, less resistance to change, and more productivity. It was based on the premise that the person who understands the work best, the worker who does the work every day, should be the person who is responsible for quality (Irving, 1979).

The first quality circle was recognized in Japan in 1962 (Konz, 1979). Since that time they have flourished. Gryna (1981) reported that, in 1980, there were about 1,000,000 officially recognized quality circles in Japan. Dr. Juran, in addressing the Japanese Union of Scientists and Engineers, summarized the results thus far in Japan:

- Collectively, the millions of improvement projects have saved billions of dollars.
- The reliability and salability of Japanese products has increased, thus, increasing Japanese market share, thereby making the jobs of the Japanese workers more secure.
- Worker motivation has improved.
- Workers have an increased sense of participation.
- Quality circle experience makes workers better prepared to assume positions in management (Gryna, 1981, p. 12).

Quality Circles in America?

If quality circles have been around since 1962, and if they are indeed a significant part of a productivity improvement program, along with training, leadership, and so on, the question must be raised, ''Why hasn't American industry rushed in to adopt the program?'' The most frequently given answer to this question reflects Frederick Taylor's scientific management theory, which has served as the basis for the organization and management of our industrial system. Our system, using Taylor's Scientific Management, was designed to aid mass production by relieving the worker of all but the most elementary decision-making responsibilities. The quality circle concept, however, cannot work in this type of system because no motivational technique can work properly in such an environment. Obviously then, quality circles are not for everyone.

Gregerman (1979) listed the conditions that should be present within an organization that will give quality circles the best chance of succeeding:

1. The management should not be authoritarian. It should be a management that stresses participation and that shows a real interest in improving conditions.
2. There should be good communication between labor and management. If there is a labor union, the union leaders must be included at the earliest stage of planning for quality circles.
3. Management must include the first-line supervision in the communication process. First-line supervisors are the key to the program.

Establishing a Quality Circle Program

Presuming that an organization is serious about using quality circles, our next question should be, "How are they organized?" Gryna (1981) offered the following "road map" to designing a quality circle program. He cautioned, however, that, "the road is not always straight and it can be rocky" (p. 41).

Select a Coordinator The first step in establishing a program is to select a coordinator. The coordinator is responsible for the project and is charged with implementing the rest of the program. It is felt that the program will work best when the coordinator is a member of the line rather than of the staff organization. The line organization will ultimately either make or break the program.

Organize a Steering Committee The first task the coordinator should perform is to organize a steering committee. This committee, quite simply, oversees and directs the program within the organization. The prime task this committee initially faces is setting objectives for the program. Once objectives are set, the committee will set the policies that govern the operation of the circles. The steering committee, in effect, gives the circles their authority to act. It also provides the physical resources necessary for the program to succeed. As the program develops, the steering committee can provide advice and a continuing sense of direction. This committee, which must have the full support of top management, can also ease the way for quality circles to obtain needed data.

Establish Objectives As was mentioned earlier, one of the main tasks initially facing the steering committee is defining objectives of the quality circle program. Although the objectives will not be the same for every organization, it will be illustrative to look at one company's quality circle objectives. Dewar (1976) reported the following as some of the objectives at Lockheed:

1. Promote job involvement and employee motivation
2. Enhance quality of products and services
3. Promote more effective teamwork
4. Promote training in problem-solving
5. Promote employee decision-making
6. Provide job enrichment opportunities
7. Further communications within the organization
8. Improve the professional posture of hourly workers
9. Improve management-employee relations.

Establish Policies After the objectives are established, and it is important to note that the objectives should be set for each organization based on that organization's needs, the policies to govern the operation of the quality circles must be established. Included in the areas where policies should be developed are the following (Gryna, 1981, p. 47):

1. Voluntary vs. compulsory circle membership
2. Scope of projects for circles
3. Types of management control for circles
4. Size of circles
5. Selection of circle members
6. Selection of circle leaders
7. Personnel-to-manage circle program
8. Frequency of circle meetings
9. Training provided to circles
10. Conflict resolution
11. Noncircle/member relations
12. Evaluation of circles.

There are, undoubtedly, many more areas where quality circle policies must be developed. However, this listing is certainly illustrative of the type of situations that might require direct action by the chief without consulting with everybody under the sun.

Plan a Pilot Run Once these preliminaries are completed, the steering committee would next plan a pilot run. The following activities are necessarily a part of any pilot run.

A name should be selected that reflects the objectives of the group. The Japanese selected "quality control circles" and many Americans have adopted "quality circles." However, depending on the company's objectives, the name may employ a different purpose. Squires (1981) cautioned about the proper selection and use of a name for a quality circle. If it is not handled properly, the circle is viewed as merely another management "gimmick" to reduce costs and the choice of name can determine the image that is projected.

Concurrent with naming the group is explaining its function and purpose, as well as the policies that govern it, to all levels of the organization's management. Management's understanding and support are required for the success of the program. The same type of explanation, naturally, should be given to the union officials, and the rest of the employees. If educational or training material is needed, it should be developed for trial use in the pilot program. The quality circle, especially in the pilot program, must be led until it knows where it is going.

A very important part of the pilot program is the selection of facilitators. In quality circles, "the facilitator helps a circle get started, sometimes acts as a technical consultant, and generally oversees the circle to help it overcome obstacles" (8-53). Coortwright (1979) of Hughes Aircraft listed some of the characteristics that Hughes looks for in selecting facilitators:

1. The facilitator likes people.
2. The facilitator is a leader.
3. The facilitator is a trainer.
4. The facilitator can carry responsibility.
5. The facilitator is available to meet with and work with the quality circle in its activities (p. 686).

After the facilitator is chosen, the circle leader must be selected. The circle leader conducts the meeting and coordinates the circle's activities. Usually, the circle leader is the circle members' regular supervisor. He or she has been trained in problem-solving skills. When conducting a circle meeting, the leader must temporarily forget that he or she is the "boss" and work with the members in joint problem-solving exercises. Not surprisingly, this change is difficult. As Gryna (1981) reported:

> "Private discussions were held with supervisors who were circle leaders. All were worried and doubtful when the role of leader was first proposed to them. In time most became comfortable with the rule. They learned that the circle could make their department run more effectively" (p. 55).

A typical organization chart for a quality control circle is shown in Figure 12-2.

Once leaders are selected, the facilitators must train them both in the conduct of a quality circle meeting and in developing their problem-solving skills. Topics such as the following are part of this training:

- Nature and purpose of quality circles
- Brainstorming
- Data summarization techniques
- Making presentations.

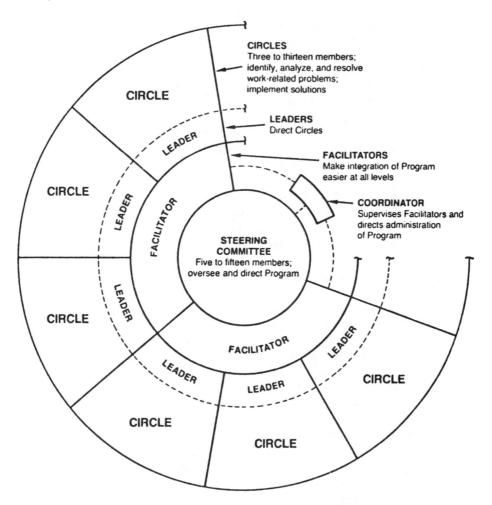

Figure 12-2 Quality Circles Organization. (From "Our Experience with Quality Circles" by J. Hanley. *Quality Progress,* February 1980, p. 24. Reprinted with permission.)

Generally, circle meetings are regularly scheduled for one hour per week. Formal minutes are kept to provide a permanent record of items discussed, responses generated, and future commitments. The methodology employed by a typical quality circle, as suggested by Schleicher (1981) and reprinted with permission of *Quality* magazine, is as follows:

1. The circle identifies problems. The identification may come from members who've been concerned with it on an individual basis. It may be suggested by the circle leader, or a member of management, or another department. There are always enough problems to tackle.

2. The circle selects the problem. The selection of a problem for solution is a prerogative of circle members.

3. It analyzes the cause. Tremendous contributions have been made by circles in analyzing causes of problems. Because of their daily contact with a specific type of work, circle members are in an excellent position to perform the analysis function. With technical help from supervisors and/or technical support personnel, causes are uncovered to help eliminate the problem.

4. The solution is developed. A solution is agreed upon through the analytical problem-solving method or through brainstorming.

5. The solution is presented to management (p. 30).

Once all of this framework is in place, the pilot program is implemented. Results are carefully monitored and the system refined or tailored for the specific company. After the bugs are worked out, the program is expanded following the policies developed. At that time, the program is producing results.

Quality Circle Success Stories

There have been, obviously, many successes in this country with quality circles. There also have been some failures. First, however, I will describe some success stories. Some results that have been reported are simply amazing. Annual productivity increases of up to 10 or 20 percent, a 60-percent decrease in defects, a 50-percent drop in scrap, and so on, have all been among the successes reported (Dewar, 1979). Federal Products Corporation of Providence, Rhode Island, reported increases in productivity, higher employee morale, and more job satisfaction (Hanley, 1980). Lockheed Missiles in Sunnyvale, California, one of the first U.S. firms to use the quality circle concept, reported total savings, since 1974, or more than 3 million dollars (*Wall Street Journal,* February 21, 1980). An assembly line quality circle at the solar turbine division of International Harvester found a way to boost both quality and productivity on the line producing turbine compression discs. Several production steps were eliminated and $8,700 a year were saved (*Wall Street Journal,* February 21, 1980). A quality circle at the American Airlines maintenance engineering center in Tulsa, Oklahoma, came up with a savings of $100,000 a year by recommending that old hand-grinders be replaced with new, efficient tools. A "machine shop circle was concerned about abnormal scrap rates caused by damage incurred in machining thin, flat test samples. They designed and built a unique vacuum chuck that held the samples securely so that the necessary machining could be accomplished without damage. Estimated savings were 2625 dollars per year" (Gregerman, 1979, p. 94). Going back to Lockheed Missiles for a moment—not only has Lockheed Missiles' use of quality circles or the "team" approach, as they call it, saved millions of dollars, the following advantages have also been cited:

1. Quality circles developed a team atmosphere.
2. Quality circles provided for an easy exchange of ideas.
3. The circles helped inter-group cooperation and relationships.

4. Defects were reduced after circles were started.

5. Workers had a better understanding of their job requirements.

6. Worker self-confidence was improved.

7. Problems were identified sooner.

8. Better communications developed between workers and supervisors.

9. Workers were provided with an opportunity to interact with support groups.

10. Circles provided an opportunity for frank and open discussion (Gregerman, 1979).

> In Cairo, Georgia, where the Torrington Company employs 450 people to manufacture precision anti-friction bearings used in machinery and automobiles, quality circles have been operating for a year . . . In addressing the problems of a leaky hydraulic system, members of one circle recommended that the company maintain a supply of hydraulic fittings, install a trouble shooting checklist on each machine, filter the coolant for grinder machines, and plan a preventative maintenance schedule for each of the hydraulic department's 90 machines (Rubin, 1981, p. 1c).

In addition to straightforward productivity benefits, there can be some other motivational or morale gains resulting from the use of quality control circles. The quality circles at J. C. Penney's Atlanta catalog distribution center make presentations to top management, "celebrations with corsages, punch, and cake all around. . . . (workers) don't have the opportunity to talk to upper management very often, so they feel important, and rightly so, because they are, says a J. C. Penney official" (Rubin, 1981, p. 3c).

Problems with Quality Circles

There are some problems and difficulties encountered with trying to establish a quality circle program. The problem of selecting the name is really an important one. Witness what happened in the steel industry. As part of the basic contractual agreement between organized labor, as represented by the United Steel Workers Union, and 10 major steel-producing companies, labor-management committees were formed at many plants to work on projects to improve productivity, provide training opportunities for employees, help redesign jobs, and do all of the other things that a quality control circle normally would do. As Mills (1978) reported, these committees initially met with some degree of resistance.

> The progress of the co-operative relationship has not been smooth. Originally the committees were called simply productivity committees. There was such opposition from some local unions to the committees that at the 1972 convention of the Steelworkers, the union decided to seek certain changes, both in the names and functions of the committees. And the President of the Steelworkers, I. W. Abel, though a strong supporter of the co-operative effort, felt compelled to oppose management's alleged use of the productivity committees as a camouflage for a speedup. Nevertheless, the Steelworkers' convention endorsed the continuation of the program, and it was included in the 1974 basic steel agreement under the new name of employment security and plant productivity committees (p. 268).

Quality control circles are so much more than mere attempts to improve quality, increase productivity, or reduce costs. They are intended to be part of a total QWL program. However, these problems exist and must be dealt with. Middle and upper management may feel threatened and try to limit the use of the circles.

Quality control circles are not always successful. The following discussion describes a program that failed. The reasons for the failure should be obvious. An industrial plant, employing about 300 people, located in a small industrial city, operated three shifts a day, five days a week. Employees in each of the production departments, via the quality circles, were responsible not only for their own quality, but also had responsibility for the entire production process. The teams had complete say-so as to how operations were to be performed and how setups were to be made. These teams had the authority to shut down assembly lines and have a meeting if problems warranted it.

And shut down the production line they did. It was not at all uncommon to see an entire department completely stopped because its workers were having a team meeting. These almost daily meetings, ostensibly held to discuss production problems, would usually last an hour. After about 20 minutes, however, these meetings would turn into name-calling contests, even though management representatives were present to try to help. Rather than showing increases in quality, productivity, teamwork, or communication, there was a significant decrease in all of these areas. According to an interview with one first-hand observer, "less and less was getting accomplished at these meetings until it seemed as though a meeting would be called so everyone could take a break."

Why did the concept fail? It would be easy to assert that quality circles failed because the workers took advantage of the system. While that statement is true, there are always two sides to these situations. Management knew what was happening but they wouldn't take hard, corrective steps to rectify the situation. Management was afraid that if they got the employees mad, a union would be voted in. Lack of proper training of the group leaders was another obvious problem.

The company, after losses at this once-profitable plant hit a million dollars a year under the quality circles program, scrapped the plan. Top management at the plant was replaced and traditional motivational procedures were returned. After about two years, the company returned to profitability.

Quality control circles fail. Schleicher (1981) cited the following as the most common reasons for circles not meeting their objectives.

> Inadequate top management support; forcing workers to join the circles through intimidation, arm twisting, or the like; falling into the trap of making people feel used instead of making people feel that they have contributed; lack of proper training; and emphasis on the individual's contribution rather than the group's efforts (p. 31).

Quality circles are not a magical cure-all.

> The research suggests that quality circles *can* make a major contribution to an enlightened industrial worker. What is necessary, first, is the courage to admit that the long standing roles of managers, production supervisors, and workers may need to be changed. We also need the

wisdom to help those who have long been guided by another philosophy to understand and become comfortable with an expanded role for workers (Gryna, 1981, p. 91).

If quality control circles are considered an easy way to make a buck without spending much time or money, forget it. The effort will fail. If there is a more serious commitment, but cost reductions and quality improvements are primarily in mind, the final success is decidedly like betting on a horse race. If the objectives of starting (the quality circles program) are employee development and training, and encouraging involvement in the growth of an organization, and if the effort is supported even though it leads automatically to training middle managers to be supportive and less authoritarian, then bevies of golden quail will be hatched (Gryna, 1981, p. 28).

ALTERNATIVE WORK PATTERNS—FLEXITIME

As part of the organizational development of this country, we have, as noted in this chapter, been tied to rigid structuring primarily as the result of following Taylor's philosophy. The entire concept of quality of work life, whether represented by formal QWL programs or quality circles, requires major attempts at breaking with tradition and helping the worker overcome alienation with the job. One relatively new program is sweeping Europe and claims to provide not only reduced worker alienation toward the job but also promises to increase productivity.

This program is flexitime. Simply defined, *flexitime* permits workers to set their own hours. As adopted in this country, there are some limitations, such as a certain time when all workers must be present, known as *core hours,* and a minimum number of hours required at work each week. Basically, it is a system that gives workers almost complete freedom and autonomy about when they work. "It sharply reduces worker alienation, raises productivity, weakens the tyranny of numerous middle managers, opens opportunity to working mothers and students and elderly people . . ." (Nathan, 1977, p. 36).

Flexitime has arrived and seems to be spreading like wildfire. It has also become a significant topic of learned discussion. For example, the American Institute of Industrial Engineers held a conference on alternative work patterns in 1980. Thousands of companies in this country, and even more in Europe, have adopted flexitime. Companies, for quick and easy fixes to employee morale, are implementing flexitime instead of more sophisticated cures. Other motivational cures such as job enrichment, work modules, and four-day weeks are being eclipsed by flexitime (Nathan, 1977).

The results of permitting workers this degree of self-determination with their jobs consistently reduces tardiness, virtually eliminates short-term absenteeism, and reduces overtime. However, its use is not without some drawbacks. When workers set their own hours, which may be different from their supervisors' hours, there is a natural tendency for the supervisors to think the workers are cheating. As a result, some firms use computerized time recording devices that make employees feel that they are not trusted. These problems can be overcome, however. As with other changes in management style, it is a matter of management being willing to make a change.

The bottom line can make this plan worthwhile. Companies have reported gains in productivity of 12 percent when flexitime is employed (Nathan, 1977). The same workers, getting paid the same amount of money, produce more and they produce more because they are given greater freedom to make decisions about when they work.

OTHER MOTIVATIONAL APPROACHES TO PRODUCTIVITY IMPROVEMENT

Motivational literature abounds with methods that have held promise of great increases in productivity. A sampling of some of these references follows. As with the extended discussions of quality circles, flexitime, and even QWL, the reader is urged to investigate further before attempting to apply these techniques to his or her own organization.

Job Enrichment

Job enrichment, originally tried by IBM and Detroit Edison, is the process of redesigning a worker's job, in effect making it larger and giving the worker more to do. "It encourages the acceptance of responsibility at the bottom of the organization; it provides opportunities for satisfying needs" (McGregor, 1966, p. 29). According to Herzberg (cited in Schultz, 1978) job enrichment can be performed through any or all of the following:

1. Give employees more responsibility for their own work.
2. Give employees complete units of work rather than just isolated components. Completing entire products gives more of a sense of satisfaction to the worker.
3. Give employees more freedom to do their jobs as they see fit. As long as responsibility for completing the work goes along with it, this method is certain to increase worker satisfaction.
4. Give production reports directly to workers rather than to the supervisor.
5. Give workers the encouragement to attempt new and more difficult tasks.
6. Give the worker a chance to become an expert on a particular task. Being the best in a particular area is a very satisfying feeling (p. 268).

To some degree, proper implementation of these suggestions can result in a more satisfied worker, reduced turnover, reduced absenteeism, and increased productivity. The literature has many examples of success stories. Take, for example, the use of job enrichment at Polaroid Corporation. An application at that company involved operators and assistant operators who worked on a major film assembly machine. The workers felt their work was repetitive and that they were, in effect, "chained to their machines." They also resented being told by the inspection and quality control departments when the film they produced didn't meet standards.

As a result of the workers expressing their frustrations, the operation was changed so that they are now responsible for their own quality. They continually run quality checks and, if they discover that something is not performing properly, they can make immediate changes. The results have been impressive—quality has increased while unit cost has decreased. Several inspection jobs were eliminated with the inspectors being transferred to other jobs in the company (*Personnel Management Bulletin,* May 15, 1962). This example shows productivity improvement—increased output with less input.

Job Rotation

Another technique some companies use to attempt to motivate workers and increase productivity is job rotation. This practice provides variety, lets the employee see more of the total operation, and gives workers a chance to learn additional skills. "Anything that will give the worker a chance to change his pace when he works will make him feel less like a robot and more like a human being—will lend variety to his work . . . if workers are permitted to change their pace they can build 'banks' and thus obtain visible evidence of accomplishment" (Strauss and Sayles, 1972). Job rotation, or giving a worker the opportunity to do something different every day can be quite a motivator in itself. The author's first industrial job was as an assembly line worker in a television assembly plant. The job was fairly simple and was mastered after about an hour. The job was also highly repetitive, being repeated 82 times an hour. After just a few hours on the job, the boredom, coupled with the knowledge that the work would never change, placed a gloom and depression over the author that killed his motivation, enthusiasm, and even his concern about the quality of the product.

Contrast this example with a company that practiced job rotation. Each production employee was trained on 10 different jobs required in the manufacture of shipping containers. Every four hours, the entire work force rotated or changed jobs. No worker performed the same job more than one-half day per week. The company always had mentally fresh employees performing the jobs. By the time the workers could get set in a routine, their stint was almost over. The motivational advantage and relative enjoyment provided by the work at this second company were obvious, even to the production workers at the two factories.

Giving workers a sense of accomplishment, a sense of completing something, is another commonly-used tool of the industrial relations specialist. Maier (1955) reported on how telephone maintenance work is redefined as accomplishable tasks:

> There is no challenge of diagnosing trouble, and the job is very confining since a man can work for hours within the space of a few feet. There is never any real experience of progress. When the job is finished the worker starts all over again. In one office, the frames in which the men worked were subdivided by means of chalk lines. Each block required between one and a half and two hours to complete. The worker made his choice of unit . . . The benefits of work were immediately apparent . . . once a man completed a unit he took a smoke or a stretch. Even lunch and quitting time found no untagged units. The men liked the plan and the supervisors reported that complaining decreased and the trouble with meeting work schedules was eliminated (p. 489).

SUMMARY

As the productivity improvement techniques described in this chapter show, when the individual is given control over his or her job, then job satisfaction, quality, and productivity improve. It is time to acknowledge that people will work better when they have more control over their work environment. All of the motivational techniques have one element in common—increased worker participation. As McGregor (1966) said, "We cannot tell industrial management how to apply this new knowledge in simple, economic ways. We know it will require years of exploration, much costly development research, and a creative imagination on the part of management to discover how to apply this growing knowledge to the organization of human effort in industry" (p. 5). As the years go by undoubtedly the same concept will continue to reappear under a variety of names. However, the fact will remain that the person who knows the most about a job and the person who is capable of finding the way to achieve maximum productivity out of an existing job is the person who performs that job on a regular basis.

> The workers in both the private and public sector are a virtually untapped natural resource of ingenuity and enthusiasm. That ingenuity can be put to work in a directly observable way when they are able to participate in decisions . . . that participation has to extend to the ways to improve the just treatment of working people . . . American workers are not morons, and they resent being treated with less simple respect than the machines they operate (Donohue, 1980, p. 15).

We can make significant strides toward increasing productivity. After all, we are all in this problem together whether we are management, an industrial engineer, or a production worker. We must continue to work to improve productivity. We must use every avenue available to us, including involving the worker in the decision-making process and using traditional industrial engineering procedures of productivity improvement. We will be successful only when we all work together.

REVIEW QUESTIONS

1. What is meant by the term *quality of work life?*
2. Describe the characteristics of a QWL program.
3. How does the Scanlon plan relate to QWL programs?
4. What is a *participatory management program?* What is this style of management?
5. What is a *quality circle?*
6. What is *job enrichment?*
7. What is *flexitime?*
8. What is *job rotation?*
9. What factors do all of the alternative patterns of productivity management have in common?
10. What are the problems encountered in adopting any of the QWL programs? How can these problems be overcome?

APPENDIX A

Computer Data
Analysis Programs

INTRODUCTION

The following three computer programs are written in BASIC. These programs should assist in the analysis and development of data for creating time standards.

The first program takes continuous stopwatch time study data and calculates the time standard. The second program calculates the coefficient of correlation. The third program calculates the necessary coefficients for linear multivariate regression analysis.

Once the programs are loaded correctly they are user-friendly and request data via a series of prompting statements.

Program 1—BASIC Program to Determine Time Standard

```
00100 MARGIN 120
00110 PRINT ''THIS PROGRAM WILL CALCULATE TIME STANDARDS BASED ON CONTIN-
            UOUS TIME STUDY DATA.''
00120 PRINT ''THE PROGRAM WILL HANDLE 15 ELEMENTS AND UP TO 500 CYCLES''
00130 PRINT
00140 PRINT ''ENTER DATA STARTING ON LINE 1000 WITH DATA STATEMENTS. DATA
            SHOULD BE ENTERED CONTINUOUSLY.''
00150 PRINT
00160 PRINT TAB(10);''ENTER MISSED READINGS OR FOREIGN ELEMENTS WITH A
            '9999'''
00170 PRINT
00180 PRINT TAB(10);''FOR CYCLES WHERE NON-CYCLICAL ELEMENTS OCCUR, SHOW
            THE TIMES.''
```

```
00190 PRINT TAB(10); ``OTHERWISE, USE `9999' ''
00200 PRINT
00210 PRINT TAB(10); ``MAXIMUM VALUE OF A READING IS `2999' ''
00220 PRINT
00230 PRINT TAB(10); ``DO NOT ENTER DECIMAL POINTS''
00240 PRINT
00250 PRINT ``HAS DATA BEEN ENTERED, YES OR NO'';
00260 INPUT A$
00270 IF A$ = ``NO'' THEN 00960
00280 PRINT
00290 DIM T(15,500),A(15),B(15)
00300 PRINT ``HOW MANY ELEMENTS'';
00310 INPUT E
00320 PRINT ``HOW MANY CYCLES'';
00330 INPUT C
00340 PRINT ``RATING FACTOR AS A DECIMAL'';
00350 INPUT R
00360 PRINT
00370 PRINT ``PFD ALLOWANCE AS A DECIMAL'';
00380 INPUT P
00390 PRINT ``RELATIVE FREQUENCY AS A PERCENT FOR''
00400 FOR I=1 TO E
00410 PRINT ``ELEMENT'';I;
00420 INPUT B(I)
00430 NEXT I
00440 PRINT
00450 FOR J=1 TO C
00460 FOR I=1 TO E
00470 READ T(I,J)
00480 NEXT I
00490 NEXT J
00500 T=T(1,1)
00510 N=1
00520 FOR J=2 TO C
00530 T1=T(E,J-1)
00540 T2=T(1,J)
00550 IF T1=9999 THEN 00630
00560 IF T2=9999 THEN 00630
00570 IF T1 < T2 THEN 00600
00580 T=T2+(3000-T1)+T
00590 GO TO 00610
00600 T=(T2-T1)+T
00610 N=N+1
00620 A(1)=T/N
00630 NEXT J
00640 T=0
00650 N=0
00660 FOR I=2 TO E
```

```
00670 FOR J=1 TO C
00680 T1=T(I-1,J)
00690 T2=T(I,J)
00700 IF T1=9999 THEN 00780
00710 IF T2=9999 THEN 00780
00720 IF T1 < T2 THEN 00750
00730 T=T2+(3000-T1)+T
00740 GO TO 00760
00750 T=(T2-T1)+T
00760 N=N+1
00770 A(I)=T/N
00780 NEXT J
00790 T=0
00800 N=0
00810 NEXT I
00820 FOR I=1 TO E
00830 S=S+A(I)*B(I)/100
00840 NEXT I
00850 S1=(S*R)*(1+P)
00860 S1=S1/100
00870 PRINT``THE STANDARD IS``;S1;``MINUTES``
00880 PRINT
00890 PRINT``THE ELEMENTAL TIMES FOLLOW``
00900 PRINT
00910 PRINT``ELEMENT``;TAB(15);``TIME``
00920 PRINT
00930 FOR I=1 TO E
00940 PRINT TAB(3);I;TAB(15);(A(I)/100)*R*(1+P)
00950 NEXT I
00960 STOP
01000 DATA 10,25,50,60,75,100,9999,300,315,350
READY.
```

Program 2—BASIC Program to Compute Correlation Coefficients

```
010 DIM R(15,100)
015 PRINT ``CORRELATION OF UP TO 100 READINGS OF 15 VARIABLES``
016 PRINT ``MINIMUM OF 4 VALUES FOR EACH VARIABLE REQUIRED``
017 PRINT ``DATA ARE READ THROUGH THE USE OF DATA STATEMENTS BEGINNING
          WITH``
018 PRINT ``STATEMENT 1000. LIST ALL OF THE DATA FOR EACH VARIABLE BEFORE``
019 PRINT ``STARTING THE DATA FOR THE NEXT VARIABLE.``
020 PRINT ``DID YOU PREPARE YOUR DATA STATEMENTS``;
021 INPUT B$
022 IF B$=``NO`` THEN 24
023 GO TO 27
024 PRINT ``THE PROGRAM WILL STOP NOW. PLEASE PREPARE YOUR DATA AND RUN
          THEM``
```

```
025 PRINT ``ONE MORE TIME.''
026 GO TO 999
027 PRINT ``HOW MANY VARIABLES'';
030 INPUT N1
031 PRINT ``YOUR VARIABLES WILL BE NUMBERED 1 THROUGH'';N1
040 PRINT ``HOW MANY READINGS'';
050 INPUT N2
060 FOR I=1 TO N1
070 FOR J=1 TO N2
080 READ R(I,J)
090 NEXT J
100 NEXT I
200 PRINT ``WHICH VARIABLES DO YOU WANT CORRELATED'';
201 T1=0
202 T5=0
203 T6=0
204 T7=0
205 T2=0
206 PRINT
210 INPUT A,B
220 PRINT ``WHAT ALPHA VALUE DO YOU WANT FOR SIGNIFICANCE, .01,.05, OR
        .10'';
230 INPUT C
240 FOR J=1 TO N2
250 T1=T1+R(A,J)
260 T2=T2+R(B,J)
290 T5=T5+R(A,J)**2
300 T6=T6+R(B,J)**2
310 T7=T7+R(A,J)*R(B,J)
330 NEXT J
340 S1=N2*T5-(T1**2)
350 S2=N2*T6-(T2**2)
360 S3=N2*T7-(T1*T2)
400 R=S3/(S1*S2)**.5
410 Z=(((N2-3)**.5)/2)*(LOG((1+R)/(1-R)))
420 IF C=.01 THEN 445
425 IF C=.05 THEN 440
430 Z1=1.645
431 Z2=90
435 GO TO 450
440 Z1=1.96
441 Z2=95
442 GO TO 450
445 Z1=2.576
446 Z2=99
450 IF ABS(Z)Z1 THEN 480
455 PRINT ``CORRELATION BETWEEN VARIABLES'';A;``AND'';B;``IS'';R
456 PRINT ``AT THE'';Z2;``PERCENT CONFIDENCE LEVEL THIS IS SIGNIFICANT.''
```

```
470 GO TO 500
480 PRINT ``CORRELATION BETWEEN VARIABLES'';A;``AND'';B;``IS'';R
481 PRINT ``AT THE ``;Z2;``PERCENT CONFIDENCE LEVEL THIS IS NOT SIGNIFI-
        CANT.''
500 PRINT ``DO YOU WANT TO DETERMINE R FOR ANOTHER PAIR OF VARIABLES'';
510 INPUT A$
520 IF A$=``YES'' THEN 200
999 STOP
```

Program 3—BASIC Program for Multivariate Regression

```
0010 MARGIN 100
0100 DIM Y(5,5),G(5),E(5),Q(50,5),A(10),C(5,5),X(5),R(5,5),Z(5,5)
0102 DIM L(1,5)
0110 PRINT``MULTIPLE REGRESSION PROGRAM''
0120 PRINT``MAXIMUM 50 OBS. AND 5 VARIABLES''
0140 PRINT ``HOW MANY OBSERVATIONS'';
0150 INPUT N
0160 PRINT ``HOW MANY OBSERVATIONS'';
0170 INPUT M
0180 PRINT``THIS ANALYSIS HAS'';N;``OBS. AND'';M;``VARIABLES''
0190 MAT Q=ZER(N,M)
0230 PRINT ``TYPE IN'';N;``BY'';M;``MATRIX, ALL DATA IN A STRING ON A ROW
        BY ROW BASIS''
0231 MAT X=ZER(M)
0232 MAT Y=ZER(M,M)
0233 MAT R=ZER(M,M)
0234 MAT C=ZER(M,M)
0235 MAT E=ZER(M)
0236 MAT L=ZER(1,M)
0240 MAT INPUT Q
0250 PRINT``LIST OF INPUT DATA? IF YES, TYPE 1; ELSE 0'';
0260 INPUT M9
0270 IF M9=0 THEN 450
0280 REM GOSUB 1660
0290 PRINT``INPUT DATA''
0300 MAT PRINT Q
0310 REM GOSUB 1660
0450 FOR K1=1 TO N
0460 FOR I=1 TO M
0470 0(I)=Q(K1,I)
0480 NEXT I
0490 FOR 1=1 TO M
0500 X(I)=X(I)+A(I)
0510 FOR J=1 TO M
0520 (I,J)=Y(I,J)+A(I)*A(J)
0530 NEXT J
0540 NEXT I
```

```
0550 NEXT K1
0560 GOSUB 1660
0572 FOR I=1 TO M
0574 L(1,I)=X(I)
0576 NEXT I
0590 FOR I=1 TO M
0600 O=X(1)/N
0610 L(1,I)=O
0620 FOR J=1 TO M
0630 C(T,J)=Y(I,J)-O*X(J)
0640 NEXT J
0650 NEXT T
0660 REM GOSUB 1660
0690 REM GOSUB 1660
0720 REM GOSUB 1660
0750 REM GOSUB 1660
0770 FOR T=1 TO M
0780 X(I)=X(1)/N
0790 FOR J=1 TO M
0800 (I,J)=C(I,J)/SQR(C(I,I)*C(J,J))
0810 NEXT J
0820 NEXT I
0840 REM GOSUB 1660
0850 PRINT``HOW MANY VARIABLES (INCLUDES A DEPENDENT VARIABLE)''
0870 INPUT R4
0880 IF R4M THEN 1080
0890 IF R4=0 THEN 1650
0900 R5=R4-1
0910 MAT Y=ZER(R5,R5)
0912 MAT E=ZER(R5)
0913 MAT G=ZER(R5)
0914 MAT R=ZER(R5,R5)
0915 MAT Z=ZER(R5,R5)
0920 PRINT ``WHAT ARE THEY, (NAME THEM 1 THROUGH 5 LIST THE DEPENDENT VARI-
          ABLE LOST)'';
0930 MAT A=ZER(R4)
0932 MAT L=ZER(1,R4)
0940 MAT INPUT L
0960 PRINT `` VARIABLES ARE'';
0970 FOR I=1 TO R4
0972 A(I)=L(1,I)
0980 PRINT A(I);
0990 IF A(I)>M THEN 1050
1000 IF A(I)=0 THEN 1050
1010 NEXT I
1020 T=A(R4)
1030 X1=X(T)
1040 GO TO 1100
```

```
1050 PRINT''YOUR VARIABLE IS OUT OF THE VARIABLE RANGE''
1060 PRINT''TRY AGAIN''
1070 GO TO 920
1080 PRINT''YOU ASK MORE THAN OUR ORIGINAL VARIABLE NUMBER SIZE''
1090 GO TO 850
1100 GOSUB 1660
1112 MAT L=ZER(1,R5)
1120 FOR I=1 TO R5
1130 I1=A(I)
1140 H(I)=X(I1)
1150 E(I)=C(T,I1)
1160 L(1,I)=E(I)
1170 FOR J=1 TO R5
1180 I2=A(J)
1190 B=C(I1,I2)
1200 Y(I,J)=B
1210 Y(J,I)=B
1220 NEXT J
1230 NEXT I
1260 REM GOSUB 1660
1290 REM GOSUB 1660
1300 MAT R=INV(Y)
1330 REM GOSUB 1660
1340 T=C(T,T)
1342 FOR I=1 TO R5
1343 G(I)=0
1344 FOR J=1 TO R5
1345 G(I)=G(I)+R(I,J)*E(J)
1346 NEXT J
1347 NEXT I
1360 FOR I=1 TO R5
1370 PRINT A(I);''REG COEF'';G(I)
1380 NEXT I
1390 X2=0
1400 FOR I=1 TO R5
1410 X1=X1-G(I)*H(I)
1420 X2=X2+G(I)*E(I)
1430 NEXT I
1440 PRINT'' INTERCEPT'';X1
1450 N1=N-1
1460 GOSUB 1660
1500 R=T-X2
1510 I=N1-R5
1520 C1=R/I
1530 C2=X2/R5
1540 C3=C2/C1
1570 R1=SQR(X2/T)
1580 GOSUB 1660
```

```
1600 R1=R1*R1
1620 GOSUB 1660
1630 GO TO 850
1650 STOP
1660 PRINT
1670 PRINT
1680 RETURN
READY.
```

Military Standard for Work Measurement

Mil-Std 1567 (USAF)
30 June 1975

The purpose of this standard is to assist in achieving increased discipline in contractors' work measurement programs with the objectives of improved productivity and efficiency in contractor-industrial operations. Experience has shown that excess manpower and lost time can be identified, reduced, and continued method improvements made regularly where work measurement programs have been implemented and conscientiously pursued.

Active support of the program by all affected levels of management, based on an appreciation of work measurement and its objectives, is vitally important. Work measurement and the reporting of labor performance is not considered an end in itself but a means to more effective management. Understanding the implications inherent in the objectives of the work measurement program will promote realization of its full value. It is important that objectives be presented and clearly demonstrated to all personnel who will be closely associated with the program.

The following are benefits which can accrue as a result of employing a work measurement program:

(a) Achieving greater output from a given amount of resources

(b) Obtaining lower unit cost at all levels of production because production is more efficient

(c) Reducing the amount of waste time in performing operations

(d) Reducing extra operations and the extra equipment needed to perform these operations

(e) Encouraging continued attention to methods and process analysis because of the necessity for achieving improved performance

(f) Improving the budgeting process and providing a basis for price estimating

(g) Acting as a basis for planning for long-term manpower, equipment, and capital requirements

(h) Improving production control activities and delivery time estimation

(i) Focusing continual attention on cost reduction and control

(j) Helping solve layout and materials handling problems by providing accurate figures for planning and utilization of such equipment.

WORK MEASUREMENT

1. SCOPE.

1.1 Purpose. This standard requires the application of a disciplined work measurement program as a management tool to improve productivity on those contracts to which it is applied. It establishes criteria which must be met by the contractor's work measurement program and provides guidance for implementation of these techniques and their use in assuring cost effective development and production of systems and equipment.

1.2 Applicability. This standard is applicable to new/follow-on procurements for:
a. Full-scale developments which exceed $100 million.
b. Production and/or major system modifications which exceed $20 million annually or $100 million dollars cumulatively. It shall not be applied to contacts or subcontracts for: construction, facilities, off-the-shelf commodities, time and materials, research, study, developments which are not connected with an acquisition program, or to firm fixed price contracts.

1.2.1 *Subcontracting.* The application of this standard will be required on all subcontracts, exclusive of firm fixed price which exceed $1 million annually or $5 million cumulatively. If it is determined that such application is not cost-effective or inappropriate for other reasons, the contractor may request the Government to waive the specific application. Requests for waivers shall be supported with the data used to make the determination.

1.3 Contractual Intent. This standard requires the application of a documented work measurement system. This standard further requires that the contractor apply procedures to maintain and audit the work measurement system. It is not the intent of this standard to prescribe or imply organization, structure, management methodology, or the details of implementation procedures.

1.4 Corrective Actions. When surveillance by the contractor or the Government discloses that the work measurement program does not meet the requirements of this standard, a plan will be initiated to expeditiously assure that corrective measures shall be implemented, demonstrated and documented. The contractor's system is subject to disapproval by the Government whenever it does not meet the requirements of this standard.

1.5 Documentation. The work measurement program shall include sufficient documentation to assure effective operation of the program and to provide for internal audits as required by paragraph 5.13. Documentation will specify organizational responsibilities, state policies, and provide operational procedures and instructions. The results of contractor system audits and plans for corrective actions shall be made readily available to the Government for review.

2. REFERENCED DOCUMENTS. (None)

3. DEFINITIONS.

3.1 Definitions of terms used in this standard are:

3.1.1 *Earned Hours.* The time in standard hours credited to a workman or group of workmen as the result of successfully completing a given task or group of tasks: usually calculated by adding all of the products of applicable standard times multipled by the completed work units.

3.1.2 *Labor Efficiency.* The ratio of earned hours to actual hours spent on a prescribed task during a reporting period. When earned hours equal actual hours, the efficiency equals 100%.

3.1.3 *Methods Engineering.* The technique that subjects each operation of a given piece of work to close analysis to eliminate every unnecessary element or operation and to approach the quickest and best method of performing each necessary element or operation. It includes the improvement and standardization of methods, equipment, and working conditions; operator training; and the determination of standard times.

3.1.4 *Predetermined Time System.* An organizd body of information, procedures and techniques employed in the study and evaluation of manual work elements. The system is expressed in terms of the motions used, their general and specific natures, the conditions under which they occur, and their previously determined performance times. The term Predetermined Motion Time System is used synonymously.

3.1.5 *Realization Factor.* A calculated factor (exclusive of personal, fatigue and delay (PF&D) allowances) by which labor standards are modified when developing actual manhour requirements.

3.1.6 *Subcontract.* A contract between the prime contractor and a third party to produce parts, components or assemblies in accordance with the prime contractor's designs, specifications or directions and applicable only to the prime contract.

3.1.7 *Touch Labor.* Production labor which can be reasonably and consistently related directly to a unit of work being manufactured, processed, or tested. It involves work affecting the composition, condition, or production of a product; it may also be referred to as "hands-on-labor" or "factory labor." NOTE: As used in this standard touch labor *DOES NOT INCLUDE* the functions of engineering, production control, and production planning. It includes such functions as machining, welding, fabricating, cleaning, painting, assembling, and functionally testing production articles.

3.1.8 *Touch Labor Standard.* A standard time set on a touch labor operation.

3.1.9 *Type I Engineered Labor Standards.* These standards are established using a recognized technique such as time study, work sampling, standard data, or a recognized predetermined time system to derive at least 90% of the total time associated with the labor effort covered by the standard.

3.1.10 *Type II Labor Standards.* All labor standards not meeting the criteria established in paragraph 3.1.9, above.

3.1.11 *Standard Time Data.* A compilation of all elements used for performing a given class of work with normal elemental time values for each element. The data are used as a basis for determining time standards on work similar to that from which the data were determined without making actual time studies.

4. REQUIREMENTS.

4.1 General. Minimum requirements which must be met in the implementation of an acceptable work measurement program are:

a. A work measurement plan and supporting procedures

b. A clear designation of the organization and personnel responsible for the execution of the system

c. A plan to establish and maintain engineered labor standards of known accuracy

d. A plan of continued improved work methods in connection with the established labor standards

e. A defined plan for the use of labor standards as an input to budgeting, estimating production planning, and "touch labor" performance evaluation.

5. SPECIFIC REQUIREMENTS.

5.1 Type I Engineered Labor Standards. Type I standards shall be backed up by sufficient data to statistically support an accuracy of plus or minus 25%, with at least a 90% confidence level.

5.2 Standard Data. The contractor shall take full advantage of available standard time data of known accuracy and traceability. The contractor shall not charge costs directly to the contract for the development of basic or general purpose data.

5.3 Labor Standards Coverage. The contractor shall develop and implement a Work Measurement Coverage Plan which provides a time-phased schedule for achieving 80% coverage of all categories of touch labor by Type I standards.

5.3.1 The Work Measurement Coverage Plan shall be based on cost trade-off analyses which relate savings to be accrued through improved productivity and simplification of work methods to the cost of attaining Type I standards coverage.

5.3.2 Type II Standards are acceptable for initial coverage. All Type II standards shall be approved by the organization responsible for establishing and implementing work measurement standards.

5.3.3 The Work Measurement Coverage Plan shall provide a schedule for converting Type II to Type I Standards.

5.4 Leveling/Performance Rating. All time studies shall be rated using recognized techniques.

5.5 Allowances. Allowances for personal, fatigue, and unavoidable delays shall be developed and included as part of the labor standard. Allowances should not be excessive or inconsistent with those normally allowed for like work and conditions.

5.6 Estimating. The contractor's estimating procedures shall describe how touch labor standards are utilized to develop price proposals.

5.7 Use of Labor Standards. Labor standarads shall be used:
a. As an input to developing budgets, plans, and schedules
b. As a basis for estimating touch labor hours when issuing proposals for changes to contracts and as a basis for estimating the prices of initial spares and replenishment spares
c. As a basis for measuring touch labor performance.

5.8 Realization Factor. When labor standards have been modified by realization factors, each element which contributed to the total factor shall be identified. The analysis supporting each element will be available to the Government for review.

5.9 Labor Efficiency. A forecast of anticipated touch labor efficiency shall be used in manpower planning, both on a long-range and current scheduling basis.

5.10 Revisions. Labor standards shall be reviewed for accuracy when changes occur to:
a. Methods or procedures
b. Tools, jigs, and fixtures
c. Workplace and work layout
d. Specified materials
e. Work content of the job.

5.11 Production Count. Work units shall be clearly and discretely defined so as to cause accurate measurement of the work completed and shall be expressed in terms of completed:
a. End items
b. Operations
c. Lots or batches of end items.

5.11.1 *Partial Credit.* In those cases where partial production credit is appropriate, the work measurement procedures will define the method to be used to permit a timely and current production measure.

5.12 Labor Performance Reporting. The contractor's work measurement program shall provide for periodic reporting of labor performance. The report shall be prepared at least weekly for each work center and be summarized at each appropriate management level; it will indicate labor efficiency and compare current results with pre-established contractor goals.

5.12.1 *Variance Analysis.* The labor performance reports shall be analyzed by supervisory and staff support functions on an exception basis as a minimum. The variance analyses will: be written, identify causes and corrective actions necessary, and indicate actions taken to attain established performance goals.

5.12.2 *Report Retention.* Performance reports and related variance trend analyses will be retained for a six-month period.

5.13 System Audit. The contractor shall use an internal review process to monitor the work measurement system. This process shall be so designed that weaknesses or failures of the system are identified and brought to the attention of management to enable timely corrective action. Written procedures will describe the audit techniques to be used in evaluating system compliance.

5.13.1 *Scope of Audit.* The audit shall cover compliance with the requirements of this standard at least annually. The audit, based upon a representative sample of all active labor standards and work measurement activities, shall determine:

    a. The accuracy of the labor standard time values and the validity of the prescribed method

    b. Percent of coverage by Type I and Type II labor standards

    c. Effectiveness of the use of labor standards for planning, estimating, budgeting, and scheduling

    d. The timeliness, accuracy and traceability of production count reporting

    e. The accuracy of labor performance reports

    f. The reasonableness of efficiency goals established

    g. The effectiveness of corrective actions resulting from variance analyses.

5.13.2 *Audit Reports.* A copy of the audit findings will be retained in company files for at least a two-year period and will be made available to the Government for review upon request.

Participative Management— Challenge to Competition

Jerry L. Schmidt, Quality Emphasis Coordinator
Buick Motor Division
Flint, Michigan 48550

ABSTRACT

Using the basic concepts of people building and individual recognition, Buick is making use of the creativity of its people in small groups on a voluntary basis. The process begins with the staff of an area and continues as participative teams are formed within middle management. The first line supervisors and hourly employees contribute through the process of Q.C. Circles.

The Buick organization challenges American industry to improve quality, efficiency, and schedule with the use of participative management at all levels of the organization.

TEXT

I appreciate this opportunity to share with you a presentation on Participative Management—Buick's Challenge to Competition. I will relate to you our experience with the Employee Participation Program which we feel is an integral part of the quality of work life activity at Buick Motor Division.

First, a brief introduction to Buick. Located in Flint, Michigan—Buick is the largest General Motors manufacturing complex in North America, employing 19,000 people.

The Buick Motor Division complex is over two miles long with over 100 buildings on 325 acres of land consisting of 7 different manufacturing and assembly complexes in addition to an engineering center. All of the major components that make up the Buick automobile are produced here.

Buick Motor Division is 77 years old this year. Since its beginning in 1903, we have expanded, and today Buick, along with other Flint area G.M. divisions, offers 75,000 jobs to the community.

Buick assembles 1,600 cars daily in two adjacent car assembly plants working two eight hour shifts . . . that's approximately one car per minute on each assembly line.

In addition to car assembly, Buick manufactures transmissions, engines, coil springs, trailing axles, and engine cradles which are used on the General Motors front wheel drive ''X'' cars.

In our Foundry, we produce castings for engine blocks.

Buick's Metal Fabricating and Plastics Plant produces major components for all Buicks. Located adjacent to the car assembly plants, the hoods and fenders are formed and shipped to the assembly line with a minimum of handling.

Buick's new paint facility is one of the most modern in the world, automatically painting front end sheet metal and applying corrosion protection to critical areas of the fenders.

Cars are assembled in two assembly plants located side by side. In one plant we assemble the Century and Regal models, while the other plant assembles the LeSabre and Electra. Each car is custom ordered by a customer or dealer and is assembled to that car order.

We, at Buick Motor Division, realize that if we are to remain competitive on the world market, we must utilize the creative thinking ability of each person, from the hourly person to the general manager.

The process of involving each person to the extent of his ability is *Participative Management.*

Participative Management at Buick has taken three basic forms:

- Upper management participating in off-sites
- Operating management problem solving
- First-line supervision and hourly people participating in Quality Control Circles

I would now like to review each of these participation forms in detail.

Off-sites are necessary because one of the hardest things for a company to do is to commit a large time block to plan strategy for meeting objectives. Buick has found that an effective way to do this is to schedule several days at a nearby hotel to provide an atmosphere conducive to creative planning.

Usually these ''off-sites'' have an established goal or objective, and all management personnel who can play a part in meeting the goal are present. Once all of the actions necessary to meet the goal have been clearly identified and understood by the entire group, commitments to perform those actions are solicited on a voluntary basis. These commitments are then reported to higher management by the participants.

Follow-up meetings are then held on a regular basis to review progress toward commitments made at the previous meeting. This concept of making commitments to peer level managers with follow-up has proven to be very effective at Buick.

Problem solving for quality improvement with management at the operating levels takes on a similar format. Many times upper management will identify a chronic problem which they wish resolved and will call appropriate management together to resolve the problem. Following the format previously discussed, commitments are made to resolve the problem. An additional benefit is that the participants develop a team relationship which they carry back into their working environment.

The third form of participative management is the *Quality Control Circle* process which involves supervisors and hourly people in the management of the business.

Our union representatives encourage us to utilize the ideas of the hourly work force on quality related items. Union and management have built a team relationship during the past several years and are working together to reach common goals. During one team building session at Black Lake, (The United Auto Workers Union Training Facility), our local union president suggested that we initiate the concept of Quality Control Circles with our hourly people in the car assembly plants.

To better understand the concept of Quality Control Circles, our Final Assembly Plant manager hired a consultant to train us in a process which has been successful in Japan, the United States, and other parts of the world.

As soon as leaders and facilitators had completed their training course, we organized the Employee Participation Program in Final Assembly. We were one of the first auto assembly plants in North America to organize Q.C. Circles. Why do we call them Employee Participation Groups rather than Quality Control Circles? To the average worker at Buick, the words, "Quality Control" mean inspection. The employees participating in the program wanted to be sure that the process involved all employees, not just inspectors.

An *Employee Participation Group* is a group of four to ten people, working in the same department or having similar jobs, meeting weekly for one hour to select and solve work related problems.

All particpation is voluntary from the plant manager to the hourly workers. It has been proved many times that volunteers respond better than non-volunteers.

Training is provided to build the individual's skills. The small investment required to train volunteers in basic problem solving techniques is returned many times as problems are systematically solved individually, and in small groups.

In participative groups, each person concentrates on one project at a time even though the project does not affect them or they don't fully understand it. Sometimes the best ideas come from people who don't know the problem can't be solved.

Creativity is encouraged. Many of the training techniques are designed to encourage people to stretch their imaginations for solutions to problems.

No criticism is allowed. This concept is basic to keeping the attention on problem solutions rather than on persons contributing to problems.

Each person who volunteers to be part of a problem solving group wants to contribute during meetings. Sometimes a strong personality will dominate the meeting. Therefore, equal participation is encouraged by the leader by going from person to person around the room.

Usually the objectives of the problem solving groups are specified by higher management. The members of the group are given the freedom to choose the projects and procedures they will use to meet those objectives.

When the concept of participative management is used with hourly personnel and elected union officials, our experience has shown that all participants work together effectively on projects to meet common objectives. Union representatives participate to insure that collective bargaining issues are not discussed.

An important by-product of Participative Management is that it provides an established mechanism by which information is accurately communicated to all levels of the organization.

Participative groups are recognized for their achievements. Individual and group recognition is the most prominent reason that people volunteer to give of themselves to help the company meet the objectives of improved quality, on schedule, and at lower cost.

Since October, 1977, dozens of groups in all areas of our car assembly plants have been organized. Each group chooses a project which is high on its priority list to resolve. Projects undertaken by Q.C. Circles are of special benefit to Buick because many of the projects do not have the priority to be resolved by any other management team. Also, some of the projects would be difficult to complete any other way because they require commitment from many assembly line operators.

One of the first projects was the United Way Campaign which is successfully run each year by a committee of hourly people.

Other major projects given the hourly employees are open houses for employees, families and friends. The hourly employees have decorated the factories and completely organized several open houses involving thousands of visitors.

To better communicate with our employees, a closed circuit T.V. system transmits quality and efficiency information to all employees. Staffed by hourly personnel, this system broadcasts information on many subjects including news and sports.

Another widely accepted form of communication is our in-plant newspaper which is also staffed by hourly workers; many give hours of their own time to assure that accurate and interesting articles are published on a timely basis.

A Quality Awareness Program was organized to communicate quality performance daily. Each employee is given an opportunity to discuss the quality of his workmanship at a daily quality review session.

Recreation projects such as ping-pong, basketball free throw, and card tournaments are organized by the employees.

A program was organized under the leadership of hourly employees to eliminate mutilations which occur on paint and sheet metal surfaces.

One project resulted in a 72% improvement in the paint quality of steering columns which we paint in the assembly plant. Improvements were made in paint quality, maintenance of the paint booth, and operator painting techniques. These improvements took place only because of the persistance of each member as they worked with each department involved.

One group in frame assembly worked on the inconvenience this small assembly aid washer caused when it allowed the bolt to fall thru before the assembly could be completed.

The problem of improving the body drop hoist operation was addressed.

Another group worked on improving the housekeeping by magnetically picking up small production parts which fall on the shop floor.

The problem of providing clean aprons and gloves to each person was tackled. The group designed a new cabinet and requested that each person in the department return their gloves and aprons at the end of the day. Now each person has clean gloves and aprons when they need them.

Members work on improving the locations of machines and conveyors to make operations more efficient and easier.

One group was so proud of their accomplishments that they designed a four by eight foot sign showing each person at their job and how the employee interacts with the department's activities. This sign was painted and assembled by two of the members in their homes on their own time.

After successfully completing a project, the group members usually make a presentation to the plant management on their accomplishments and items they need help with.

A management presentation is the highpoint of the employee participation process. Members spend much of their own time preparing and perfecting these presentations which are attended by the plant manager. Strong support by the plant manager and his staff is the key to a successful Employee Participation Program.

The success of the groups is monitored in terms of attitude changes, feelings about the job and company, and the extra effort and care that a person puts into the job. Monetary savings can also be calculated, and Buick can testify to one problem solution which more than paid for the cost of the entire program.

In conclusion, most people want to make a contribution to the organization they are associated with. People have shown they are cooperative and competent in effectively resolving problems, and we find they gain recognition through participative management.

This form of participative management allows for communication to flow to all levels of the organization. Buick's experience has shown that union and management people will cooperate to achieve their common objectives; and through a common commitment, union and management will assure the future success of the Employee Participation Activity which will improve the quality, costs, and schedule at Buick Motor Division. Participative Management is Buick's answer to the challenge of our competition . . . both at the present time and in the future.

Productivity Monitoring of Indirect Labor

Maurice E. Chapman, S.E.T.
Kennestone Hospital System
Marietta, Georgia

INTRODUCTION

In the last few years, the health care industry has experienced such a large increase in the cost of administering medical care that management has been challenged to find new and better ways to reduce costs. One potential source of cost reduction that has been virtually untapped is improvinmg the productivity of the indirect labor force. *Indirect labor* in the health care industry is described as labor functions that are not directly applied to the patient. A nurse placing a bandage on a patient would be considered direct labor, while the activities of purchasing, storing, and transporting the bandage to the patient would be indirect labor. Other indirect labor activities include management, maintenance, materials handling, engineering, accounting, housekeeping, and so forth.

The importance of better management of indirect labor has been realized by many institutions. Many of these institutions have experienced growth to a point of inefficiency in the flow of materials, linens, food, and medical care products. In addition, their growth has also required a proportional increase of support personnel such as materials handlers, housekeepers, maintenance mechanics, and food preparers. All activities of the indirect labor force not only present a substantial potential for saving the direct cost of the indirect labor payroll, but also overall cost, due to increased efficiency of the medical care facility. Generally speaking, the cost savings potential will be directly proportional to the size of the organization's indirect labor staff.

Note: This appendix was prepared for use in this book as an illustration of the application of the principles described within the text.

PRODUCTIVITY MONITORING SYSTEM

The indirect labor activities have generally been avoided by methods analysts due to the higher degree of difficulty and higher costs involved in their improvement, standardization, and measurement. The difficulties are attributed mainly to the greater variability of conditions, work content, and thinking/decision process time found in indirect labor activities (Krick, 1962, p. 417).

Because of extreme differences in the basic nature of various types of indirect labor activities in a hospital, it is difficult to generalize the design of productivity monitoring systems. Therefore, the remainder of this appendix will focus on one of the most difficult areas—the maintenance and repair activities within a hospital environment. The system described was established by the author for the maintenance and engineering department at Kennestone Hospital in Marietta, Georgia.

Maintenance has some routine activities that can be measured and standardized using conventional direct labor methods. These activities include preventive maintenance, scheduled repairs, and renovation projects.

PREVENTIVE MAINTENANCE

Preventive maintenance (PM) is prescheduled work with repetitive activities that are conducive to conventional direct labor measurement and control techniques. It requires the assignment of standard man hours to each task and a system of scheduling the appropriate discipline for the job. Standard man hours can be determined in a number of different ways. Since many of the hospital's preventive maintenance functions are unique, they may require individual time studies. Some activities, such as medical equipment PMs, have published "recommended man hours required" such as those found in *Medical Equipment Management in Hospitals,* published by the American Society for Hospital Engineers. A third alternative is to use historic "time required" standards. That is, after the tasks have been performed enough times for the "time required" data to be statistically significant, an average time can be used as the standard. If the hospital maintenance department employees are represented by a union, it would be advisable to have the union representative approve the method of establishing the standards, but not the standards themselves, before beginning.

The evaluation process can begin only after the standard times have been established for each task. As with any new program, a review system should be set up to periodically evaluate the engineered standards and adjust them as necessary. If the PM mechanic's actual "time required" is consistently and substantially less or greater than the engineered standard, it would indicate that additional evaluation is needed.

The following are productivity monitoring documents that are a part of the preventive maintenance program. Included are the task description sheet, the PM work order, and the PM record card.

Task Description Sheet

A task description form is used to describe the tasks, the frequency, and the standard man hours assigned for each frequency. Figure D-1 shows an example of this form. The tasks are grouped into weekly, monthly, quarterly, semi-annual, and annual frequency periods. In order to simplify the record-keeping, statistical analysis, and assignment of standard times, the man hours required for each individual task should be accumulated for a specific frequency. Therefore, if a weekly PM requires 0.5 man hours, a monthly PM requires 1.3 man hours, and a quarterly requires 0.2 man hours, then a monthly PM would have 1.8 man hours assigned to the PM work order, accounting for the weekly and the monthly times. Similarly, when a quarterly PM is due, the assigned "time required" would be 2.0 man hours, including the weekly, monthly, and quarterly times.

Preventive Maintenance Work Order

The PM work order is a form used to communicate information to the mechanic and includes such data as: date work order was issued; frequencies due; PM control number; date PM is scheduled; man hours required; equipment description, location, and ID number. Figure D-2 shows an example of a typical form.

   The mechanic is given the PM work order and task description sheet and is instructed to fill in the time required to complete each task in the appropriate column. In addition, the PM mechanic describes work performed, repairs needed, and total time required per frequency on the preventive maintenance work order form. Both the PM work order and the task sheet are turned in daily to the PM supervisor who reviews the recorded information and evaluates the productivity by comparing the actual time required (with complications considered) with the standard man hours. These "time required" totals for each frequency are then recorded on the equipment PM maintenance record card for fast and easy future reference.

Preventive Maintenance Record Card

The PM record card is a part of the master recordkeeping system. This card serves as a permanent record of when the PMs were done, who did the work, work order number, man hours required to complete the task, and other information not relevant to productivity monitoring. An example of such a form is shown in Figure D-3.

   Scheduled PM is relatively routine, but there are exceptions to the man hour requirements. Any complications or unforeseen problems that require additional time should be noted on the completed work order during the supervisor's review. Also, any productivity accounting should take these exceptions into consideration for a more accurate picture of an individual's productivity.

   In addition to increasing productivity and efficiency through these controls on PM activity, other savings such as less equipment down time, and lower emergency repair (overtime) costs will be realized. As mentioned before, the cost savings will not only be found in the indirect labor payroll. Figure D-4 shows a graph of monthly repair costs

P.M. CONTROL NO. 003442 EQUIP. I.D. NO. I-N11-001

DATE ISSUED 12-7-81

PREVENTIVE MAINTENANCE	FREQ. CODE	STD. MHS
EQUIPMENT: CONTRA-FLO WASHER	W - WEEKLY	0.5
LOCATION: LAUNDRY	M - MONTHLY	1.0
	Q - QUARTERLY	2.0
	S-SEMI-ANNUAL	0.5
	A - ANNUAL	

ITEM No.	TASK DESCRIPTION	FREQ	MECH	TIME REQD.
1.	Lubricate drive shaft roller bearings, timing chain and bronze rushing; and open gears.	W	JH	0.1
2.	Check water pump grease and service, as necessary.	W	JH	0.05
3.	Check and service air lubricator as necessary. Drain water from air bowl.	W	JH	0.1
4.	Check drive chains for proper tension and wear. Stretch should not exceed 1/8" at 10" measurement.	W	JH	0.05
5.	Check for noise such as thumping, scraping, squeaking, etc. Investigate and correct as necessary.	W	JH	0.1
6.	Check solution pumps for operation and leaks.	W	JH	0.1
7.	Lubricate drive motor.	M	JH	0.1
8.	Check brake for proper adjustment.	M	JH	0.1
9.	Lubricate roller chains and bearings for conveyor.	M	JH	0.1
10.	Check conveyor belt for condition and alignment.	M	JH	0.1
11.	Lubricate hopper door pivot points, latches, etc.	M	JH	0.1
12.	Adjust roller tension as necessary.	M	JH	0.1
13.	Vacuum all electrical motors.	M	JH	0.3
14.	Remove and check support rollers and bearings for condition.	Q	JH	0.5
15.	Check support roller jig fixture for security.	Q	JH	0.2
16.	Check all nuts and bolts for tightness. Particular attention should be given to those associated with the cylinder.	Q	JH	0.2

REMARKS:

Figure D-1 Task Description Sheet. (Reprinted from *Preventative Maintenance Management for Hospitals* by M. D. Chapman. Internal Communication of Kennestone Hospital, Marietta, Georgia.)

PREVENTIVE MAINTENANCE
WORK ORDER

DATE: *12-4-81*

✓	Weekly
✓	Monthly
✓	Quarterly
	Semi-Annual
	Annual

P.M. CONTROL NUMBER *3442*

DATE SCHEDULED *11-23-81*

HOURS ASSIGNED *3.5*

EQUIPMENT DESCRIPTION   LAUNDRY - Contra Flo Washer - I.D. #I-N11-001

DESCRIPTION OF WORK PERFORMED & REPAIRS NEEDED	REPAIR/REPLACEMENT RE: SHOP CONTROL NO.	DOLLAR COST
Weekly P.M. .5 hr.		
Monthly P.M. .9 hr.		
Quarterly P.M. 1.7 hr.		
TOTAL 3.1 hr.		
Drive Chains Need Replacing	*12-15-07*	*$350.95*

DATE COMPLETED *11-24-81*

HOURS REQUIRED *3.1*

HOURS REQUIRED _____

REVIEWED BY *Bob Sharp*
            (Supervisor)

Total Repair Cost $ _____

MECHANIC *Red Schneider*
            (Signature)

MECHANIC _____
            (Signature)

Figure D-2 Preventive Maintenance Work Order. (Reprinted from *Preventative Maintenance Management for Hospitals* by M. E. Chapman. Internal Communication of Kennestone Hospital, Marietta, Georgia.)

PREVENTIVE MAINTENANCE RECORD

EQUIPMENT / DESCRIPTION	LOCATION	I D NUMBER
Contra-Flo Washer TDS #25	Laundry	I-N11-001

MODEL	MANUFACTURER	SERIAL NO
379-189A Gr 1	Ametek, Inc.	ZA-45542

WEEKLY ✓ (0.5) MONTHLY ✓ (1.0) 4th WK (1.0) QUARTERLY MDS. 12,2,5,8 (2.0) SEMI-ANNUAL MDS. 12,2,5,8 (2.0) ANNUAL MDS. 11,5 (0.5) ANNUAL

DATE	MECHANIC	WORK ORDER NO	HRS
11-25-80	SCHNEIDER	0050 (W,M,Q,S)	13.2
12-10-80	SCHNEIDER	0170	.7
12-16-80	SCHNEIDER	0280 (W)	1.5
12-23-80	SCHNEIDER	0294 (W,M)	2.3
12-30-80	SCHNEIDER	0303 (W)	1.5
1-7-81	SCHNEIDER	0403 (W)	.8
1-13-81	SCHNEIDER	0410 (W)	.7
1-19-81	SCHNEIDER	0520 (W)	.5
1-26-81	SCHNEIDER	0666 (W,M)	1.0
2-2-81	SCHNEIDER	0735	1.0
2-9-81	SCHNEIDER	0758 (W)	.5
2-17-81	SCHNEIDER	0902 (W)	1.0
2-23-81	SCHNEIDER	0957 (B,W,M)	1.5
3-2-81	SCHNEIDER	1036 (W)	.5
3-11-81	SCHNEIDER	1111	1.0

DATE	MECHANIC	WORK ORDER NO	HRS
3-16-81	SCHNEIDER	1240 (W)	13.2
3-23-81	SCHNEIDER	1327 (W,M)	.7
3-30-81	SCHNEIDER	1415	1.5
4-6-81	SCHNEIDER	1519 (W)	2.3
4-16-81	SCHNEIDER	1613 (W)	1.5
4-24-81	REEVES	1651	.8
4-29-81	REEVES	1748 (W,M)	.7
5-7-81	SCHNEIDER	1836	.5
5-13-81	SCHNEIDER	1932 (W)	1.0
5-21-81	SCHNEIDER	1981 (W,M,a)	1.0
6-2-81	SCHNEIDER	2119 (W,M,a,S)	.5
6-6-81	SCHNEIDER	2203	1.0
6-16-81	SCHNEIDER	2195 (W)	.5
6-24-81	SCHNEIDER	1284 (W,M)	1.0

DATE	MECHANIC	WORK ORDER NO	HRS
6-29-81	SCHNEIDER	0369 (D)	.5
7-7-81	SCHNEIDER	0473 (W)	1.0
7-20-81	SCHNEIDER	0518 (W)	.5
7-24-81	SCHNEIDER	0782 (W,M)	1.0
8-6-81	SCHNEIDER	0856	.5
8-11-81	SCHNEIDER	1031	1.8
8-22-81	SCHNEIDER	1111	2.7
8-24-81	SCHNEIDER	1184 (W,M,Q,S)	.5
9-3-81	SCHNEIDER	1282	1.0
9-8-81	SCHNEIDER	1353 (W)	1.0
9-14-81	SCHNEIDER	1570 (W)	.5
9-23-81	SCHNEIDER	1722 (W)	.5
10-1-81	SCHNEIDER	1850 (W,M)	1.9
10-9-81	SCHNEIDER	1974	.5
10-13-81	SCHNEIDER	2094 (W,M)	1.5

DATE	MECHANIC	WORK ORDER NO	HRS
10-20-81	SCHNEIDER	2275 (W)	.5
11-7-81	COLE	2560 (W,M)	2.9
11-12-81	COLE	2813	1.5
11-17-81	REEVES	2146 (W)	.5
12-8-81	REEVES	3096 (W,M,a,S)	3.0
12-8-81	SCHNEIDER	3222 (W)	.5
12-18-81	SCHNEIDER	3348	.5
12-28-81	SCHNEIDER	2829 (W,M)	1.5
1-6-82	SCHNEIDER	3967 (W)	.5

WEEKLY CONTROL 1 2 3 4

MONTHLY INSPECTION CONTROL

JAN	FEB	MAR	APR	MAY	JUN	JUL	AUG	SEP	OCT	NOV	DEC

I D NUMBER	DESCRIPTION
I-N11-001	Contra-Flo Washer

Figure D-3 Preventive Maintenance Record Card. (Courtesy of Acme Visible Records, Inc., Crozet, Virginia, 1970.)

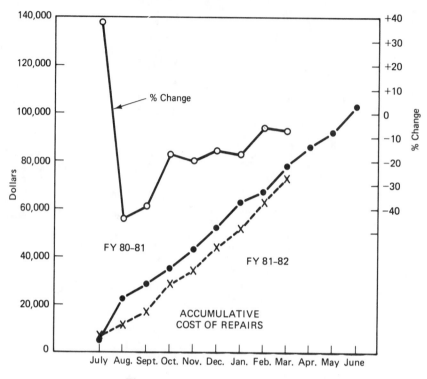

Figure D-4   Cost of Repairs Graph.

showing Kennestone Hospital's expenditures for fiscal year 1980–81 and 1981–82. Kennestone Hospital started implementing its new preventive maintenance program in November of 1980. Since that time, the maintenance department's work force has been *reduced by 15 percent* and year-to-date cost of equipment and building repairs has been *reduced by 20 percent.*

Work Request Control

The second area of maintenance activities that can be scheduled is the written work request system. Once the request is received in the maintenance office it is logged in, assigned a priority, and given to the appropriate supervisor. The supervisor evaluates the work needed and estimates the time required to complete the work. This estimated time is then used when planning each mechanic's daily work schedule and computing work backlog. This type of estimating is dependent solely on the supervisor's experience and rarely is repetitive enough to make use of the established man hour standards. This method of assigning "time required" is crude and has much room for improvement, but it does give a benchmark to measure from.

Daily Planning

Using the above referenced "time required" standards and other standard man hour data, the maintenance supervisors plan the following day's activities for each person in the section using the daily work assignment form shown in Figure D-5. Use of this form not only requires the supervisor to plan each mechanic's day, which in itself increases productivity, but also gives the supervisor a vehicle to evaluate his estimating skills.

The supervisor also fills out a "Mechanic's Daily Assignment" form for each mechanic. Figure D-6 shows a sample of this form. This form, and the related work orders, are given to each mechanic as the schedule for the following day. The activities are listed in order of priority and include estimated times for completion. Normally, the supervisor will schedule more than eight hours of estimated work in order to compensate for underestimated times required. Each mechanic completes the actual time required information, description of assigned work, completion status of each work order, and lists any unscheduled activity performed. The completed form is turned in at the end of the day and becomes a permanent record of each mechanic's daily activities.

Monitoring Department Performance

In order to monitor the department's performance it is necessary to keep a master log of all mechanic man hours spent on PMs, work requests, and unscheduled activities. At the end of each month, the actual man hours required totals are added together and divided by actual man hours available less a "non-work order man hour constant" to arrive at a ratio that represents the efficiency of the department. This ratio may not be relevant to any conventional productivity monitoring figures, but it gives an indicator useful in comparing the effectiveness of one period to another.

SUMMARY

The methods described are simple but effective. They represent an awareness of a need, information to evaluate the work system, appraisal of the methods, and an organized system of measurement and control. Indirect labor methods engineering presents a wide open and potentially substantial opportunity for management engineers to exercise creativity, ingenuity, and industrial engineering analysis skills.

MECHANIC'S DAILY ASSIGNMENT

NAME: _____

DATE: _____

TIME: _____

WORK ORDER NO.	PRIORITY	EST. TIME REQD.	TIME REQD.	COMPLETE	DESCRIPTION OF WORK INCLUDING ANY COMPLICATIONS

MECHANIC'S SIGNATURE

Figure D-5   Mechanic's Daily Assignment Form.

SUPERVISOR'S DAILY WORK ASSIGNMENT

SUPERVISOR: _____   DATE: _____

TECHNICIAN'S NAME	SHIFT	WORK ORDER CONTROL NO.	PRIOR-ITY	ESTIMATED TIME REQUIRED (MANHOURS)	ACTUAL TIME REQUIRED (MANHOURS)	REMARKS

Figure D-6 Supervisor's Daily Work Assignment Form.

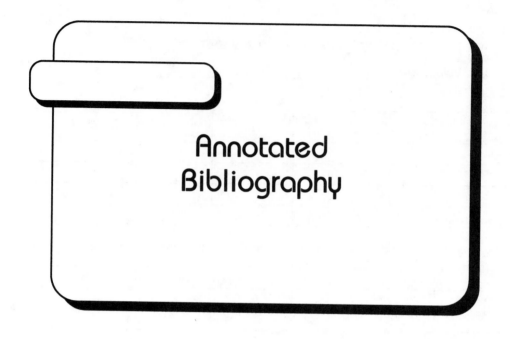

Annotated
Bibliography

INTRODUCTION

This book has explained and illustrated a number of procedures for improving productivity by using traditional industrial engineering methodologies. While the book has used numerous examples, the following few pages describe and reference full-scale applications of some of these techniques. Following these references are some additional suggestions regarding the overall implementation of productivity improvement and work measurement programs are summarized.

APPLICATIONS

American Productivity Center. "Reward Systems and Productivity." Maritz Corporation, 1991.

America's greatest national resource is its people. It has been generally recognized that our human productivity problem is not the work ethic. The American worker has as strong a work ethic as at any time in our history. The problem has been our inability to pool our knowledge and experience to find an effective way to tap that resource. This report shows how a new approach to the concept of reward systems can benefit productivity. Key among the new approaches is the sharing of information with employees. Employees specifically need more and better information about their jobs, business conditions, customers, and competitors.

Coppey, J. M., and Wyskida, R. M. "The Supervisor and Flexible Working Hours." *Industrial Engineering,* November 1977.

"A better working climate that is a great advantage to supervisors is reported repeatedly by companies operating under flexible working hours." Supervisors are the workers' major contact with the company and their attitude can help to increase the workers' productivity. A number of benefits, including the following, can result when the supervisor is concerned more with the work than the hours: individuals learn to manage their own time, supervisors learn to delegate, supervisors develop their own abilities in addition to their supervisory skills, and supervisors are more humanistic in their approach to their workers. All of these factors can lead to an increase in overall productivity.

Gibson, D. "Work Sampling Monitors Job Shop Productivity." *Industrial Engineering,* June 1970.

"Systematic sampling of work in 16 maintenance shops provided accurate work standards as well as valuable information for management." The traditional variability of maintenance tasks provided a suitable atmosphere for setting performance standards based on random sampling. A productivity increase of 15 percent was attributed to the standards-setting procedure.

Lanphier, T. "Flow Charting Speeds Hospitl Work." *Industrial Engineering,* November 1970.

"Logical use of flow process charts and redesign of simple equipment simplifies the vital job of blood specimen collection." The use of the flow process chart in this activity reduced the transportations required from 24 to 16, a 33 percent savings. The reduced transportation speeded patient care and reduced the overall time and cost for the treatment.

Leavens, J. M., and Nanda, R. "Standard Data Aids Air Cargo Terminal Planning." *Industrial Engineering,* May 1971.

"Predetermined time systems are used to develop standard data for work analysis of proposed air cargo terminals." The principles of standard data are implemented to aid management in planning air cargo operations. By preparing manuals based on the analysis, industrial engineering was able to predict the expected performance and standards for a number of jobs. Additionally, these jobs were able to be changed as the jobs' contents changed.

Meck, F. S. "Work Sampling Study of Scattered Maintenance Workers." *Industrial Engineering,* January 1972.

When maintenance workers perform their tasks in a variety of locations around a large manufacturing plant even the observation of their activities poses special problems when work sampling is performed. ''The IE's of this plant solved the problem with a cleverly designed route and random observation plan for a work sampling study.'' The study was conducted to determine the effectiveness of the maintenance department, develop recommendations to help improve the work, and to check on the previously developed standards.

O'Dell, Carla. *People, Performance, and Pay.* American Productivity Center, Houston, 1987.

This report documents what organizations are doing to respond to the reward and human resource challenges facing them, why they are responding the way they are, and the results they are achieving.

The report indicates that there has been a striking growth in the number of firms adopting non-traditional reward systems during the last five years. It further goes on to indicate that there will be an increasing trend to adopt these non-traditional systems.

Ottinger, L. V. ''Robotics for the IE, Terminology, Types of Robots.'' *Industrial Engineering,* November 1981.

The future of many typical production tasks seems tied to the development of industrial robots. ''Today's industrial engineer is well equipped to be a robot project manager.'' Robots can be specified using traditional industrial engineering skills. Therblig symbols, chronocyclegraphs, principles of motion economy, and so on, can be used to specify the robotic characteristics needed for a job. In a series of articles beginning in November 1981, *Industrial Engineering* discussed the use of robots in production.

Patton, J. A. ''Wage Incentives: From Failure to Success.'' *Industrial Engineering,* June 1974.

''Having installed or revised several hundred wage incentive plans during a thirty year period, the author writes with authority about the causes of ineffectiveness and failure of incentives. In addition, he describes the ingredients of successful applications.'' For a wage incentive system to be successful, management must provide a fair and equitable system—one in which all employees have an equal opportunity to earn a bonus for work in excess of standard. To ensure this equality, the company must have competent supervision, qualified industrial engineers, and a maintenance system for the standards used in the system. The workers must understand the incentive system and accept the fact that above average performance will assure above average pay.

Ricks, L. ''MTM and Hospital Nursing.'' *Proceedings,* AIIE Annual Spring Conference, 1980.

''The major purpose of the Nursing Study was the development of a labor standard, or to measure the nursing service workload in nursing hours per patient day, for budgetary considerations." Various versions of MTM, such as MTM-2 and MTM-C, were used to establish performance standards. The productivity of individual staffing units, as well as the need for increased staffing or reduced staffing, was documented.

PHILOSOPHY

In addition to articles describing the specifics of how productivity was improved, how work was measured, or both, there are a number of generally interesting and useful discussions of productivity improvement and work measurement. Following is a brief summary of some representative articles.

Burnham, D. C. "Three Steps to Productivity Improvement." *Industrial Engineering,* September 1972.

"When the Chairman of one our largest corporations (Westinghouse Electric) takes the time to discuss productivity with you, you can be sure it's a top priority problem—for you as well as him. Further, he proposes a three prong Federal Government program to improve U.S. productivity growth." These three steps are research, removing roadblocks, and positive encouragement for productivity improvement. Based on this philosophy private industry can make strides in overcoming the "productivity problem." The commitment and results Westinghouse has attained are described in detail and attributed to following this policy.

Cooling, W. C. "Production Roundtable—How to Increase Production." *Proceedings,* AIIE Annual Spring Conference, 1980.

"If IE should not develop standards, issue the standards to the floor, and walk away." Instead, the IE should ensure that the supervisor is able to do his or her job, including training, administration, leadership, and helping the employee do his job.

Frohman, Mark. "Lower Level Management: The Internal Customers of Change." *Industry Week,* November 5, 1990.

Any improvement to any operating system is destined to succeed or fail based on the response of the managers and supervisors who must implement the change. This article provides a succinct recipe of actions that will bring about either success or failure for those actions.

Failure-guaranteeing actions such as mandating change from the top without providing training are discussed in-depth. Similarly, suggestions for successful implementation are presented in enough detail to raise appropriate levels of awareness.

Gottlieb, B. "A Fair Day's Work is Anything You Want It to Be." *Journal of Industrial Engineering,* December 1968.

"A fair day's work for a fair day's pay is unreal. An acceptable level of work for an acceptable level of pay is real." The problems associated with defining what is meant by *fair* lead the author to this conclusion. The many factors that influence productivity— technological, management, financial, labor, governmental, economic, and natural— make a uniform definition difficult if not impossible to state. Equal problems surround the concept of standard performance. The author suggests a way to deal with these difficulties without abdicating the job that industrial engineering is charged with in work measurement.

James, C. F. "Incentives for Machine-paced Operations." *Industrial Engineering,* September 1975.

"An incentive system for machine or line paced operations is developed by establishing a standard based on reducing downtime." It is possible to reward people, not for the work they do, but for the problems they prevent. This perspective to incentive pay is explored through the use of two examples showing how incentive earnings are calculated.

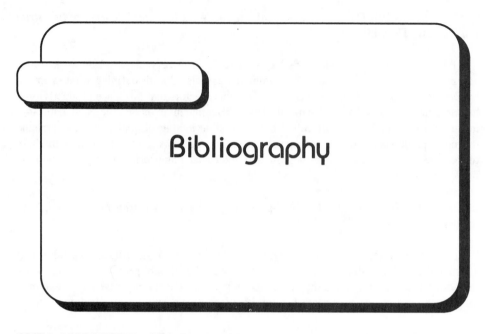

Bibliography

PART I: PRODUCTIVITY—AN INTRODUCTION

ALLEN, S. "Unionized Workers are More Productive." *Research Report for the Center to Protect Workers' Rights.* Washington, D.C., November 1979.

"Better Ways to Solve Plant Problems." *Factory,* May 1956.

BIRN, S. A. "Productivity Slump: Broader IE's Needed." *Industrial Engineering,* February 1979.

BUTCHER. "Closing Our Productivity Gap: Key to U.S. Economic Health." *Industrial Engineering,* December 1979.

CAMP, LANE GARDNER. "IEs Evaluate Productivity Improvement Efforts in Own Organizations and Across U.S." *IE,* January 1985.

CHUNG, K. H., AND GRAY, M. A. "Can We Adopt the Japanese Methods of Human Resources Mangement?" *Personnel Administrator,* May 1982.

COHEN, W. A. *Principles of Technical Management.* New York: Amacom, 1980.

DEWITT. "Productivity and the Industrial Engineer." *Industrial Engineering,* January 1976.

"Discipline Called Japan's Key to High Production." *Atlanta Journal-Constitution,* January 17, 1981.

DONOHUE, T. R. "The Human Factor in Productivity." *Federationist,* December 1980.

DRUCKER, P. "Behind Japan's Success." *Harvard Business Review,* January-February 1981.

DUMOLIEN, W. J., AND SANTEN, W. P. "Cellular Manufacturing Becomes Philosophy of Management at Components Facility." *IE,* November 1983.

FALLOWS, J. "American Industry, What Ails It, How to Save It." *Atlantic Monthly,* September 1980.

FEIGENBAUM, A. *Total Quality Control.* New York: McGraw Hill, 1983.

"Frank Talk About Productivity." *The Collegiate Forum,* Fall 1980.

GIESEL, C. E., AND NADLER, G. "The 'Best' Method is Not Good Enough." *Proceedings of the AIIE Fall Industrial Engineering Conference*, 1978.

GORDON, M. E., AND FITZGIBBONS, W. J. "Unions and Psychology—An Updated View." *Federalionist*, September 1979.

HAMLIN. "Productivity Improvement: An Organized Effort." *Proceedings of the AIIE Annual Spring Conference*, 1978.

HAYES, J. L. "Problem Decision Making." *Quality*, December 1981.

HAYES, R. H. "Why Japanese Factories Work." *Harvard Business Review*, July-August 1981.

HENRICI, S. B. "How Deadly is the Productivity Disease?" *Harvard Business Review*, November-December 1981.

HICKS, P. "The IE and Productivity." Paper presented at the AIIE Region IV Conference, San Juan, Puerto Rico, November 1980.

HINES, W. H. "Guidelines for Implementing Productivity Measurement." *Industrial Engineering*, June 1976.

"IBM's Automated Factory, A Giant Step Forward." *Modern Materials Handling*, March 1985.

"IE's Offer Views on Reasons Behind and Possible Cures for Declining U.S. Productivity." *Industrial Engineering*, October 1981.

KEHLBECK, J. "Productivity, an International Contest." *Industrial Engineering*, January 1978.

KHALIL, T. M., AND AYOUB, M. A. "Applied Ergonomics" *Industrial Engineering*, April 1976.

KIRKLAND, L. "Productivity: A Labor View." *Proceedings of the American Productivity Congress*, 1982.

KRICK, E. *Methods Engineering*. New York: Wiley, 1962. [1969]

KUPER. "Action for Productivity Improvement: Role of the IE." *Proceedings of the AIIE Annual Spring Conference*, 1978.

"Labor Letter." *Wall Street Journal*, January 6, 1981.

LIKER, J. K., AND HANCOCK, WALTON. "Work Environment Survey Generates Ideas on Increasing White Collar Productivity." *IE*, July 1984.

METZ, D. L., AND KLEIN, R. E. *Man and the Technological Society*. Englewood Cliffs, NJ: Prentice-Hall, 1973.

NIEHOFF, M., AND ROMANS, M. J. "Needs Assessment as Step One Towards Enhancing Productivity." *Personnel Administrator*, May 1982.

"No Miracle. Just Good Management." *Quality*, August 1980.

OSBORN, A. *Applied Imagination*, New York: Scribner's, 1963.

OTIS, I. "Industrial Work Standards and Productivity." *Proceedings of the AIIE Annual Spring Conference*, 1978.

"Productivity for Profit." *Industrial Distribution*, February 1981.

"Productivity: How We Can Put the Pieces Together." *Modern Materials Handling*, January 20, 1982.

"Productivity's Link to Economic Growth." *Federationist*, March 1978.

"Productivity . . . What's in a Word." *Viewpoint*, Spring 1980.

"Quality and Productivity . . . The Japanese Point of View." *Quality*, June 1981.

RICCIO, L. J. "Productivity in the Public Sector, Its Measurement and Improvement." *Proceedings of the AIIE Systems Engineering Conference*, 1976.

Ross, J. E. *Productivity, People, and Profits.* Reston VA: Reston Publishing Company, 1981.

Schlick. "Productivity Improvement at the Naval Weapons Center, Crane." *Proceedings of the AIIE Annual Spring Conference,* 1980.

Schoneberger, Richard, J. "Production Workers Bear Major Quality Responsibility in Japanese Industry." *IE,* December 1982.

Sedan, S. M. "Engineering Next in Line for Productivity Push?" *Industrial Engineering,* June 1972.

Sellie, C. "Better Use of Better Tools Should Make Work Measurement Increasingly Valuable in Future." *IE,* July 1984.

Sirota. "The Conflict Between IE and Behavioral Science." *Industrial Engineering,* June 1972.

Vaughn, R. L. "Productivity: The Need for Innovative Management." *Manufacturing Engineering,* March 1982.

Von Fange, E. K. *Professional Creativity.* Englewood Cliffs, NJ: Prentice-Hall, 1959.

Vough, C. F., and Asbell, B. *Productivity: A Practical Program for Improving Productivity.* New York: Amacom, 1979.

Walsh. "At Least We Gave Jimmy Four Years." *Quality,* December 1980.

"When Twain Meet." *Wall Street Journal,* October 10, 1978.

"Worker/Management Program Boosts Productivity at Nissan." *Modern Materials Handling,* March 1985.

PART II: WORK ANALYSIS

Alexander, D., and Smith L. "Designing Industrial Jobs." *Industrial Engineering,* June 1982.

Barnes, R. *Motion and Timestudy.* New York: Wiley, 1980.

Heimstrand, N., and Ellingstad, V. *Human Behavior: A Systems Approach.* Monterey, CA: Brooks/Cole, 1972.

Konz, S. *Work Design.* Columbus, OH: Grid, 1979. [1990]

Krick, E. *Methods Engineering.* New York: Wiley, 1962. [1969]

Kroemer, K. H. E., and Price, D. L. "The Office Work Place." *Industrial Engineering,* July, 1982.

Meyers, H. *Human Engineering.* New York: Harper, 1932.

Mundel, M. *Motion and Timestudy.* Englewood Cliffs, NJ: Prentice-Hall, 1978.

Neibel, B. *Motion and Timestudy.* Homewood, IL: Irwin, 1976.

Rice, R. S. "Survey of Work Measurement and Wage Incentives." *Industrial Engineering,* July 1977.

Roy, D. E. "Banana Time—Job Satisfaction and Informal Interaction." *Human Organization,* 1960.

Smith, J., and Ramsey, J. "Designing Physically Demanding Tasks." *Industrial Engineering,* May 1982.

Smith, P., Armstrong, T., and Lizza, G. "IE's and Job Designing for the Handicapped Worker." *Industrial Engineering,* April 1982.

Weiss, W. H. "Human Engineering Goals, Minimum Injuries, Maximum Productivity." *Production Engineering,* May 1982.

PART III: MEASURING PRODUCTIVITY

ABRUZZI. "Developing Standard Data for Predictive Purposes." *Journal of Industrial Engineering*, November 1952.

ANDREWS, R. "The Relationship Between Measures of Heart Rate and Energy Expenditure." *AIIE Transactions*, January 1969.

AYOUB, M. A. "Pre-employment Screening Programs That Match Job Demands with Work Abilities." *Industrial Engineering*, March 1982.

BAILEY, G. B., AND PRESGRAVE, R. *Basic Motion Timestudy.* New York: McGraw-Hill, 1948.

BARNES, R. *Work Sampling.* New York: Chapman and Hall, 1957.

BARNES, R. *Motion and Timestudy.* New York: Wiley, 1980.

BECK. "Work Sampling Sets Direct Labor Standards." *Industrial Engineering.* February 1969.

BRISLEY, C., AND DOSSETT. "Computer Use and Non-Direct Labor Measurement Will Transform Profession in the Next Decade." *Industrial Engineering,* August 1980.

BRISLEY, C., AND EIELDER, W. F. "Reliability and Accuracy in Work Measurement." *Industrial Engineering,* May 1982.

BROUHA. *Physiology in Industry.* New York: Pergamon, 1960.

CANDY, W. L. "Maintenance Groups Make and Implement Own Systems for Improving Productivity." *Industrial Engineering,* February 1982.

CARROLL. *Timestudy for Cost Control.* New York: McGraw-Hill, 1954.

DAVIDSON, H. O. "Work Sampling—Eleven Fallacies." *Journal of Industrial Engineering,* September 1960.

DAVIS, H. L., FAULKNER, T. W., AND MILLER, C.I. "Evaluating Work Performance." *The American Journal of Clinical Nutrition,* September 1971.

DAVIS, H. L., AND MILLER. "Human Productivity and Work Design." *Industrial Engineering Handbook* (3rd ed.). New York: McGraw-Hill, 1971.

DORING. "Why Timestudy Standards Cost Too Much." *Journal of Industrial Engineering,* September 1954.

DUNCAN, J. H., QUICK, J. H., AND MALCOLM, J. A. *Work Factor Time Standards.* New York: McGraw-Hill, 1962.

EDHOLM, O. G. *The Biology of Work.* New York: McGraw-Hill, 1976.

FERGUSON. "Measuring Productivity Changes: A Case Study." *Proceedings of the AIIE 1978 Annual Spring Conference,* 1978.

GOMBERG. "What are the Specifications for an Effective System of Standard Data?" *Journal of Industrial Engineering,* July 1954.

HEILAND, R. E., AND RICHARDSON, W. J. *Work Sampling.* New York: McGraw-Hill, 1957.

KARGER, D. W., AND BAYHA, F. W. *Engineered Work Measurement.* New York: Industrial Press.

KONZ, S., AND CAHILL. "Variability of the Interbeat Interval of the EKG as an Index of Mental Effort." *Kansas State University Bulletin,* September 1977.

KONZ, S. *Work Design.* Columbus, OH: Grid, 1979. [1990]

KRICK, E. *Methods Engineering.* New York: Wiley, 1962. [1969]

LANE, R. E., AND STETLER, L. "Selection and Implementation of a Predetermined Time System." *Assembly Engineering,* October 1972.

LOWRY, S. M., MAYNARD, H. B., AND STEGEMERTEN, G. J. *Time and Motion Study.* New York: McGraw-Hill, 1940.

MAYNARD, H. B. (ED.). *IE Handbook* (3rd ed.). New York: McGraw-Hill, 1971.

MAYNARD, H. B., SCHWAB, J. L., AND STEGEMERTEN, G. J. *Methods Time Measurement.* New York: McGraw-Hill, 1948.

MUNDEL, M. *Motion and Timestudy.* Englewood Cliffs, NJ: Prentice-Hall, 1978.

MUNDEL, M. "Predetermined Time Standards in the Army Ordinance Corp." *Journal of Industrial Engineering,* November 1954.

NEIBEL, B. *Motion and Timestudy.* Homewood, IL: Irwin, 1976.

NOF, S. Y., AND LECHTMAN, H. "The RTM Method of Analyzing Robot Work." *Industrial Engineering,* April 1982.

RICE, R. "Survey of Work Measurement and Wage Incentives." *Industrial Engineering,* July 1977.

RICHARDSON, W. J. *Cost Improvement, Work Sampling, and Short Interval Scheduling.* Reston, VA: Reston Publishing Co., 1976.

SALVENDY, G., AND STEWART, K. "The Prediction of Operator Performance on the Basis of Performance Tests and Biological Measures." *AIIE Transactions.* January 1969.

SINGLETON, W. T., ET AL. (EDS.). *Measurement of Man at Work.* New York: Van Nostrand Reinhold, 1971.

STEWART, K. "A Participative Approach to Productivity Measurement in a Manufacturing Environment." *Proceedings of the AIIE Annual Spring Conference,* 1978.

WRIGHT, M., "Analyzing Word Processing Productivity Improvements." *Industrial Engineering,* July 1982.

PART IV: IMPROVING PRODUCTIVITY

AYOUB, M. A. "Work Place Design and Posture." *Human Factors,* 1973, 15(3).

BARNES, R. *Motion and Timestudy.* New York: Wiley, 1980.

BLUESTONE, I. "How Quality of Work Life Projects Work for United Auto Workers." *Monthly Labor Review,* 1980, 103(7).

BREDIN, H. "Unmanned Manufacturing." *Mechanical Engineering,* February 1982.

BROOKE, K. A. "Management Readiness to Support QC Circle." *Industrial Engineering,* January 1982.

COX, J. M. "Labor Incentives—A Failure and a Success," *Journal of Industrial Engineering,* March-April 1959.

COX, M., AND BROWN, J. "Quality of Worklife: Another Fad or Real Benefit?" *Personnel Administrator,* May 1982.

COORTRIGHT, W. E. "Quality Circles at Hughes Aircraft." *Transactions, ASQC Technical Conference,* 1979.

DEWAR, D. L. "A Human Approach to Motivation and Productivity." *AIIE Proceedings,* May 1979.

DEWAR, D. L. "Measurement of Results—Lockheed QC Circles." *Quality Circles: Applications Tools and Theory.* ASQC, 1976.

DONOHUE, T. R. "The Human Factor in Productivity." *American Federationist,* December 1980.

DWORTZAN, B. "The ABC's of Incentive Programs." *Personnel Journal,* June 1982.

"Ergonomics: The Scientific Aproach to Making Work Human." *International Labour Review,* January 1961.

FEIN, M. "Work Measurement and Wage Incentives." *Industrial Engineering.* September 1973.

FEIN, M. "Job Enrichment: A Reevaluation." *Sloan Management Review,* Winter 1974.

FEIN, M. "Wage Incentive Plans." Publication Number 2 in the Monograph Series of the Work Measurement and Methods Engineering Division of the American Institute of Industrial Engineers (AIIE), Norcross, Georgia. (Copyright in 3rd edition of *IE Handbook,* New York: McGraw-Hill, 1971.)

FILLEY, R. "Productivity Case Studies Focus on Manufacturing Operations." *Industrial Engineering,* May 1982.

FITCH, J., AND BRYCE, W. "Introducing Automation into the Manufacturing Process." *Industrial Engineering,* November 1981.

GERWIN, D. "Do's and Don'ts of Computerized Manufacturing." *Harvard Business Review,* March-April 1982.

GREGERMAN, I. B. "Introduction to Quality Circles—An Approach to Participative Problem Solving." *Industrial Management,* September-October 1979.

GRYNA, F. M. *Quality Circles, A Team Approach to Problem Solving.* New York: Amacom, 1981.

GUEST, R. H. "Quality Work Life." *Vital Speeches,* March 1, 1980.

HANLEY, J. "Our Experiences with Quality Circles." *Quality Progress,* February 1980.

HANSSEN, G. M. "Productivity Gains from the Shopfloor Up." *Production Engineering,* March 1982.

Harvard Business Review, September 1969.

HEATH. "The Hawthorne Studies—50 Years Ago and Today." *Industrial Engineering,* November 1974.

HENDERSON, R. *Compensation Management.* Reston, VA: Reston Publishing Co., 1980. [1989]

HERZBERG, F. *The Motivation to Work.* New York: Wiley, 1959.

HERZOG. "New Problems, New Solutions in Human Factors Engineering." *Machine Design,* March 7, 1974.

HOWE, R. J., AND MINDELL, M. G. "Motivating the Contemporary Employee." *Management Review,* September 1979.

HUTCHINSON, R. D. *New Horizons in Human Factors.* New York: McGraw-Hill, 1981.

"Involved People Making the Difference in Manufacturing." *Modern Materials Handling,* January 20, 1982.

IRVING, R. R. "QC Payoff Attracts Top Management." *Iron Age,* August 20, 1979.

"Job Enlargement and Job Rotation." *Personnel Management Bulletin,* May 15, 1962.

KHALIL, T. M. "The Role of Ergonomics in Increasing Productivity." *Proceedings of the AIIE Annual Conference,* 1976.

KHALIL, T. M. AND AYOUB, M. M. "Applied Ergonomics." *Industrial Engineering,* April 1976.

KONZ, S. "Quality Circles: Japanese Success Story." *Industrial Engineering,* October 1979.

KRICK, E. *Methods Engineering.* New York: Wiley, 1962. [1969]

LINKERT. *The Human Organization: Its Management and Value.* New York: McGraw-Hill, 1967.

LITTLEJOHN, R. F. "Team Management: A How-to Approach to Improved Productivity, Higher Moral, and Longer Lasting Job Satisfaction." *Management Review,* January 1982.

LOKIEC, M. "Incentives and the Garment Industry." *Journal of Industrial Engineering,* June 1966.

MACY, B. A. "The Quality of Work Life Project at Bolivar: An Assessment." *Monthly Labor Review,* 1966, 103 (7).

MAIER, F. *Psychology in Industry.* Boston: Houghton Mifflin, 1955.

MAYNARD, H. B. (ED.). *IE Handbook.* New York: McGraw-Hill, 1971.

McCORMICK, E. J., AND SANDERS, M. S. *Human Factors in Engineering Design.* New York: McGraw-Hill, 1982. [1986]

McGREGOR, D. *Leadership and Motivation.* Cambridge, MA: MIT Press, 1966.

MILLS, D. O. *Labor Management Relations.* New York: McGraw-Hill, 1978. [1988]

"Mr. Rooney Goes to Work." CBS Television News, 1977.

MURRELL, K. F. H. *Ergonomics.* London: Chapman and Hall, 1969.

NATHAN, R. S. "The Scheme That's Killing the Rat Race Blues." *New York Magazine,* July 18, 1977.

"New Industrial Relations." *Business Week,* May 11, 1981.

OZLEY, L., AND BALL, J. "Quality to Work Life: Initiating Successful Efforts in Labor-Management Organizations." *Personnel Administrator,* May 1982.

PRESGRAVE, R. *The Dynamics of Timestudy.* Toronto: University of Toronto Press, 1944.

RICKER, W. S. "Quality Control Circles—Tapping the Creative Power of the Workforce." *AIIE Proceedings,* December 1976.

"Robots Steal the Show." *Production Engineering,* May 1982.

ROETHLISBERGER AND DICKSON. *Management and the Worker.* Cambridge, MA: Harvard University, 1939.

ROSENTHAL. "Application of Human Engineering Principles and Techniques in the Design of Electronics Production Equipment." *Human Factors,* April 1973.

RUBIN, E. F. "These Circle Meetings Get Right Down to Business." *Atlanta Journal-Constitution,* October 22, 1981.

SCHEUCH, R. *Labor in the American Economy.* New York: Harper and Row, 1981.

SCHLEICHER, W. F. "Quality Circles—The Participative Team Approach." *Quality,* October 1981.

SCHULTZ, D. P. *Psychology and Industry Today.* New York: Macmillan, 1978. [1990]

SLOANE, A., AND WHITNEY, F. *Labor Relations.* Englewood Cliffs, NJ: Prentice-Hall, 1977. [1988]

SMITH, L, AND SMITH, J. "Justifying an Ergonomics Program." *Industrial Engineering,* February 1982.

SQUIRES, F. "A New Role for the Quality Circle." *Quality,* August 1981.

STRAUSS, G., AND SAYLES, L. R. *Personnel.* Englewood Cliffs, NJ: Prentice-Hall, 1972.

"Stunning Turnaround at Tarrytown." *Time,* May 5, 1980.

SWAIN, A. D. "Design of Industrial Jobs a Worker Can and Will Do." *Human Factors,* 1973, 15 (3).

"U.S. Firms, Worried by Productivity Lag, Copy Japan in Seeking Employees' Advice." *Wall Street Journal,* February 21, 1980.

VAN WELY, P. *Design and Disease.* Special Report Number 86. Kansas Engineering Experiment Station, Kansas State University. Manhattan, KS, 1969.

WATMOUGH, E. B. "The Case Against Incentives." *Journal of Industrial Engineering,* November-December 1965.

WITFELDT, J. R. "The IE's Role in a Quality Control Circle." *Industrial Engineering,* January 1982.

YODER, T. A. *Some Applications of Human Factors in Industry.* Report of Environmental Health and Safety Department, Eli Lilly Company, Indianapolis, IN, 1972.

Index

Methods analysis, 15, 21
Methods improvement, 94–96
Methods Time Measurement, 243
Miller, 282
Mills, 337
Mil-Std 1567, 204, 352
Mindell, 326
Missed elements, 156
MODAPTS, 242
Motorola, 11
MTM, 221
MTM Blue Card, 222
MTM-1, 254
MTM-2, 254
MTM-3, 254
Multiple activity chart, 59–76, 77
Multiple machine responsi-
    bility, 165
Murrell, 296
Myers, 121

Nadler, 15
Nathan, 339
Neibel, 96
Non-cyclical elements, 156
Normal pace, 131
Normal, defined, 132

Olsen, 243
Operation process chart, 26–32, 76
Otis, 3, 7, 8, 10, 12
Out of sequence elements, 156
Oxygen consumption, 280, 282

Participatory management, 328
Pfd allowances, 133–134
Physiological measures of
    work, 279
Physiological measurements, 292
Physiological work measure-
    ment, 280:
    uses, 282–284
Piece rate, 309
Predetermined time standards, 220
    advantages, 240
    limitations, 241
    purpose, 220
    selection, 241
    use, 221
Principles of motion economy, 95,
    105–112
Problem definition, 16
Product modification, 20

Productivity, 3, 4, 5, 9, 291
Productivity and wage incen-
    tives, 304
Productivity growth, 12, 15
Productivity improvement, 25, 325
Productivity sharing, 308, 316
Productivity tools, 7

Quasar, 11
Quick, 243
Quality circles, 325, 330:
    establishing, 332
    problems, 337
    successes, 336
Quality control, 5
Quality of work life, 321, 325, 326
Quality of work life, implementa-
    tion, 328
Questions, 25

Reich, 5
Riccio, 4, 9
Rockway, 290
Rosenthal, 109
Ross, 20
Rubin, 337
Random, 258
Rating, 132–133
Recovery, 280
Recovery curve, 281
Regression, 214–216
Relative frequency, 141, 156
Resistance to change, 117–121

Safety, 291
Salter, 11
Sample size, 152
Sample size estimation, 275
Sanders, 294
Santen, 9
Scanlon plan, 317, 329
Schleicher, 335, 338
Schonberger, 5
Schultz, 340
Scientific management, 332
Scientific method, 16, 19
Selling solutions, 117
Singleton, 279
Snapback method, 149
Solution implementation, 18
Speeds, 137
Standard, 130
Standard calculation, 161

Standard data:
    advantages, 209
    construction, 183–196
    limitations, 211
    methodology, 196
    purpose, 182
Standard hours plan, 311
Standard time, 131
Standard time, calculation, 134–135
Standards, work, 9, 10
Standard, defined, 132
Standard, uses, 131
Static work, 282
Steady state, 280
Stopwatch, 135
Supervisor, 10, 137
Swain, 301

Taylor, Frederick, 9, 331
Taylor plan, 312
Therbligs, 95, 97–105
Time standard (see also standard
    time), 279
Time standards, 304
Time study:
    conduct, 135, 140
    data analysis, 159
    procedure, 135
Training, 10

Unions and incentives, 305
Universal Standard Data, 254

Van Wely, 295
Volume, 166
Vough, 4

Wage incentives, 304
Wage incentives and produc-
    tivity, 304
Walsh, 6
WF, 221
Work Factor, 242
Work sampling:
    advantages, 272
    definition, 257
    limitations, 272
    method, 260–267
Workplace design, 293

Yellow Fever, 20
Yoder, 295